Solutions Manual to Accompany
MODERN DIGITAL DESIGN

RICHARD S. SANDIGE

Department of Electrical Engineering
The University of Wyoming

McGraw-Hill Publishing Company
New York St. Louis San Francisco Auckland Bogotá Caracas
Hamburg Lisbon London Madrid Mexico Milan
Montreal New Delhi Oklahoma City Paris San Juan
São Paulo Singapore Sydney Tokyo Toronto

Solutions Manual to Accompany
MODERN DIGITAL DESIGN
Copyright ©1990 by McGraw-Hill, Inc. All rights reserved.
Printed in the United States of America. The contents, or
parts thereof, may be reproduced for use with
MODERN DIGITAL DESIGN
by Richard S. Sandige
provided such reproductions bear copyright notice. but may not
be reproduced in any form for any other purpose without
permission of the publisher

ISBN 0-07-054858-7

1 2 3 4 5 6 7 8 9 0   W H T   W H T   8 9 8 3 2 1 0

Contents

If you care to share with the author any comments you may have concerning the text *Modern Digital Design*, either the software package or the *PLDesigner User's Manual (Student Version)*, or this *Solutions Manual*, please direct your comments to the author Dr. Richard S. Sandige in care of the McGraw-Hill Publishing Company.

PART
I

SOLUTIONS

Section 1-2 A Mathematical Model

1-1.

1-2. $f_1 = 1/T_1 = 1/(83.33$ ns$) = 12$ MHz, $f_2 = 1/T_2 = 1/(62.5$ ns$) = 16$ MHz. Computer 2 with a shorter period (or faster clock speed) is faster than computer 1.

Section 1-3 The Algebra of Logic

1-3. $X + 0 = X$ is true in ordinary algebra. $X \cdot 1 = X$, the dual of $X + 0 = X$, is also true in ordinary algebra.

1-4. $X \cdot Y = Y \cdot X$ is true in ordinary algebra. Commutative property.

1-5. Given an expression, interchange the elements 0 and 1 and the binary operators $+$ and $\cdot$. The result is the dual of the original expression.

1-6. $X \cdot (Y + Z) = X \cdot Y + X \cdot Z$, $X + (Y \cdot Z) = (X + Y) \cdot (X + Z)$. No, addition is not distributive with respect to multiplication in ordinary algebra. Only one case is necessary to show this. Let $X = 2$, $Y = 3$, and $Z = 4$
$$X + (Y \cdot Z) = 2 + 12 = 14$$
but, $(X + Y) \cdot (X + Z) = (2 + 3) \cdot (2 + 4) = 5 \cdot 6 = 30$
and, 14 is not equal to 30

1-7. P5a: $X + \overline{X} = 1$ P5b: $X \cdot \overline{X} = 0$
$$X + (\neg X) = 1 \qquad X \cdot (\neg X) = 0$$
$$X + (\sim X) = 1 \qquad X \cdot (\sim X) = 0$$
$$X + X' = 1 \qquad\quad X \cdot X' = 0$$

1-8.

A	B	A·B
0	0	0
0	1	0
1	0	0
1	1	1

1-9.

Z	$\overline{Z}$
0	1
1	0

1-10. 0 + 0 = 0
0 + 1 = 1
1 + 0 = 1
1 + 1 = 1

1-11. highest order, Complement
second highest, AND
third highest, OR

For X = 1, Y = 0, and Z = 1,
(a) X + X·Y = 1 + 1·0 = 1 + 0 = 1
(b) ~(~X + Y) = ~(~1 + 0) = ~(0 + 0) = ~(0) = 1
In this case parenthesis are used to establish
the desired order of precedence.
(c) X' + Y + Z·Y' = 0' + 0 + 1·0' = 1 + 0 + 1·1 = 1 +
0 + 1 = 1

Section 1-4 Digital Logic Functions

1-12.

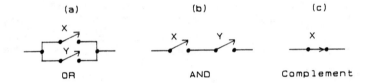

    (a)   (b)   (c)

    OR   AND   Complement

1-13.

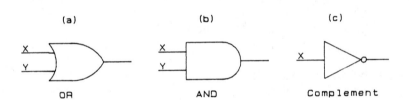

    (a)   (b)   (c)

    OR   AND   Complement

1-14.

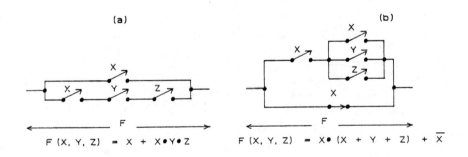

(a)

$$F(X, Y, Z) = X + X·Y·Z$$

(b)

$$F(X, Y, Z) = X·(X + Y + Z) + \overline{X}$$

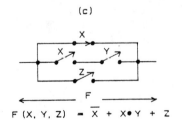

(c)

$$F(X, Y, Z) = \overline{X} + X·Y + Z$$

2

1-15. (a)

$$F(X, Y, Z) = X + X \bullet Y \bullet Z$$

(b)

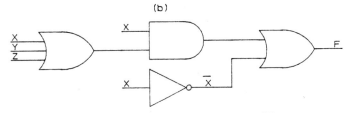

$$F(X, Y, Z) = X \bullet (X + Y + Z) + \overline{X}$$

(c)

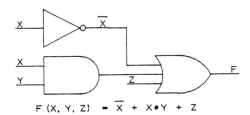

$$F(X, Y, Z) = \overline{X} + X \bullet Y + Z$$

1-16. (a) $F = (\overline{X} + Y) \bullet \overline{Z} + Z$

(b) $F = (W \bullet X \bullet \overline{Y} + Z) \bullet \overline{X}$

(c) $F = ((W + X + Y) \bullet Z) + \overline{W}$

1-17. (a) $F = X \bullet Y + Z$

(b) $F = (W + X) \bullet (Y + Z)$

(c) $F = (X \bullet Y) \bullet \overline{Z} + X$

1-18.

(a)

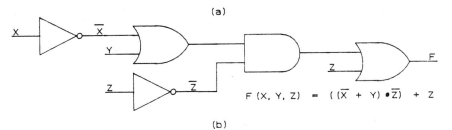

$$F(X, Y, Z) = ((\overline{X} + Y) \bullet \overline{Z}) + Z$$

(b)

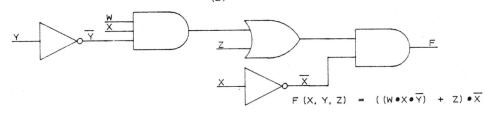

$$F(X, Y, Z) = ((W \bullet X \bullet \overline{Y}) + Z) \bullet \overline{X}$$

(c)

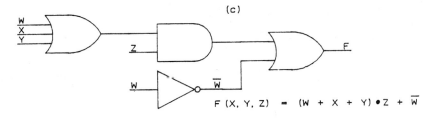

$$F(X, Y, Z) = (W + X + Y) \bullet Z + \overline{W}$$

1-19. (a) $F(X,Y) = X + X \cdot Y$ (b) $F(X,Y,Z) = X \cdot Y + X \cdot Z$

X	Y	F
0	0	0
0	1	0
1	0	1
1	1	1

X	Y	Z	F
0	0	0	0
0	0	1	0
0	1	0	0
0	1	1	0
1	0	0	0
1	0	1	1
1	1	0	1
1	1	1	1

(c) $F(X,Y,Z) = X \cdot Y \cdot \overline{Z} + \overline{X} \cdot Y + Z$

X	Y	Z	F
0	0	0	0
0	0	1	1
0	1	0	1
0	1	1	1
1	0	0	0
1	0	1	1
1	1	0	1
1	1	1	1

Section 1-5   Introduction to Logic Symbols

1-20.

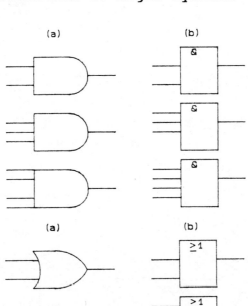

1-21.

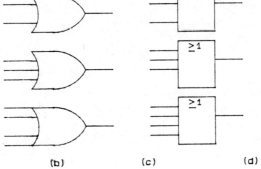

1-22.

Section 1-6 Boolean Algebra Theorems

1-23. $X \cdot X = X$ Idempotency Theorem

$X = X \cdot 1$ by P2b
$X = X \cdot (X + \overline{X})$ by P5a

$X = X \cdot X + X \cdot \overline{X}$ by P4b
$X = X \cdot X + 0$ by P5b
$X = X \cdot X$ by P2a

1-24. $X \cdot 0 = 0$ Identity Element Theorem

$X \cdot 0 = X \cdot 0 + 0$ by P2a

$X \cdot 0 = X \cdot 0 + X \cdot \overline{X}$ by P5b

$X \cdot 0 = X \cdot (0 + \overline{X})$ by P4b

$X \cdot 0 = X \cdot \overline{X}$ by P2a
$X \cdot 0 = 0$ by P5b

1-25. $X \cdot (X + Y) = X$ Absorption Theorem

$X \cdot (X + Y) = (X + 0) \cdot (X + Y)$ by P2a
$X \cdot (X + Y) = X + 0 \cdot Y$ by P4a
$X \cdot (X + Y) = X + Y \cdot 0$ by P3b

but, $Y \cdot 0 = Y \cdot 0 + 0$ by P2a

$Y \cdot 0 = Y \cdot 0 + Y \cdot \overline{Y}$ by P5b

$Y \cdot 0 = Y \cdot (0 + \overline{Y})$ by P4b

$Y \cdot 0 = Y \cdot (\overline{Y})$ by P2a

$Y \cdot 0 = 0$ by P5b

therefore,

$X \cdot (X + Y) = X + 0$ by intermediate proof
$X \cdot (X + Y) = X$ by P2a

1-26. $\overline{X \cdot Y} = \overline{X} + \overline{Y}$ DeMorgan's Theorem

Since the postulates do not have a complement over multiple variables, the proof is obtained indirectly by proving the variables satisfy P5a and P5b as follows.

First prove $W + \overline{W} = 1$, where $W = X \cdot Y$ and $\overline{W} = \overline{X} + \overline{Y}$

$X \cdot Y + \overline{X} + \overline{Y} = 1$ by P5a

5

$$X \cdot Y + X \cdot Y + \overline{X} + \overline{Y} = 1 \qquad \text{by T2a (Idempotency Theorem)}$$

$$\overline{X} + X \cdot Y + \overline{Y} + X \cdot Y = 1 \qquad \text{by P3a}$$

$$\overline{X} + Y + \overline{Y} + X = 1 \qquad \text{by T9a (Simplification Theorem)}$$

$$\overline{X} + X + \overline{Y} + Y = 1 \qquad \text{by P3a}$$

$$1 + 1 = 1 \qquad \text{by P5a}$$

$$1 = 1 \qquad \text{by the OR property or T3a with X = 1}$$

Second prove $W \cdot \overline{W} = 0$, where $W = X \cdot Y$ and $\overline{W} = \overline{X} + \overline{Y}$

$$(X \cdot Y) \cdot (\overline{X} + \overline{Y}) = 0 \qquad \text{by P5a}$$

$$X \cdot Y \cdot \overline{X} + X \cdot Y \cdot \overline{Y} = 0 \qquad \text{by P4b}$$

$$X \cdot \overline{X} \cdot Y + X \cdot Y \cdot \overline{Y} = 0 \qquad \text{by P3b}$$

$$0 \cdot Y + X \cdot 0 = 0 \qquad \text{by P5b}$$

$$0 + 0 = 0 \qquad \text{by T3b (Identity Element Theorem)}$$

$$0 = 0 \qquad \text{by the OR property or P2a with X = 0}$$

1-27. $X + (Y + Z) = (X + Y) + Z$ Associative Theorem

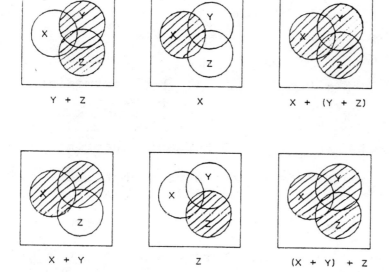

Therefore, $X + (Y + Z) = (X + Y) + Z$ by graphical observation.

6

1-28. $X + X \cdot Y = X$ Absorption Theorem

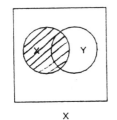 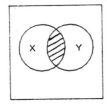 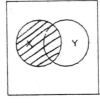

| X | X•Y | X + X•Y |

Therefore, $X + X \cdot Y = X$ by graphical observation.

1-29. $(X + Y) \cdot (X + \overline{Y}) = X$ Adjacency Theorem

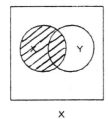 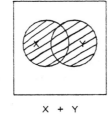 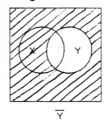

| X | X + Y | $\overline{Y}$ |

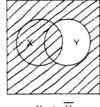 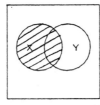

| $X + \overline{Y}$ | $(X + Y) \bullet (X + \overline{Y})$ |

Therefore, $(X + Y) \bullet (X + \overline{Y}) = X$ by graphical observation.

1-30. $X \cdot (X + Y) = X$ Absorption Theorem

X Y	X + Y	X·(X + Y)
0 0	0	0
0 1	1	0
1 0	1	1
1 1	1	1

Columns $X \cdot (X + Y)$ and X are identical; therefore, the two expressions are equal by perfect induction.

1-31. $X \cdot (Y \cdot Z) = (X \cdot Y) \cdot Z$ Associative Theorem

X Y Z	Y·Z	X·(Y·Z)	X·Y	(X·Y)·Z
0 0 0	0	0	0	0
0 0 1	0	0	0	0
0 1 0	0	0	0	0
0 1 1	1	0	0	0
1 0 0	0	0	0	0
1 0 1	0	0	0	0
1 1 0	0	0	1	0
1 1 1	1	1	1	1

7

Columns X·(Y·Z) and (X·Y)·Z are identical; therefore, the two expressions are equal by perfect induction.

1-32. $(X + Y) \cdot (\overline{X} + Z) \cdot (Y + Z) = (X + Y) \cdot (\overline{X} + Z)$ Consensus Theorem

X Y Z	X+Y	$\overline{X}$	$\overline{X}$+Z	Y+Z	$(X+Y) \cdot (\overline{X}+Z)$	$(X+Y) \cdot (\overline{X}+Z) \cdot (Y+Z)$
0 0 0	0	1	1	0	0	0
0 0 1	0	1	1	1	0	0
0 1 0	1	1	1	1	1	1
0 1 1	1	1	1	1	1	1
1 0 0	1	0	0	0	0	0
1 0 1	1	0	1	1	1	1
1 1 0	1	0	0	1	0	0
1 1 1	1	0	1	1	1	1

Columns $(X + Y) \cdot (\overline{X} + Z)$ and $(X + Y) \cdot (\overline{X} + Z) \cdot (Y + Z)$ are identical; therefore, the two expressions are equal by perfect induction.

1-33. $X + \overline{X} \cdot Y = X + Y$ Simplification Theorem

X Y	$\overline{X}$	$\overline{X} \cdot Y$	$X + \overline{X} \cdot Y$	X + Y
0 0	1	0	0	0
0 1	1	1	1	1
1 0	0	0	1	1
1 1	0	0	1	1

Columns $X + \overline{X} \cdot Y$ and $X + Y$ are identical; therefore, the two expressions are equal by perfect induction.

1-34. $\overline{X} \cdot Y + X \cdot \overline{Y} = 1$ Is this expression true?

X Y	$\overline{X}$	$\overline{X} \cdot Y$	$\overline{Y}$	$X \cdot \overline{Y}$	$\overline{X} \cdot Y + X \cdot \overline{Y}$	1
0 0	1	0	1	0	0	1
0 1	1	1	0	0	1	1
1 0	0	0	1	1	1	1
1 1	0	0	0	0	0	1

Columns $\overline{X} \cdot Y + X \cdot \overline{Y}$ and 1 are not identical; therefore, the two expressions are not equal by perfect induction.

1-35. $Y \cdot Z + X \cdot Y + \overline{X} \cdot Y = Y$ Is this expression true?

X Y Z	Y·Z	X·Y	$\overline{X}$	$\overline{X} \cdot Y$	$Y \cdot Z + X \cdot Y + \overline{X} \cdot Y$
0 0 0	0	0	1	0	0
0 0 1	0	0	1	0	0
0 1 0	0	0	1	1	1
0 1 1	1	0	1	1	1
1 0 0	0	0	0	0	0
1 0 1	0	0	0	0	0
1 1 0	0	1	0	0	1
1 1 1	1	1	0	0	1

Columns $Y \cdot Z + X \cdot Y + \overline{X} \cdot Y$ and Y are identical; therefore, the two expressions are equal by perfect induction.

Section 1-7 Minimizing Boolean Functions Algebraically

1-36. (a) $F(A,B) = A \cdot B + A \cdot \overline{B}$
$$= A \qquad \text{by Adjacency Theorem T7a}$$

(b) $F(A,B,C) = A \cdot \overline{B} \cdot C + B \cdot C + B \cdot \overline{C}$

$$= A \cdot \overline{B} \cdot C + B \quad \text{by Adjacency Theorem T7a}$$

$$= A \cdot C + B \qquad \text{by Simplification Theorem T9a}$$

(c) $F(X,Y,Z) = \overline{Y} \cdot \overline{Z} + \overline{X} \cdot Y + \overline{X} \cdot Y \cdot Z + X \cdot Y \cdot \overline{Z}$

$$= \overline{Z} \cdot (\overline{Y} + X \cdot Y) + \overline{X} \cdot Y \cdot (1 + Z)$$
$$\text{by Postulates P3a and P4b}$$

$$= \overline{Z} \cdot (\overline{Y} + X \cdot Y) + \overline{X} \cdot Y \quad \text{by Identity Element Theorem T3a and P2b}$$

$$= \overline{Z} \cdot (\overline{Y} + X) + \overline{X} \cdot Y \quad \text{by Simplification Theorem T9a}$$

(This is not a minimum form.)

$$= \overline{Z} \cdot \overline{(Y \cdot \overline{X})} + \overline{X} \cdot Y \quad \text{by DeMorgan's Theorem T6b}$$

$$= \overline{Z} + \overline{X} \cdot Y \qquad \text{by Simplification Theorem T9a}$$

(This is a minimum form.)

1-37. (a) $F(X,Y,Z) = (\overline{Y} + \overline{Z}) \cdot (\overline{X} + Y) \cdot (\overline{X} + Y + Z)$

$$\cdot (X + Y + \overline{Z})$$

$$= (\overline{Z} + \overline{Y} \cdot (X + Y)) \cdot (\overline{X} + Y \cdot (Y + Z))$$
$$\text{by Postulates P3b and P4a}$$

$$= (\overline{Z} + \overline{Y} \cdot X + \overline{Y} \cdot Y) \cdot (\overline{X} + Y \cdot Y + Y \cdot Z)$$
$$\text{by Postulate P4b}$$

$$= (\overline{Z} + \overline{Y} \cdot X) \cdot (\overline{X} + Y \cdot Y + Y \cdot Z)$$
$$\text{by Postulate P5b}$$

$$= (\overline{Z} + \overline{Y} \cdot X) \cdot (\overline{X} + Y + Y \cdot Z)$$
$$\text{by Idempotency Theorem T2b}$$

$$= (\overline{Z} + \overline{Y} \cdot X) \cdot (\overline{X} + Y)$$
by Absorption Theorem T4a

$$= (\overline{Z} + X \cdot \overline{Y}) \cdot (\overline{X} + Y)$$
by Postulate P3b

$$= (\overline{Z} + \overline{\overline{X} + Y}) \cdot (\overline{X} + Y)$$
by DeMorgan's Theorem T6a

$$= \overline{Z} \cdot (\overline{X} + Y)$$
by Simplification Theorem T9b

(b) $F(A,B,C) = (A + B) \cdot (C + B) \cdot (A + \overline{B} + \overline{C})$

$$= (A + B + C) \cdot (A + B + \overline{C}) \cdot (A + C + B)$$
$$\cdot (\overline{A} + C + B) \cdot (A + \overline{B} + \overline{C})$$
by Adjacency Theorem T7b

$$= (A + B + C) \cdot (\overline{A} + B + C) \cdot (A + B + \overline{C})$$
$$\cdot (A + \overline{B} + \overline{C})$$
by Postulates P3a and b, and Idempotency Theorem T2b

$$= (B + C) \cdot (A + \overline{C})$$
by Adjacency Theorem T7b

(c) $F(A,B,C) = (A + \overline{B} + C) \cdot (B + C) \cdot (B + \overline{C})$

$$= (A + \overline{B} + C) \cdot B$$
by Adjacency Theorem T7b

$$= (A + C) \cdot B$$
by Simplification Theorem T9b

1-38. (a) $F(X,Y,Z) = Y + \overline{X} \cdot \overline{Z}$

$$= X \cdot Y + \overline{X} \cdot Y + \overline{X} \cdot Y \cdot \overline{Z} + \overline{X} \cdot \overline{Y} \cdot \overline{Z}$$
by Adjacency Theorem T7a

$$= X \cdot Y \cdot Z + X \cdot Y \cdot \overline{X} + \overline{X} \cdot Y \cdot Z + \overline{X} \cdot Y \cdot \overline{Z}$$
$$+ \overline{X} \cdot Y \cdot \overline{Z} + \overline{X} \cdot \overline{Y} \cdot \overline{Z}$$
by Adjacency Theorem T7a

(b) $F(A,B,C) = B + C$

$$= A \cdot B + \overline{A} \cdot B + A \cdot C + \overline{A} \cdot C$$
by Adjacency Theorem T7a

$$= A \cdot B \cdot C + A \cdot B \cdot \overline{C} + \overline{A} \cdot B \cdot C + \overline{A} \cdot B \cdot \overline{C}$$

$$+ A \cdot B \cdot C + A \cdot \overline{B} \cdot C + \overline{A} \cdot B \cdot C + \overline{A} \cdot \overline{B} \cdot C$$

by Adjacency Theorem T7a

(c) $F(M,R,S) = \overline{S} + \overline{M} \cdot R$

$$= M \cdot \overline{S} + \overline{M} \cdot \overline{S} + \overline{M} \cdot R \cdot S + \overline{M} \cdot R \cdot \overline{S}$$

by Adjacency Theorem T7a

$$= M \cdot R \cdot \overline{S} + M \cdot \overline{R} \cdot \overline{S} + \overline{M} \cdot R \cdot \overline{S} + \overline{M} \cdot \overline{R} \cdot \overline{S}$$

$$+ \overline{M} \cdot R \cdot S + \overline{M} \cdot R \cdot \overline{S}$$

by Adjacency Theorem T7a

1-39. (a) $F(X,Y,Z) = (Y) \cdot (\overline{X} + \overline{Z})$

$$= (X + Y) \cdot (\overline{X} + Y) \cdot (\overline{X} + Y + \overline{Z})$$

$$\cdot (\overline{X} + \overline{Y} + \overline{Z})$$

by Adjacency Theorem T7b

$$= (X + Y + Z) \cdot (X + Y + \overline{Z}) \cdot (\overline{X} + Y + Z)$$

$$\cdot (\overline{X} + Y + \overline{Z}) \cdot (\overline{X} + Y + \overline{Z}) \cdot (\overline{X} + \overline{Y} + \overline{Z})$$

by Adjacency Theorem T7b

(b) $F(A,B,C) = (B) \cdot (C)$

$$= (A + B) \cdot (\overline{A} + B) \cdot (A + C) \cdot (A + \overline{C})$$

by Adjacency Theorem T7b

$$= (A + B + C) \cdot (A + B + \overline{C}) \cdot (\overline{A} + B + C)$$

$$\cdot (\overline{A} + B + \overline{C}) \cdot (A + B + C) \cdot (A + \overline{B} + C)$$

$$\cdot (\overline{A} + B + C) \cdot (\overline{A} + \overline{B} + C)$$

by Adjacency Theorem T7b

(c) $F(M,R,S) = \overline{S} \cdot (\overline{M} + R)$

$$= (M + \overline{S}) \cdot (\overline{M} + \overline{S}) \cdot (\overline{M} + R + S)$$

$$\cdot (\overline{M} + R + \overline{S})$$

by Adjacency Theorem T7b

$$= (M + R + \overline{S}) \cdot (M + \overline{R} + \overline{S}) \cdot (\overline{M} + R + \overline{S})$$

$$\cdot (\overline{M} + \overline{R} + \overline{S}) \cdot (\overline{M} + R + S) \cdot (\overline{M} + R + \overline{S})$$

by Adjacency Theorem T7b

1-40. (a) $F(W,X,Y,Z) = \overline{W} \cdot \overline{X} \cdot Y \cdot Z + X \cdot Y \cdot Z + W \cdot X \cdot Z$

$= \overline{W} \cdot \overline{X} \cdot Y \cdot Z + W \cdot X \cdot Y \cdot Z + \overline{W} \cdot X \cdot Y \cdot Z$

$+ W \cdot X \cdot Y \cdot Z + W \cdot X \cdot \overline{Y} \cdot Z$
   by Adjacency Theorem T7a

$= \overline{W} \cdot Y \cdot Z + W \cdot X \cdot Z$
   by Idempotency Theorem T2b, P3a,
   and Adjacency Theorem T7a

(b) $F(A,B,C,D) = A \cdot \overline{B} \cdot \overline{D} + A \cdot B \cdot C \cdot \overline{D} + \overline{A} \cdot C \cdot \overline{D}$

$= A \cdot \overline{B} \cdot C \cdot \overline{D} + A \cdot \overline{B} \cdot \overline{C} \cdot \overline{D} + A \cdot B \cdot C \cdot \overline{D}$

$+ \overline{A} \cdot B \cdot C \cdot \overline{D} + \overline{A} \cdot \overline{B} \cdot C \cdot \overline{D}$
   by Adjacency Theorem T7a

$= A \cdot \overline{B} \cdot C \cdot \overline{D} + A \cdot \overline{B} \cdot C \cdot \overline{D} + A \cdot \overline{B} \cdot \overline{C} \cdot \overline{D}$

$+ A \cdot B \cdot C \cdot \overline{D} + \overline{A} \cdot B \cdot C \cdot \overline{D} + \overline{A} \cdot \overline{B} \cdot C \cdot \overline{D}$
   by Idempotency Theorem T2a

$= A \cdot \overline{B} \cdot C \cdot \overline{D} + A \cdot \overline{B} \cdot \overline{C} \cdot \overline{D} + A \cdot \overline{B} \cdot C \cdot \overline{D}$

$+ A \cdot B \cdot C \cdot \overline{D} + \overline{A} \cdot B \cdot C \cdot \overline{D} + \overline{A} \cdot \overline{B} \cdot C \cdot \overline{D}$
   by Postulate P3a

$= A \cdot \overline{B} \cdot \overline{D} + (A \cdot (\overline{B} + B) + \overline{A} \cdot (B + \overline{B})) \cdot C \cdot \overline{D}$
   by Adjacency Theorem T7a and P4b

$= A \cdot \overline{B} \cdot \overline{D} + (A \cdot 1 + \overline{A} \cdot 1) \cdot C \cdot \overline{D}$
   by Postulate P5a

$= A \cdot \overline{B} \cdot \overline{D} + (A + \overline{A}) \cdot C \cdot \overline{D}$
   by Postulate P2b

$= A \cdot \overline{B} \cdot \overline{D} + 1 \cdot C \cdot \overline{D}$
   by Postulate P5a

$= A \cdot \overline{B} \cdot \overline{D} + C \cdot \overline{D}$
   by Postulate P2b

(c) $F(W,X,Y,Z) = W \cdot \overline{X} \cdot \overline{Y} \cdot \overline{Z} + W \cdot X \cdot Y \cdot Z + \overline{W} \cdot \overline{X} \cdot Y \cdot \overline{Z}$
   This is a minimum form and cannot be reduced further.

Section 1-8   Canonical or Standard Forms for Boolean Functions

1-41. $F1(X,Y,Z) = \Sigma\, m(3,4,5,7)$
   $F2(X,Y,Z) = \Sigma\, m(1,3,4,6)$
   $F3(X,Y,Z) = \Sigma\, m(0,1,2,5)$
   $F4(X,Y,Z) = \Sigma\, m(2,3,4,5,6)$

12

1-42. (a) $F(X,Y) = \Sigma\, m(1,2,3)$
$$= m_1 + m_2 + m_3$$
$$= \overline{X}\cdot Y + X\cdot\overline{Y} + X\cdot Y$$

(b) $F(X,Y,Z) = \Sigma\, m(0,5,6,7)$
$$= m_0 + m_5 + m_6 + m_7$$
$$= \overline{X}\cdot\overline{Y}\cdot\overline{Z} + X\cdot\overline{Y}\cdot Z + X\cdot Y\cdot\overline{Z} + X\cdot Y\cdot Z$$

(c) $F(W,X,Y,Z) = \Sigma\, m(7,10,12,14,15)$
$$= m_7 + m_{10} + m_{12} + m_{14} + m_{15}$$
$$= \overline{W}\cdot X\cdot Y\cdot Z + W\cdot\overline{X}\cdot Y\cdot\overline{Z} + W\cdot X\cdot\overline{Y}\cdot\overline{Z}$$
$$+ W\cdot X\cdot Y\cdot\overline{Z} + W\cdot X\cdot Y\cdot Z$$

(d) $F(A,B,C,D) = \Sigma\, m(3,6,9,11,14)$
$$= m_3 + m_6 + m_9 + m_{11} + m_{14}$$
$$= \overline{A}\cdot\overline{B}\cdot C\cdot D + \overline{A}\cdot B\cdot C\cdot\overline{D} + A\cdot\overline{B}\cdot\overline{C}\cdot D$$
$$= A\cdot\overline{B}\cdot C\cdot D + A\cdot B\cdot C\cdot\overline{D}$$

1-43. $F1(X,Y,Z) = \pi\, M(0,1,2,6)$
$F2(X,Y,Z) = \pi\, M(0,2,5,7)$
$F3(X,Y,Z) = \pi\, M(3,4,6,7)$
$F4(X,Y,Z) = \pi\, M(0,1,7)$

1-44. (a) $F(X,Y) = \pi\, M(0)$
(b) $F(X,Y,Z) = \pi\, M(1,2,3,4)$
(c) $F(W,X,Y,Z) = \pi\, M(0,1,2,3,4,5,6,8,9,11,13)$
(d) $F(A,B,C,D) = \pi\, M(0,1,2,4,5,7,8,10,12,13,15)$

1-45. (a) $F(X,Y) = \pi\, M(0,2)$
$$= M_0\cdot M_2$$
$$= (X + Y)\cdot(\overline{X} + Y)$$

(b) $F(X,Y,Z) = \pi\, M(3,5,7)$
$$= M_3\cdot M_5\cdot M_7$$
$$= (X + \overline{Y} + \overline{Z})\cdot(\overline{X} + Y + \overline{Z})\cdot(\overline{X} + \overline{Y} + \overline{Z})$$

(c) $F(W,X,Y,Z) = \pi\, M(2,4,6,8,10)$
$$= M_2\cdot M_4\cdot M_6\cdot M_8\cdot M_{10}$$
$$= (W + X + \overline{Y} + Z)\cdot(W + \overline{X} + Y + Z)$$
$$\cdot(W + \overline{X} + \overline{Y} + Z)\cdot(\overline{W} + X + Y + Z)$$
$$\cdot(\overline{W} + X + \overline{Y} + Z)$$

(d) $F(A,B,C,D) = \pi M(1,3,5,6,8,9,13,14)$
$= M_1 \cdot M_3 \cdot M_5 \cdot M_6 \cdot M_8 \cdot M_9 \cdot M_{13} \cdot M_{14}$

$= (A + B + C + \overline{D}) \cdot (A + B + \overline{C} + \overline{D})$

$\cdot (A + \overline{B} + C + \overline{D}) \cdot (A + \overline{B} + \overline{C} + D)$

$\cdot (\overline{A} + B + C + D) \cdot (\overline{A} + B + C + \overline{D})$

$\cdot (\overline{A} + \overline{B} + C + \overline{D}) \cdot (A + B + C + \overline{D})$

1-46.

W X Y Z	Minterms	Maxterms
0 0 0 0	$\overline{W} \cdot \overline{X} \cdot \overline{Y} \cdot \overline{Z}$	$W + X + Y + Z$
0 0 0 1	$\overline{W} \cdot \overline{X} \cdot \overline{Y} \cdot Z$	$W + X + Y + \overline{Z}$
0 0 1 0	$\overline{W} \cdot \overline{X} \cdot Y \cdot \overline{Z}$	$W + X + \overline{Y} + Z$
0 0 1 1	$\overline{W} \cdot \overline{X} \cdot Y \cdot Z$	$W + X + \overline{Y} + \overline{Z}$
0 1 0 0	$\overline{W} \cdot X \cdot \overline{Y} \cdot \overline{Z}$	$W + \overline{X} + Y + Z$
0 1 0 1	$\overline{W} \cdot X \cdot \overline{Y} \cdot Z$	$W + \overline{X} + Y + \overline{Z}$
0 1 1 0	$\overline{W} \cdot X \cdot Y \cdot \overline{Z}$	$W + \overline{X} + \overline{Y} + Z$
0 1 1 1	$\overline{W} \cdot X \cdot Y \cdot Z$	$W + \overline{X} + \overline{Y} + \overline{Z}$
1 0 0 0	$W \cdot \overline{X} \cdot \overline{Y} \cdot \overline{Z}$	$\overline{W} + X + Y + Z$
1 0 0 1	$W \cdot \overline{X} \cdot \overline{Y} \cdot Z$	$\overline{W} + X + Y + \overline{Z}$
1 0 1 0	$W \cdot \overline{X} \cdot Y \cdot \overline{Z}$	$\overline{W} + X + \overline{Y} + Z$
1 0 1 1	$W \cdot \overline{X} \cdot Y \cdot Z$	$\overline{W} + X + \overline{Y} + \overline{Z}$
1 1 0 0	$W \cdot X \cdot \overline{Y} \cdot \overline{Z}$	$\overline{W} + \overline{X} + Y + Z$
1 1 0 1	$W \cdot X \cdot \overline{Y} \cdot Z$	$\overline{\overline{W}} + \overline{X} + Y + \overline{Z}$
1 1 1 0	$W \cdot X \cdot Y \cdot \overline{Z}$	$\overline{W} + \overline{X} + \overline{Y} + Z$
1 1 1 1	$W \cdot X \cdot Y \cdot Z$	$\overline{W} + \overline{X} + \overline{Y} + \overline{Z}$

1-47. (a) $\overline{F}(P,Q,R,S) = \Sigma m(0,5,9,13)$
$= m_0 + m_5 + m_9 + m_{13}$

$F(P,Q,R,S) = \overline{m_0} \cdot \overline{m_5} \cdot \overline{m_9} \cdot \overline{m_{13}}$
$= M_0 \cdot M_5 \cdot M_9 \cdot M_{13}$
$= \pi M(0,5,9,13)$
Standard POS form of F
$F(P,Q,R,S) = \Sigma m(1,2,3,4,6,7,8,10,11,12,14,15)$
Standard SOP form of F

(b) $\overline{F}(D,C,B,A) = \Sigma\ m(3,6,8,14)$
$= m_3 + m_6 + m_8 + m_{14}$

$F(D,C,B,A) = \overline{m_3} \cdot \overline{m_6} \cdot \overline{m_8} \cdot \overline{m_{14}}$
$= M_3 \cdot M_6 \cdot M_8 \cdot M_{14}$
$= \pi\ M(3,6,8,14)$
              Standard POS form of F
$F(D,C,B,A) = \Sigma\ m(0,1,2,4,5,7,9,10,11,12,13,15)$
              Standard SOP form of F

(c) $\overline{F}(W,X,Y,Z) = \Sigma\ m(4,5,6,7)$
$= m_4 + m_5 + m_6 + m_7$

$F(W,X,Y,Z) = \overline{m_4} \cdot \overline{m_5} \cdot \overline{m_6} \cdot \overline{m_7}$
$= M_4 \cdot M_5 \cdot M_6 \cdot M_7$
$= \pi\ M(4,5,6,7)$
              Standard POS form of F
$F(W,X,Y,Z) = \Sigma\ m(0,1,2,3,8,9,10,11,12,13,14,15)$
              Standard SOP form of F

(d) $\overline{F}(P,Q,R,S) = \Sigma\ m(12,13,14,15)$
$F(P,Q,R,S) = \pi\ M(12,13,14,15)$
              Standard POS form of F
$F(P,Q,R,S) = \Sigma\ m(0,1,2,3,4,5,6,7,8,9,10,11)$
              Standard SOP form of F

1-48. (a) $\overline{F}(J,K,L,M) = \pi\ M(7,9,12,15)$
$= M_7 \cdot M_9 \cdot M_{12} \cdot M_{15}$

$F(J,K,L,M) = \overline{M_7} + \overline{M_9} + \overline{M_{12}} + \overline{M_{15}}$
$= m_7 + m_9 + m_{12} + m_{15}$
$= \Sigma\ m(7,9,12,15)$
              Standard SOP form of F
$F(J,K,L,M) = \pi\ M(0,1,2,3,4,5,6,8,10,11,13,14)$
              Standard POS form of F

(b) $\overline{F}(B,J,M,Z) = \pi\ M(3,6,8,14)$
$= M_3 \cdot M_6 \cdot M_8 \cdot M_{14}$

$F(B,J,M,Z) = \overline{M_3} + \overline{M_6} + \overline{M_8} + \overline{M_{14}}$
$= m_3 + m_6 + m_8 + m_{14}$
$= \Sigma\ m(3,6,8,14)$
              Standard SOP form of F
$F(B,J,M,Z) = \pi\ M(0,1,2,4,5,7,9,10,11,12,13,15)$
              Standard POS form of F

(c) $\overline{F}(W,X,Y,Z) = \pi\ M(4,7,9,13)$
$F(W,X,Y,Z) = \Sigma\ m(4,7,9,13)$
              Standard SOP form of F
$F(W,X,Y,Z) = \pi\ M(0,1,2,3,5,6,8,10,11,12,14,15)$
              Standard POS form of F

(d) $\overline{F}(U,T,P,W) = \pi\, M(0,3,7,9,14)$

$F(U,T,P,W) = \Sigma\, m(0,3,7,9,14)$
                        Standard SOP form of F

$F(U,T,P,W) = \pi\, M(1,2,4,5,6,8,10,11,12,13,15)$
                        Standard POS form of F

1-49. (a) $F(J,K,L,M) = \pi\, M(7,9,12,15)$

(b) $\overline{F}(B,J,M,Z) = \pi\, M(3,6,8,14)$

J	K	L	M	F
0	0	0	0	1
0	0	0	1	1
0	0	1	0	1
0	0	1	1	1
0	1	0	0	1
0	1	0	1	1
0	1	1	0	1
0	1	1	1	0
1	0	0	0	1
1	0	0	1	0
1	0	1	0	1
1	0	1	1	1
1	1	0	0	0
1	1	0	1	1
1	1	1	0	1
1	1	1	1	0

B	J	M	Z	F
0	0	0	0	0
0	0	0	1	0
0	0	1	0	0
0	0	1	1	1
0	1	0	0	0
0	1	0	1	0
0	1	1	0	1
0	1	1	1	0
1	0	0	0	1
1	0	0	1	0
1	0	1	0	0
1	0	1	1	0
1	1	0	0	0
1	1	0	1	0
1	1	1	0	1
1	1	1	1	0

(c) $F(W,X,Y,Z) = \Sigma\, m(4,7,9\ 13)$

(d) $\overline{F}(U,T,P,W) = \Sigma\, m(0,3,7,9,14)$

W	X	Y	Z	F
0	0	0	0	0
0	0	0	1	0
0	0	1	0	0
0	0	1	1	0
0	1	0	0	1
0	1	0	1	0
0	1	1	0	0
0	1	1	1	1
1	0	0	0	0
1	0	0	1	1
1	0	1	0	0
1	0	1	1	0
1	1	0	0	0
1	1	0	1	1
1	1	1	0	0
1	1	1	1	0

U	T	P	W	F
0	0	0	0	0
0	0	0	1	1
0	0	1	0	1
0	0	1	1	0
0	1	0	0	1
0	1	0	1	1
0	1	1	0	1
0	1	1	1	0
1	0	0	0	1
1	0	0	1	0
1	0	1	0	1
1	0	1	1	1
1	1	0	0	1
1	1	0	1	1
1	1	1	0	0
1	1	1	1	1

1-50. (a) $F(V,W,X,Y,Z) = m_0 + m_6 + m_{23}$

$$= \overline{V} \cdot \overline{W} \cdot \overline{X} \cdot \overline{Y} \cdot \overline{Z} + \overline{V} \cdot \overline{W} \cdot X \cdot Y \cdot \overline{Z} + V \cdot \overline{W} \cdot X \cdot Y \cdot Z$$

(b) $F(U,V,W,X,Y,Z) = m_{19} + m_{22} + m_{30}$

$$= \overline{U} \cdot V \cdot \overline{W} \cdot \overline{X} \cdot Y \cdot Z + \overline{U} \cdot V \cdot \overline{W} \cdot X \cdot Y \cdot \overline{Z}$$

$$+ \overline{U} \cdot V \cdot W \cdot X \cdot Y \cdot \overline{Z}$$

(c) $F(U,V,W,X,Y,Z) = m_{25} + m_{35} + m_{47}$

$$= \overline{U} \cdot V \cdot W \cdot \overline{X} \cdot \overline{Y} \cdot Z + U \cdot \overline{V} \cdot \overline{W} \cdot \overline{X} \cdot Y \cdot Z$$

$$+ \overline{U} \cdot V \cdot \overline{W} \cdot \overline{X} \cdot \overline{Y} \cdot \overline{Z}$$

(d) $F(T,U,V,W,X,Y,Z) = m_{55} + m_{67} + m_{93}$

$$= \overline{T} \cdot U \cdot V \cdot \overline{W} \cdot X \cdot Y \cdot Z + T \cdot \overline{U} \cdot \overline{V} \cdot \overline{W} \cdot \overline{X} \cdot Y \cdot Z$$

$$+ T \cdot \overline{U} \cdot V \cdot W \cdot X \cdot \overline{Y} \cdot Z$$

1-51. (a) $F(U,V,W,X,Y,Z) = M_0 \cdot M_6 \cdot M_{23}$
$$= (U + V + W + X + Y + Z)$$

$$\cdot (U + V + W + \overline{X} + \overline{Y} + Z)$$

$$\cdot (U + \overline{V} + W + \overline{X} + \overline{Y} + \overline{Z})$$

(b) $F(U,V,W,X,Y,Z) = M_5 \cdot M_9 \cdot M_{43}$

$$= (U + V + W + \overline{X} + Y + \overline{Z})$$

$$\cdot (U + V + \overline{W} + X + Y + \overline{Z})$$

$$\cdot (\overline{U} + V + \overline{W} + X + \overline{Y} + \overline{Z})$$

(c) $F(U,V,W,X,Y,Z) = M_7 \cdot M_{26} \cdot M_{56}$

$$= (U + V + W + \overline{X} + \overline{Y} + \overline{Z})$$

$$\cdot (U + \overline{V} + \overline{W} + X + \overline{Y} + Z)$$

$$\cdot (\overline{U} + \overline{V} + \overline{W} + X + Y + Z)$$

(d) $F(U,V,W,X,Y,Z) = M_{19} \cdot M_{36} \cdot M_{62}$

$$= (U + \overline{V} + W + X + \overline{Y} + \overline{Z})$$

$$\cdot (\overline{U} + V + W + \overline{X} + Y + Z)$$

$$\cdot (\overline{U} + \overline{V} + \overline{W} + X + \overline{Y} + Z)$$

1-52. (a) $F1(X,Y) = \Sigma\, m(0) = \overline{X} \cdot \overline{Y} = \overline{\overline{\overline{X} \cdot \overline{Y}}} = \overline{X + Y}$

NOR function

(b) $F7(X,Y) = \Sigma\ m(0,1,2) = \overline{X}\cdot\overline{Y} + \overline{X}\cdot Y + X\cdot\overline{Y}$

$$= \overline{X}\cdot(\overline{Y} + Y) + X\cdot\overline{Y}$$

$$= \overline{X} + X\cdot\overline{Y} = \overline{X} + \overline{Y}$$

$$= \overline{\overline{\overline{X} + \overline{Y}}} = \overline{X\cdot Y} \quad \text{NAND function}$$

(c) $F8(X,Y) = \Sigma\ m(3) = X\cdot Y \quad \text{AND function}$

(d) $F14(X,Y) = \Sigma\ m(1,2,3) = \overline{X}\cdot Y + X\cdot\overline{Y} + X\cdot Y$

$$= \overline{X}\cdot Y + X\cdot(\overline{Y} + Y)$$

$$= \overline{X}\cdot Y + X = Y + X \quad \text{OR function}$$

1-53. $F = X\cdot Y = \overline{\overline{X\cdot Y}} = \overline{(\overline{X} + \overline{Y})}$ by Double Complementation Theorem T1 and DeMorgan's Theorem T6b

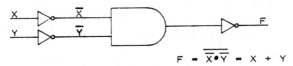

$F = \overline{\overline{X} + \overline{Y}} = X \bullet Y$

X Y	$\overline{X}$ $\overline{Y}$	$\overline{X} + \overline{Y}$	$\overline{\overline{X} + \overline{Y}}$	$X\cdot Y$
0 0	1 1	1	0	0
0 1	1 0	1	0	0
1 0	0 1	1	0	0
1 1	0 0	0	1	1

1-54. $F = X + Y = \overline{\overline{X + Y}} = \overline{(\overline{X}\cdot\overline{Y})}$ by Double Complementation Theorem T1 and DeMorgan's Theorem T6a

$F = \overline{\overline{X}\bullet\overline{Y}} = X + Y$

X Y	$\overline{X}$ $\overline{Y}$	$\overline{X}\cdot\overline{Y}$	$\overline{\overline{X}\cdot\overline{Y}}$	$X + Y$
0 0	1 1	1	0	0
0 1	1 0	0	1	1
1 0	0 1	0	1	1
1 1	0 0	0	1	1

1-55. First consider the complement operation since it is also use in both the OR and AND operations.

NOR

X Y	$\overline{X + Y}$
0 0	1
0 1	0
1 0	0
1 1	0

==>

	NOR	Complemented Input
X = Y	$\overline{X + Y}$	$\overline{X = Y}$
0	1	1
1	0	0

Next consider the OR operation

	NOR	Complemented NOR	OR
X Y	$\overline{X + Y}$	$\overline{\overline{X + Y}}$	$X + Y$
0 0	1	0	0
0 1	0	1	1
1 0	0	1	1
1 1	0	1	1

Now consider the AND operation

	Complemented Inputs		NOR	AND
X Y	$\overline{X}$	$\overline{Y}$	$\overline{\overline{X} + \overline{Y}}$	$X \cdot Y$
0 0	1	1	0	0
0 1	1	0	0	0
1 0	0	1	0	0
1 1	0	0	1	1

1-56. $F = \overline{X} = \overline{X \cdot X}$ by Idempotency Theorem T2b

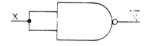

or

$F = \overline{X} = \overline{X \cdot 1}$ by Postulate P2b

Preferred method
(See Sec. 6-3)

$F = \overline{X} = \overline{X + X}$ by Idempotency Theorem T2a

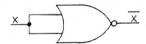

or

$F = \overline{X} = \overline{X + 0}$ by Postulate P2a

Preferred method
(See Sec. 6-3)

1-57. $F = X + Y = \overline{\overline{X + Y}} = \overline{(\overline{X} \cdot \overline{Y})}$

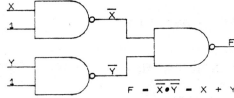

$F = \overline{\overline{X} \cdot \overline{Y}} = X + Y$

1-58. $F = X \cdot Y = \overline{\overline{X \cdot Y}} = \overline{(\overline{X \cdot Y})}$

$F = \overline{\overline{X \cdot Y}} = X \cdot Y$

Section 1-9 Specifying Designs Using Logic descriptions

1-59. (a)

X	Y	Z	F
0	0	0	0
0	0	1	0
0	1	0	0
0	1	1	1
1	0	0	0
1	0	1	1
1	1	0	1
1	1	1	0

(b) $F(X,Y,Z) = \Sigma\, m(3,5,6)$

(c) $F(X,Y,Z) = \pi\, M(0,1,2,4,7)$ Standard POS form of F

$$\overline{F}(X,Y,Z) = \Sigma\ m(0,1,2,4,7)\quad \text{Complement of the standard POS form of F}$$

1-60. (a)

X Y Z	F1	F2
0 0 0	0	1
0 0 1	0	0
0 1 0	0	0
0 1 1	0	0
1 0 0	0	0
1 0 1	0	0
1 1 0	0	0
1 1 1	1	1

(b) $F1(X,Y,Z) = \Sigma\ m(7)$
$F2(X,Y,Z) = \Sigma\ m(0,7)$

(c) $F1(X,Y,Z) = \pi\ M(0,1,2,3,4,5,6)$ Standard POS form of F1

$F2(X,Y,Z) = \pi\ M(1,2,3,4,5,6)$ Standard POS form of F2

$\overline{F1}(X,Y,Z) = \Sigma\ m(0,1,2,3,4,5,6)$ Complement of the standard POS form of F1

$\overline{F2}(X,Y,Z) = \Sigma\ m(1,2,3,4,5,6)$ Complement of the standard POS form of F2

1-61. In a computer programming if-then-else-form, the logic statement can be written as follows.

If (NOT X AND Y AND Z) OR (X AND NOT Y AND Z) OR (X AND Y AND NOT Z) then F = 1, else F = 0.

$$F = \overline{X}\cdot Y\cdot Z + X\cdot\overline{Y}\cdot Z + X\cdot Y\cdot\overline{Z}$$

1-62. $F = A + \overline{B}\cdot C\cdot\overline{D} + E$

Section 1-10 Number of Different Functions for n Independent Variables

1-63. $2^{2^{n}}$ or 2 raised to the power of 2 raised to power of n, where n is the number of independent variables. An abbreviated notation for this relationship is 2^2^n.
(a) 2^2^2 = 2^4 = 16 considering a function and its complement as separate functions, and 8 considering a function and its complement as only one function.

(b) $2^{2^3} = 2^8 = 256$ considering a function and its complement as separate functions, and 128 considering a function and its complement as only one function.

(c) $2^{2^4} = 2^{16} = 65,536$ considering a function and its complement as separate functions, and 32,768 considering a function and its complement as only one function.

(d) $2^{2^5} = 2^{32} = 4,294,967,296$ considering a function and its complement as separate functions, and 2,147,483,648 considering a function and its complement as only one function.

Section 2-2 Number systems

2-1. (a) IMICIX = -1 + 1000 - 1 + 100 - 1 + 10 = 1107
      (b) MCMLVI = 1000 - 100 + 1000 + 50 + 5 + 1 = 1956
      (c) CCMLVXII = -100 - 100 + 1000 + 50 - 5 + 10 + 1 +
          1 = 857 (A more concise way to write this number
          is CCMLVII)
      (d) MMLLCCXXII = 1000 + 1000 - 50 - 50 + 100 + 100
          + 10 + 10 + 1 + 1 = 2122 (A more concise way to
          write this number is MMCXXII)

2-2. (a) $23 = 2 \times 10^1 + 3 \times 10^0$
      (b) $4087 = 4 \times 10^3 + 0 \times 10^2 + 8 \times 10^1 + 7 \times 10^0$
      (c) $39.28 = 3 \times 10^1 + 9 \times 10^0 + 2 \times 10^{-1} + 8 \times 10^{-2}$

2-3. $(36)_8 = 3 \times 8^1 + 6 \times 8^0$
      $(E5.3)_{16} = 14 \times 16^1 + 5 \times 16^0 + 3 \times 16^{-1}$

Section 2-3 The Binary Number system

2-4. For binary number 10001001.0101,
      (a) bit 2 is 0
      (b) bit -2 is 1
      (c) bit 8 is 0 (leading 0)
      (d) bit -4 is 1
      (e) bit 4 is 0

2-5. (a) $(11011)_2 = 1 \times 2^4 + 1 \times 2^3 + 0 \times 2^2 + 1 \times 2^1 + 1 \times 2^0$
          $= 16 + 8 + 2 + 1 = (27)_{10}$
      (b) $(110110)_2 = 1 \times 2^5 + 1 \times 2^4 + 0 \times 2^3 + 1 \times 2^2 + 1 \times 2^1 +$
          $0 \times 2^0 = 32 + 16 + 4 + 2 = (54)_{10}$
      (c) $(1101101)_2 = 1 \times 2^6 + 1 \times 2^5 + 0 \times 2^4 + 1 \times 2^3 + 1 \times 2^2$
          $+ 0 \times 2^1 + 1 \times 2^0 = 64 + 32 + 8 + 4 + 1 = (109)_{10}$
      (d) $(11011.110)_2 = 1 \times 2^4 + 1 \times 2^3 + 0 \times 2^2 + 1 \times 2^1 + 1 \times 2^0$
          $+ 1 \times 2^{-1} + 1 \times 2^{-2} + 0 \times 2^{-3} = 16 + 8 + 2 + 1 + 1/2 +$
          $1/4 = (27.75)_{10}$

2-6. (a) $(1110)_2 = 8 + 4 + 2 = (14)_{10}$ (using positional
          weights for bits 3, 2, and 1)
      (b) $(1011)_2 = 8 + 2 + 1 = (11)_{10}$ (using positional
          weights for bits 3, 1, and 0)
      (c) $(10101010)_2 = 128 + 32 + 8 + 2 = (170)_{10}$ (using
          positional weights for bits 7, 5, 3, and 2)
      (d) $(00110110)_2 = 32 + 16 + 4 + 2 = (54)_{10}$
          (using positional weights for bits 5, 4, 2, and
          1)

2-7. (a) $(27431)_8 = 2 \times 8^4 + 7 \times 8^3 + 4 \times 8^2 + 3 \times 8^1 + 1 \times 8^0 =$
          $(12057)_{10}$

(b) $(476620)_8 = 4 \times 8^5 + 7 \times 8^4 + 6 \times 8^3 + 6 \times 8^2 + 2 \times 8^1 + 1 \times 8^0 = (163216)_{10}$

(c) $(1234.567)_8 = 4 \times 8^3 + 2 \times 8^2 + 3 \times 8^1 + 4 \times 8^0 + 5 \times 8^{-1} + 6 \times 8^{-2} + 7 \times 8^{-3} = (668.732421875)_{10}$

(d) $(11011110)_8 = 1 \times 8^7 + 1 \times 8^6 + 0 \times 8^5 + 1 \times 8^4 + 1 \times 8^3 + 1 \times 8^2 + 1 \times 8^1 + 1 \times 8^0 = (2363976)_{10}$

(e) $(FFFCC)_{16} = 16 \times 16^4 + 16 \times 16^3 + 16 \times 16^2 + 12 \times 16^1 + 12 \times 16^0 = (1048524)_{10}$

(f) $(123430)_{16} = 1 \times 16^5 + 2 \times 16^4 + 3 \times 16^3 + 4 \times 16^2 + 3 \times 16^1 + 0 \times 16^0 = (1193008)_{10}$

(g) $(E2B4.5E7)_{16} = 14 \times 16^3 + 2 \times 16^2 + 11 \times 16^1 + 4 \times 16^0 + 5 \times 16^{-1} + 14 \times 16^{-2} + 7 \times 16^{-3} = (58036.368896484375)_{10}$

(h) $(11011110)_{16} = 1 \times 16^7 + 1 \times 16^6 + 0 \times 16^5 + 1 \times 16^4 + 1 \times 16^3 + 1 \times 16^2 + 1 \times 16^1 + 0 \times 16^0 = (285282576)_{10}$

2-8.
(a) $(10001010)_2 = 10\ 001\ 010 = (212)_8 = 1000\ 1010 = (8A)_{16}$

(b) $(11110000)_2 = 11\ 110\ 000 = (360)_8 = 1111\ 0000 = (F0)_{16}$

(c) $(100000011.111)_2 = 100\ 000\ 011.111 = (403.7)_8 = 1\ 0000\ 0011.1110 = (103.E)_{16}$

(d) $(11001100.1)_2 = 11\ 001\ 100.100 = (314.4)_8 = 1100\ 1100.10 = 1100\ 1100.1000 = (CC.8)_{16}$

2-9.

Power of 2	Decimal	Binary	Octal	Hex
2^0	1	1	1	1
2^1	2	10	2	2
2^2	4	100	4	4
2^3	8	1000	10	8
2^4	16	10000	20	10
2^5	32	100000	40	20
2^6	164	1000000	100	40
2^7	128	10000000	200	80
2^8	256	100000000	400	100
2^9	512	1000000000	1000	200
2^{10}	1024	10000000000	2000	400
2^{11}	2048	100000000000	4000	800
2^{12}	4096	1000000000000	10000	1000
2^{13}	8192	10000000000000	20000	2000
2^{14}	16384	100000000000000	40000	4000
2^{15}	132768	1000000000000000	100000	8000

2-10.
(a) $(3451)_8 = (11\ 100\ 101\ 001)_2$

(b) $(65473)_8 = (110\ 101\ 100\ 111\ 011)_2$

(c) $(563451)_8 = (101\ 110\ 001\ 100\ 101\ 001)_2$

(d) $(7657.1100)_8 = (111\ 110\ 101\ 111.001\ 001\ 000\ 000)_2$

(e) $(BDE)_{16} = (1011\ 1101\ 1110)_2$

(f) $(13F5)_{16} = (0001\ 0011\ 1111\ 0101)_2$

(g) $(563.4512)_{16} = (0101\ 0110\ 0011.0100\ 0101\ 0001\ 0010)_2$

(h) $(1.1)_{16} = (0001.0001)_2$

2-11.

Hex	Binary	Hex	Binary
00	00000000	10	00010000
01	00000001	11	00010001
02	00000010	12	00010010
03	00000011	13	00010011
04	00000100	14	00010100
05	00000101	15	00010101
06	00000110	16	00010110
07	00000111	17	00010111
08	00001000	18	00011000
09	00001001	19	00011001
0A	00001010	1A	00011010
0B	00001011	1B	00011011
0C	00001100	1C	00011100
0D	00001101	1D	00011101
0E	00001110	1E	00011110
0F	00001111	1F	00011111
		20	00100000

Section 2-4   Converting from Decimal to Other Number Systems

2-12.   (a) $(23)_{10}$

$$23 \div 2, \ Q = 11, \ R = 1 \ \text{LSB}$$
$$11 \div 2, \ Q = 5, \ R = 1$$
$$5 \div 2, \ Q = 2, \ R = 1$$
$$2 \div 2, \ Q = 1, \ R = 0$$
$$1 \div 2, \ Q = 0, \ R = 1 \ \text{MSB}$$

$= (10111)_2$

$$23 \div 8, \ Q = 2, \ R = 7 \ \text{LSD}$$
$$2 \div 8, \ Q = 0, \ R = 2 \ \text{MSD}$$

$= (27)_8$

$$23 \div 16, \ Q = 1, \ R = 7 \ \text{LSD}$$
$$1 \div 16, \ Q = 0, \ R = 1 \ \text{MSD}$$

$= (17)_{16}$

(b) $(47)_{10}$

$$47 \div 2, \ Q = 23, \ R = 1 \ \text{LSB}$$
$$23 \div 2, \ Q = 11, \ R = 1$$
$$11 \div 2, \ Q = 5, \ R = 1$$
$$5 \div 2, \ Q = 2, \ R = 1$$
$$2 \div 2, \ Q = 1, \ R = 0$$
$$1 \div 2, \ Q = 0, \ R = 1 \ \text{MSB}$$

$= (101111)_2$

$$47 \div 8, \ Q = 5, \ R = 7 \ \text{LSD}$$
$$5 \div 8, \ Q = 0, \ R = 5 \ \text{MSD}$$

$= (57)_8$

$$47 \div 16, \ Q = 2, \ R = 15 = F \ \text{LSD}$$
$$2 \div 16, \ Q = 0, \ R = 2 \ \text{MSD}$$

$= (2F)_{16}$

(c) $(268)_{10}$

$$268 \div 2, \quad Q = 134, \quad R = 0 \text{ LSB}$$
$$134 \div 2, \quad Q = 67, \quad R = 0$$
$$67 \div 2, \quad Q = 33, \quad R = 1$$
$$33 \div 2, \quad Q = 16, \quad R = 1$$
$$16 \div 2, \quad Q = 8, \quad R = 0$$
$$8 \div 2, \quad Q = 4, \quad R = 0$$
$$4 \div 2, \quad Q = 2, \quad R = 0$$
$$2 \div 2, \quad Q = 1, \quad R = 0$$
$$1 \div 2, \quad Q = 0, \quad R = 1 \text{ MSB}$$

$= (100001100)_2$

$$268 \div 8, \quad Q = 33, \quad R = 4 \text{ LSD}$$
$$33 \div 8, \quad Q = 4, \quad R = 1$$
$$4 \div 8, \quad Q = 0, \quad R = 4 \text{ MSD}$$

$= (414)_8$

$$268 \div 16, \quad Q = 16, \quad R = 12 = C \text{ LSD}$$
$$16 \div 16, \quad Q = 1, \quad R = 0$$
$$1 \div 16, \quad Q = 0, \quad R = 1 \qquad \text{MSD}$$

$= (10C)_{16}$

2-13.   (a) $(.75)_{10}$

$$.75 \times 2, \quad I = 1 \text{ MSB}, \quad F = .5$$
$$.5 \times 2, \quad I = 1 \text{ LSB}, \quad F = 0$$

$= (.11)_2$

$$.75 \times 8, \quad I = 6 \text{ MSD}, \quad F = 0$$

$= (.6)_8$, also, $(.11)_2 = .110 = (.6)_8$

$$.75 \times 16, \quad I = 12 = C \text{ MSD}, \quad F = 0$$

$= (.C)_{16}$

also $(.11)_2 = .1100 = (C)_{16}$

(b) $(.475)_{10}$

$$.475 \times 2, \quad I = 0 \text{ MSB}, \quad F = .95$$
$$.95 \times 2, \quad I = 1, \qquad F = .9$$
$$.9 \times 2, \quad I = 1, \qquad F = .8$$
$$.8 \times 2, \quad I = 1, \qquad F = .6$$
$$.6 \times 2, \quad I = 1, \qquad F = .2$$
$$.2 \times 2, \quad I = 0, \qquad F = .4$$
$$.4 \times 2, \quad I = 0, \qquad F = .8$$
$$.8 \times 2, \quad I = 1, \qquad F = .6$$
$$.6 \times 2, \quad I = 1, \qquad F = .2$$
$$.2 \times 2, \quad I = 0, \qquad F = .4$$
$$.4 \times 2, \quad I = 0, \qquad F = .8$$
$$.8 \times 2, \quad I = 1, \qquad F = .6$$
$$.6 \times 2, \quad I = 1, \qquad F = .2$$

$$\vdots$$

$$= (.011110011001\ldots)_2 \quad \text{approximation}$$

$$(.47497\ldots)_{10}$$

```
               .475x8,  I = 3 MSD,   F = .8
                 .8x8,  I = 6,       F = .4
                 .4x8,  I = 3,       F = .2
                 .2x8,  I = 1,       F = .6
                 .6x8,  I = 4,       F = .8
                 .8x8,  I = 6,       F = .4
                                .
                                .
                                .
```

$$= (.363146\ldots)_8 \quad \text{approximation}$$

```
              .475x16,  I = 7 MSD,  F = .6
                .6x16,  I = 9,      F = .6
                .6x16,  I = 9,      F = .6
                               .
                               .
                               .
```

$$= (.799\ldots)_{16} \quad \text{approximation}$$

(c) $(.96)_{10}$

```
               .96x2,  I = 1 MSB,  F = .92
               .92x2,  I = 1,      F = .84
               .84x2,  I = 1,      F = .68
               .68x2,  I = 1,      F = .36
               .36x2,  I = 0,      F = .72
               .72x2,  I = 1,      F = .44
               .44x2,  I = 0,      F = .88
               .88x2,  I = 1,      F = .76
               .76x2,  I = 1,      F = .52
               .52x2,  I = 1,      F = .04
               .04x2,  I = 0,      F = .08
               .08x2,  I = 0,      F = .16
                               .
                               .
                               .
```

$$= (.111101011100\ldots)_2 \quad \text{approximation}$$

```
               .96x8,  I = 7 MSD,  F = .68
               .68x8,  I = 5,      F = .44
               .44x8,  I = 3,      F = .52
               .52x8,  I = 4,      F = .16
                               .
                               .
                               .
```

$$= (.7534\ldots)_8 \quad \text{approximation}$$

```
              .96x16,  I = 15 = F MSD,  F = .36
              .36x16,  I =  5,          F = .76
              .76x16,  I = 12 = C       F = .16
                               .
                               .
                               .
```

$$= (.F5C\ldots)_{16} \quad \text{approximation}$$

27

2-14. (a) Convert $(12.87)_{10}$ to a base-2 number

$$(12)_{10} = (1100)_2$$

```
.87x2, I = 1  MSB, F = .74
.74x2, I = 1,      F = .48
.48x2, I = 0,      F = .96
.96x2, I = 1,      F = .92
.92x2, I = 1,      F = .84
.84x2, I = 1,      F = .68
.68x2, I = 1,      F = .36
.36x2, I = 0,      F = .72
.72x2, I = 1,      F = .44
              .
              .
              .
```

$$= (1100.110111101\ldots)_2 \quad \text{approximation}$$

(b) Convert $(12.87)_{10}$ to a base-8 number

$$(12)_{10} = (14)_8$$

```
.87x8, I = 6  MSD, F = .96
.96x8, I = 7,      F = .68
.68x8, I = 5,      F = .44
              .
              .
              .
```

$$= (14.675\ldots)_8 \quad \text{approximation}$$

(c) Convert $(12.87)_{10}$ to a base-16 number

$$(12)_{10} = (C)_{16}$$

```
.87x16, I = 13 = D MSD, F = .92
.92x16, I = 14 = E      F = .72
.72x16, I = 11 = B      F = .52
              .
              .
              .
```

$$= (C.DEB\ldots)_{16} \quad \text{approximation}$$

2-15. To save a little work we can solve for the number in hexadecimal first.

(c) $(234.56)_{10} = (?)_{16}$

```
234÷16, Q = 14, R = 10 = A LSD
 14÷16, Q =  0, R = 14 = E MSD

.56x16, I = 8 MSD,  F = .96
.96x16, I = 15 = F, F = .36
              .
              .
              .
```

$$= (EA.8F\ldots)_{16} \quad \text{approximation for fraction}$$

(a) $(234.56)_{10} = (?)_2 = (1110\ 1010.1000$
$1111\ 0101...)_2$ from (c), approximation for
fraction

(b) $(234.56)_{10} = (?)_8 = (352.4365...)_8$ from (a),
approximation for fraction

2-16. (a) $(17)_{10}$

64	32	16	8	4	2	1
0	0	1	0	0	0	1

$= (10001)_2$

(b) $(37)_{10}$

64	32	16	8	4	2	1
0	1	0	0	1	0	1

$= (100101)_2$

(c) $(56)_{10}$

64	32	16	8	4	2	1
0	1	1	1	0	0	0

$= (111000)_2$

(d) $(72)_{10}$

64	32	16	8	4	2	1
1	0	0	1	0	0	0

$= (1001000)_2$

(d) $(475)_{10}$

512	256	128	64	32	16	8	4	2	1
0	1	1	1	0	1	1	0	1	1

$= (111011011)_2$

Section 2-5 Number Representations

2-17. (a) Decimal 10, using 6 bits SM representation =
001001

(b) Decimal 29, using 6 bits SM representation =
011101

(c) Decimal 51, using 8 bits SM representation =
00011011

(d) Decimal 75, using 8 bits SM representation =
01001011

(e) Decimal 327, using 12 bits SM representation =
000101000111

2-18. (a) +1001010 = 01001010, requires a minimum of 8 bits
in SM representation

(b) -11110000 = 111110000, requires a minimum of 9
bits in SM representation

(c) -11001100.1 = 111001100.1, requires a minimum of
10 bits in SM representation

(d) +100000011.111 = 0100000011.111, requires a
minimum of 13 bits in SM representation

2-19.   (a) 1's complement of 1010110 = 0101001 complementing
        each bit
        (b) 1's complement of 01011010011 = 10100101100
        complementing each bit
        (c) 1's complement of 010010.0101 = 101101.1010
        complementing each bit
        (d) 1's complement of 101110101.1001 = 010001010.0110
        complementing each bit

2-20.   (a) Decimal +6, 1's complement representation using
        8 bits = 00000110
        (b) Decimal -25, 1's complement representation using
        8 bits = 1's complement of 00011001 = 11100110
        (c) Decimal +125, 1's complement representation using
        8 bits = 01111101
        (d) Decimal -126, 1's complement representation using
        8 bits = 1's complement of 01111110 = 10000001

2-21.   (a) 0010110 expressed in 1's complement
        representation = decimal +22
        (b) 00011010011 expressed in 1's complement
        representation = decimal +211
        (c) 110010.0101 expressed in 1's complement
        representation = 1's complement of 001101.1010
        = decimal -13.625
        (d) 100000101.1000 expressed in 1's complement
        representation = 1's complement of 011111010.0111
        = decimal -250.4375

2-22.   (a) 2's complement of 1010110 = 0101001 + 1 = 0101010
        (2's C of N = 1's C of N + 1_{LSB})
        (b) 2's complement of 01011010011 = 10100101100 + 1 =
        10100101101
        (c) 2's complement of 010010.0101 = 101101.1010 +
        .0001 = 101101.1011
        (d) 2's complement of 101110101.1001 = 010001010.0110
        + .0001 = 010001010.0111

2-23.   (a) Decimal +9, 2's complement representation using 8
        bits = 00001001
        (b) Decimal -36, 2's complement representation using
        8 bits = 2's complement of 00100100 = 11011100
        (using quick inspection method to write the 2's
        complement)
        (c) Decimal +85, 2's complement representation using
        8 bits = 01010101
        (d) Decimal -114, 2's complement representation using
        8 bits = 2's complement of 01110010 = 10001110

2-24.   (a) 0110110 expressed in 2's complement
        representation = decimal +54
        (b) 01011010111 expressed in 2's complement
        representation = decimal +727

(c) 110001.0111 expressed in 2's complement
representation = 2's complement of 001110.1001 =
decimal -14.5625

(d) 101100101.1010 expressed in 2's complement
representation = 2's complement of 010011010.0110
= decimal -154.375

2-25. (a) Decimal +7, 1's complement and 2's complement
representations using 8 bits = 00000111 (positive
numbers are expressed the same in 1's C and 2's C
representations)

(b) Decimal -37, 1's complement representation using
8 bits = 1's complement of 00100101 (+37) =
11011010, 2's complement representation using 8
bits = 2's complement of 00100101 = 11011011

(c) Decimal +91, 1's complement and 2's complement
representations using 8 bits = 01011011

(d) Decimal -113, 1's complement representation using
8 bits = 1's complement of 01110001 (+113) =
10001110, 2's complement representation using 8
bits = 2's complement of 01110001 = 10001111

2-26. (a) Decimal +38, SM, 1's C, and 2's C representations
using 16 bits = 0000000000100110, (positive
numbers are expressed the same in all three
representations)

(b) Decimal -192, SM representation using 16 bits =
1000000011000000 (SM representation of +192 with
sign bit 15 changed to 1), 1's complement
representation using 16 bits = 1's complement of
0000000011000000 (+192) = 1111111100111111, 2's
complement representation using 16 bits =
2's complement of 0000000011000000
= 1111111101111111.

(c) Decimal +389, SM, 1's C, and 2's C
representations using 16 bits = 0000000110000101

(d) Decimal -4751, SM representation using 16 bits =
1001001010001111 (SM representation of +4751 with
sign bit 15 changed to 1), 1's complement
representation using 16 bits = 1's complement of
0001001010001111 (+4751) = 1110110101110000 , 2's
complement representation using 16 bits =
2's complement of 0001001010001111 =
1110110101110001

Section 2-6 arithmetic Operations

2-27. (a) Decimal   2's complement representation

```
          6           00000110   or   00000110
         -3          -00000011        +11111101
          3           00000011        100000011
                                     (ignore carry out of
                                      sign bit)
```

(b) Decimal   2's complement representation

```
  95        01011111
 +27       +00011011
 122        01111010
```

(c) Decimal   2's complement representation

```
 101        01100101   or    01100101
 -46       -00101110        +11010010
  55        00110111        100110111
                        (ignore carry out of
                             sign bit)
```

(d) Decimal   2's complement representation

```
  39        00100111   or    00100111
 -17       -00010001        +11101111
  22        00010110        100010110
                        (ignore carry out of
                             sign bit)
```

2-28. (a) Decimal   1's complement representation

```
  17        00010001   or    00010001
 -12       -00001100        +11110011
   5        00000101        100000100
                                  +1
                             00000101
                        (add carry out of
                        sign bit to least
                        significant bit of
                        sum)
```

(b) Decimal   1's complement representation

```
  42        00101010
 +25       +00011001
  67        01000011
```

(c) Decimal   1's complement representation

```
 123        01111011   or    01111011
 -76       -01001100        +10110011
  47        00101111        100101110
                                  +1
                             00101111
                        (add carry out of
                        sign bit to least
                        significant bit of
                        sum)
```

(d) Decimal   1's complement representation

```
   27          00011011    or      00011011
  -15         -00001111           +11110000
   12          00001100           100001011
                                        +1
                                   ──────────
                                   00001100
                              (add carry out of
                              sign bit to least
                              significant bit of
                              sum)
```

2-29. The following 6-bit numbers are expressed in 2's complement representation.

(a)
```
       N1    =   001000  =   +8
       N2    =   011111  =   +31
   N1 + N2   =   100111  =   -25     error (should be
                                              +39)
```
Two positive numbers must result in a positive number or overflow has occurred. The change in the sign bit indicates that overflow has indeed occurred. 31 is the largest positive number that will fit in the six bit word size specified by the problem.

(b)
```
       N1      =   010100  =   +20
       N2      =   101011  =   -21
      -N2      =   010101  =   +21
   N1 + (-N2)  =   101001  =   -23     error (should
                                              be +41)
```
For six bits the largest positive number that will fit in the six bit word size specified by the problem is 31. Overflow has occurred since the sign bit is negative and it should be positive when two positive numbers are added.

(c)
```
       N1    =   100011  =   -29
       N2    =   101101  =   -19
   N1 + N2   =   010000  =   +16     error (should be
                                              -48)
```
Two negative numbers must result in a negative number or overflow has occurred. The change in the sign bit indicates that overflow has indeed occurred. Also, 32 is the largest negative number that will fit in the six bit word size specified by the problem.

(d)
```
       N1      =   010000   =   +16
       N2      =   000011   =    +3
      -N2      =   111101   =    -3
   N1 + (-N2)  =   1001101  =   +13     correct
                                        (ignore carry
                                        out of
                                        sign bit)
```
Overflow does not occur when the absolute value of the result of two signed numbers being added

or subtracted is smaller than the absolute value
of the larger signed number.

2-30. The following 8-bit numbers are expressed in 2's
complement representation.

```
(a)     N1        =   00110010   =     +50
        N2        =   11111101   =      -3
      N1 + N2     = 100101111   =     +47      correct
                  (ignore carry out of sign bit)

        N1        =   00110010   =     +50
        N2        =   11111101   =      -3
       -N2        =   00000011   =      +3
     N1 + (-N2)   =   00110101   =     +53      correct


(b)     N1        =   10001110   =    -114
        N2        =   00001101   =     +13
      N1 + N2     =   10011011   =    -101      correct

        N1        =   10001110   =    -114
        N2        =   00001101   =     +13
       -N2        =   11110011   =     -13
     N1 + (-N2)   = 110000001   =    -127      correct
                  (ignore carry out of sign bit)


(c)     N1        =   11111010   =      -6
        N2        =   10010101   =    -107
      N1 + N2     = 110001111   =    -113      correct
                  (ignore carry out of sign bit)

        N1        =   11111010   =      -6
        N2        =   10010101   =    -107
       -N2        =   01101011   =    +107
     N1 + (-N2)   = 101100101   =    +101      correct
                  (ignore carry out of sign bit)


(d)     N1        =   00010010   =     +18
        N2        =   10011101   =     -99
      N1 + N2     =   10101111   =     +81      correct

        N1        =   00010010   =     +18
        N2        =   10011101   =     -99
       -N2        =   01100011   =     +99
     N1 + (-N2)   =   01110101   =    +117      correct
```

34

2-31. The following 8-bit numbers are expressed in 1's complement representation.

```
(a)      N1        =   00110010  =    +50
         N2        =   11111101  =     -2
    N1 + N2        = 100101111   =    +47
                          +1
                    _____
                     100110000  =    +48    correct
         (add carry out of
          sign bit to least
          significant bit of
          sum)
         N1        =   00110010  =    +50
         N2        =   11111101  =     -2
        -N2        =   00000010  =     +2
    N1 + (-N2)     =   00110100  =    +52    correct

(b)      N1        =   10001110  =   -113
         N2        =   00001101  =    +13
    N1 + N2        =   10011011  =   -100    correct

         N1        =   10001110  =   -113
         N2        =   00001101  =    +13
        -N2        =   11110010  =    -13
    N1 + (-N2)     = 110000000   =   -127
                          +1
                    _____
                     10000001   =   -126    correct
         (add carry out of
          sign bit to least
          significant bit of
          sum)
(c)      N1        =   11111010  =     -5
         N2        =   10010101  =   -106
    N1 + N2        = 110001111   =   -112
                          +1
                    _____
                     10010000   =   -111    correct
         (add carry out of
          sign bit to least
          significant bit of
          sum)
         N1        =   11111010  =     -5
         N2        =   10010101  =   -106
        -N2        =   01101010  =   +106
    N1 + (-N2)     = 101100100   =   +100
                          +1
                    _____
                     00110101   =   +101    correct
         (add carry out of
          sign bit to least
          significant bit of
          sum)
```

(d)   N1   = 00010010 =   +18
               N2   = 10011101 =   -98
      N1 + N2   = 10101111 =   -80   correct

               N1   = 00010010 =   +18
               N2   = 10011101 =   -98
           -N2   = 01100010 =   +98
  N1 + (-N2) = 01110100 = +116   correct

Section 2-7   The Keyboard as an Input to a Switching Circuit

2-32. (a) $2^X \geq Y$ (b) $2^X \geq Y$ (c) $2^X \geq Y$
        $2^X \geq 9$ $2^X \geq 16$ $2^X \geq 22$
        X = 4 bits   X = 4 bits   X = 5 bits
    (d) $2^X \geq Y$ (e) $2^X \geq Y$
        $2^X \geq 36$ $2^X \geq 104$
        X = 6 bits   X = 7 bits

2-33. (a) $2^X \geq Y$ (b) $2^X \geq Y$
        $2^4 \geq Y$ $2^5 \geq Y$
        Y = 16 characters   Y = 32 characters
    (c) $2^X \geq Y$ (d) $2^X \geq Y$
        $2^6 \geq Y$ $2^7 \geq Y$
        Y = 64 characters   Y = 128 characters
    (e) $2^X \geq Y$
        $2^8 \geq Y$
        Y = 256 characters

2-34.

Decimal	BCD	XS3
0	0000	0011
1	0001	0100
2	0010	0101
3	0011	0110
4	0100	0111
5	0101	1000
6	0110	1001
7	0111	1010
8	1000	1011
9	1001	1100

Section 2-8   Short Introduction: Analog Interfaces to Switching Circuits

2-35. (a) For an ADC with an analog full scale voltage range of 5 volts and 4 output bits the

$$\text{Resolution of LSB} = \frac{V_{FSR}}{2^n} = \frac{5}{2^4} = 0.3125 \text{ V.}$$

    (b) For an ADC with an analog full scale voltage range of 7.5 volts and 10 output bits the

$$\text{Resolution of LSB} = \frac{V_{FSR}}{2^n} = \frac{7.5}{2^{10}} = 0.0073 \text{ V.}$$

(c) For an ADC with an analog full scale voltage range of 20 volts and 6 output bits the

$$\text{Resolution of LSB} = \frac{V_{FSR}}{2^n} = \frac{20}{2^6} = 0.3125 \text{ V.}$$

(d) For an ADC with an analog full scale voltage range of 9 volts and 12 output bits the

$$\text{Resolution of LSB} = \frac{V_{FSR}}{2^n} = \frac{9}{2^{12}} = 0.0022 \text{ V.}$$

2-36. The minimum number of output bits for a ADC with:
(a) An output resolution no larger than 0.25 volts and a full scale voltage range of 5 volts is

$$\text{Resolution of LSB} = \frac{V_{FSR}}{2^n} ,$$

$$0.25 = \frac{5}{2^n} , \quad 2^n = \frac{5}{0.25} = 20 , \quad n = 5.$$

$$\text{Resolution of LSB} = \frac{V_{FSR}}{2^n} = \frac{5}{2^5} = 0.16 \text{ V} < 0.25 \text{ V.}$$

(b) An output resolution no larger than 70 millivolts and a full scale voltage range of 10 volts is

$$\text{Resolution of LSB} = \frac{V_{FSR}}{2^n} ,$$

$$0.07 = \frac{10}{2^n} , \quad 2^n = \frac{10}{0.07} = 142.86 , \quad n = 8.$$

$$\text{Resolution of LSB} = \frac{V_{FSR}}{2^n} = \frac{10}{2^8} = 0.04 \text{ V} < 0.07 \text{ V.}$$

(c) An output resolution no larger than 8 millivolts and a full scale voltage range of 12 volts is

$$\text{Resolution of LSB} = \frac{V_{FSR}}{2^n} ,$$

$$0.008 = \frac{12}{2^n} , \quad 2^n = \frac{12}{0.008} = 1500 , \quad n = 11.$$

$$\text{Resolution of LSB} = \frac{V_{FSR}}{2^n} = \frac{12}{2^{11}} = .006 \text{ V} < .008 \text{ V.}$$

(d) An output resolution no larger than 3/4 millivolts and a full scale voltage range of 12 volts is

$$\text{Resolution of LSB} = \frac{V_{FSR}}{2^n} ,$$

$$0.00075 = \frac{12}{2^n} , \quad 2^n = \frac{12}{0.00075} = 16000 , \quad n = 14.$$

$$\text{Resolution of LSB} = \frac{V_{FSR}}{2^n} = \frac{12}{2^{14}} = 0.00073 \text{ V} <$$

0.00075 V.

2-37. The minimum number of input bits for a DAC with:
(a) An analog output voltage with no less than 32 output steps is $2^n \geq 32$, n = 5.
(b) An analog output voltage with no less than 128 output steps is $2^n \geq 128$, n = 7.
(c) An analog output voltage with no less than 425 output steps is $2^n \geq 425$, n = 9.
(d) An analog output voltage with no less than 1024 output steps is $2^n \geq 1024$, n = 10.

2-38. The required resolution in degrees of a shaft-angle encoder with:
(a) Three output bits is $1/2^3$ of one revolution (360 degrees) = $360/2^3$ = 45 degrees.
(b) Four output bits is $1/2^4$ of one revolution = $360/2^4$ = 22.50 degrees.
(c) Five output bits is $1/2^5$ of one revolution = $360/2^5$ = 11.25 degrees.
(d) Seven output bits is $1/2^7$ of one revolution = $360/2^7$ = 2.81 degrees.

2-39. (a) X3 X2 X1 X0 binary
       X3 X3 $\oplus$ X2 X2 $\oplus$ X1 X1 $\oplus$ X0   gray code
(b) X4 X3 X2 X1 X0 binary
       X4 X4 $\oplus$ X3 X3 $\oplus$ X2 X2 $\oplus$ X1 X1 $\oplus$ X0   gray code
(c) X5 X4 X3 X2 X1 X0 binary
       X5 X5 $\oplus$ X4 X4 $\oplus$ X3 X3 $\oplus$ X2 X2 $\oplus$ X1 X1 $\oplus$ X0   gray code

2-40. (a) X3 X2 X1 X0 gray code
       X3 X3 $\oplus$ X2 X3 $\oplus$ X2 $\oplus$ X1 X3 $\oplus$ X2 $\oplus$ X1 $\oplus$ X0 binary
(b) X4 X3 X2 X1 X0 gray code
       X4 X4 $\oplus$ X3 X4 $\oplus$ X3 $\oplus$ X2 X4 $\oplus$ X3 $\oplus$ X2 $\oplus$ X1
       X4 $\oplus$ X3 $\oplus$ X2 $\oplus$ X1 $\oplus$ X0   binary
(c) X5 X4 X3 X2 X1 X0 gray code
       X5 X5 $\oplus$ X4 X5 $\oplus$ X4 $\oplus$ X3 X5 $\oplus$ X4 $\oplus$ X3 $\oplus$ X2
       X5 $\oplus$ X4 $\oplus$ X3 $\oplus$ X2 $\oplus$ X1
       X5 $\oplus$ X4 $\oplus$ X3 $\oplus$ X2 $\oplus$ X1 $\oplus$ X0   binary

2-41. (a) 01110 binary
       01001 gray code
       01110 binary (check)
(b) 10110 binary
       11101 gray code
       10110 binary (check)
(c) 01100 binary
       01010 gray code
       01100 binary (check)

(d) 11101 binary
    10011 gray code
    11101 binary (check)

2-42. X4 X3 X2 X1 X0 binary input
       X4 X4 ⊕ X3 X3 ⊕ X2 X2 ⊕ X1 X1 ⊕ X0 gray code output
                                             Y4 through Y0

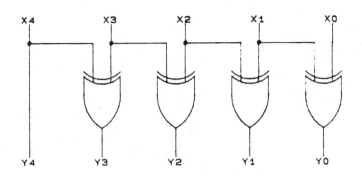

2-43. (a) 011101 gray code
           010110 binary
           011101 gray code (check)
       (b) 101111 gray code
           110101 binary
           101111 gray code (check)
       (c) 011001 gray code
           010001 binary
           011001 gray code (check)
       (d) 111010 gray code
           101100 binary
           111010 gray code (check)

2.44. X5 X4 X3 X2 X1 X0 gray code input
       X5 X5 ⊕ X4 X5 ⊕ X4 ⊕ X3 X5 ⊕ X4 ⊕ X3 ⊕ X2
       X5 ⊕ X4 ⊕ X3 ⊕ X2 ⊕ X1
       X5 ⊕ X4 ⊕ X3 ⊕ X2 ⊕ X1 ⊕ X0 binary output
                                      Y5 through Y0

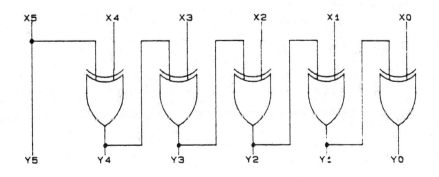

39

Section 3-2 Drawing, Filling, and Reading a Karnaugh Map

3-1.

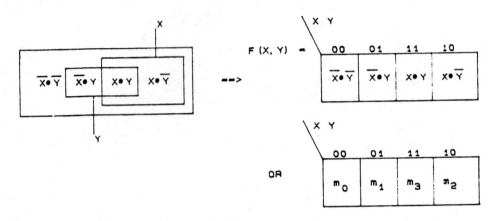

3-2.

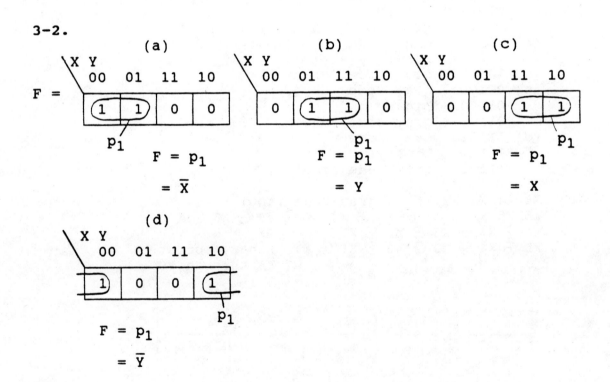

3-3.

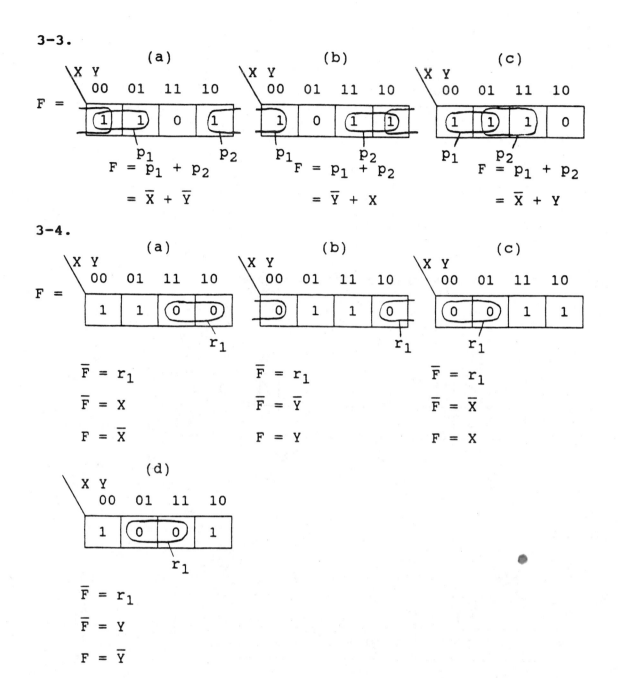

(a)

X Y

00	01	11	10

F =

(1) (1) 0 (1)

p_1 p_2

$$F = p_1 + p_2$$
$$= \overline{X} + \overline{Y}$$

(b)

X Y

00	01	11	10

(1) 0 (1) (1)

p_1 p_2

$$F = p_1 + p_2$$
$$= \overline{Y} + X$$

(c)

X Y

00	01	11	10

(1) (1) 1 0

p_1 p_2

$$F = p_1 + p_2$$
$$= \overline{X} + Y$$

3-4.

(a)

X Y

00	01	11	10

F =

1 1 (0 0)

r_1

$\overline{F} = r_1$

$\overline{F} = X$

$F = \overline{X}$

(b)

X Y

00	01	11	10

(0) 1 1 (0)

r_1

$\overline{F} = r_1$

$\overline{F} = \overline{Y}$

$F = Y$

(c)

X Y

00	01	11	10

(0 0) 1 1

r_1

$\overline{F} = r_1$

$\overline{F} = \overline{X}$

$F = X$

(d)

X Y

00	01	11	10

1 (0 0) 1

r_1

$\overline{F} = r_1$

$\overline{F} = Y$

$F = \overline{Y}$

3-5.

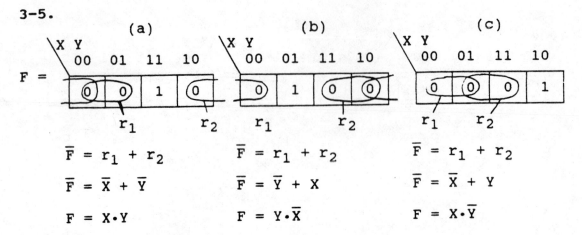

(a)	(b)	(c)
$\overline{F} = r_1 + r_2$	$\overline{F} = r_1 + r_2$	$\overline{F} = r_1 + r_2$
$\overline{F} = \overline{X} + \overline{Y}$	$\overline{F} = \overline{Y} + X$	$\overline{F} = \overline{X} + Y$
$F = X \cdot Y$	$F = Y \cdot \overline{X}$	$F = X \cdot \overline{Y}$

Section 3-3   Three-Variable Karnaugh Maps

3-6.

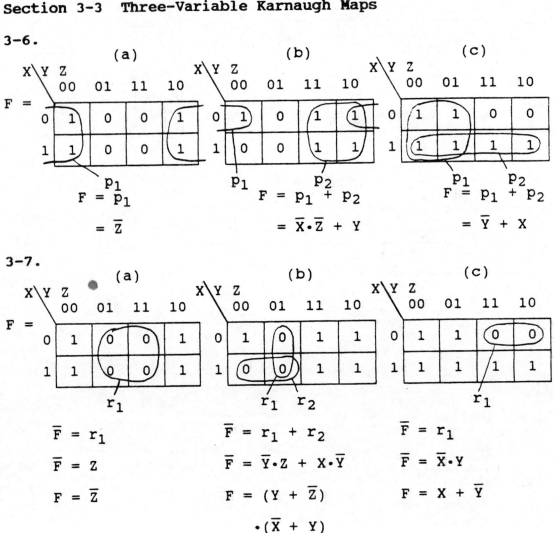

(a)	(b)	(c)
$F = p_1$	$F = p_1 + p_2$	$F = p_1 + p_2$
$= \overline{Z}$	$= \overline{X} \cdot \overline{Z} + Y$	$= \overline{Y} + X$

3-7.

(a)	(b)	(c)
$\overline{F} = r_1$	$\overline{F} = r_1 + r_2$	$\overline{F} = r_1$
$\overline{F} = Z$	$\overline{F} = \overline{Y} \cdot Z + X \cdot \overline{Y}$	$\overline{F} = \overline{X} \cdot Y$
$F = \overline{Z}$	$F = (Y + \overline{Z})$	$F = X + \overline{Y}$
	$\cdot (\overline{X} + Y)$	

3-8.

	(a)	(b)	(c)

$F(X, Y, Z) =$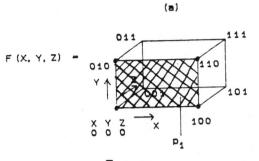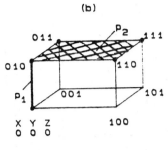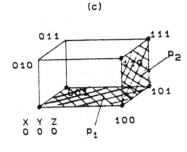

$$F = p_1$$

$$F = \overline{Z}$$
checks with
3-6a

$$F = p_1 + p_2$$

$$F = \overline{X} \cdot \overline{Z} + Y$$
checks with
3-6b

$$F = p_1 + p_2$$

$$F = \overline{Y} + X$$
checks with
3-6c

3-9. $p_1 = \overline{Y} \cdot \overline{Z}$; for $Y = 0$, $Z = 0$, $p_1 = \overline{Y} \cdot \overline{Z} = 1$
$p_2 = Y \cdot Z$; for $Y = 1$, $Z = 1$, $p_1 = Y \cdot Z = 1$
$p_3 = X \cdot Y$; for $Y = 1$, $Z = 1$, $p_1 = X \cdot Y = 1$
$F = p_1 + p_2 + p_3$

$$F = \overline{Y} \cdot \overline{Z} + Y \cdot Z + X \cdot Y$$

3-10. $r_1 = \overline{Y} \cdot Z$; for $Y = 0$, $Z = 1$, $r_1 = \overline{Y} \cdot Z = 1$

$r_2 = \overline{X} \cdot Y \cdot \overline{Z}$; for $X = 0$, $Y = 1$, $Z = 0$, $r_1 = \overline{X} \cdot Y \cdot \overline{Z} = 1$

$$\overline{F} = r_1 + r_2 = \overline{Y} \cdot X + \overline{X} \cdot Y \cdot \overline{Z}$$

3-11. $p_1 = \overline{Y}$, $p_2 = X \cdot \overline{Z}$

$$F = p_1 + p_2 = \overline{Y} + X \cdot \overline{Z}$$

3-12. $r_1 = \overline{X} \cdot Y$, $r_2 = Y \cdot Z$

$$\overline{F} = r_1 + r_2 = \overline{X} \cdot Y + Y \cdot Z$$

3-13.

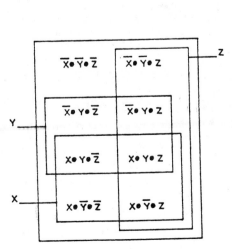

$F(X, Y, Z) =$ →

X \ Y Z	0	1
00	$\overline{X} \cdot \overline{Y} \cdot \overline{Z}$	$\overline{X} \cdot \overline{Y} \cdot Z$
01	$\overline{X} \cdot Y \cdot \overline{Z}$	$\overline{X} \cdot Y \cdot Z$
11	$X \cdot Y \cdot \overline{Z}$	$X \cdot Y \cdot Z$
10	$X \cdot \overline{Y} \cdot \overline{Z}$	$X \cdot \overline{Y} \cdot Z$

or

X \ Y Z	0	1
00	m_0	m_1
01	m_2	m_3
11	m_6	m_7
10	m_4	m_5

3-14.
&
3-15.

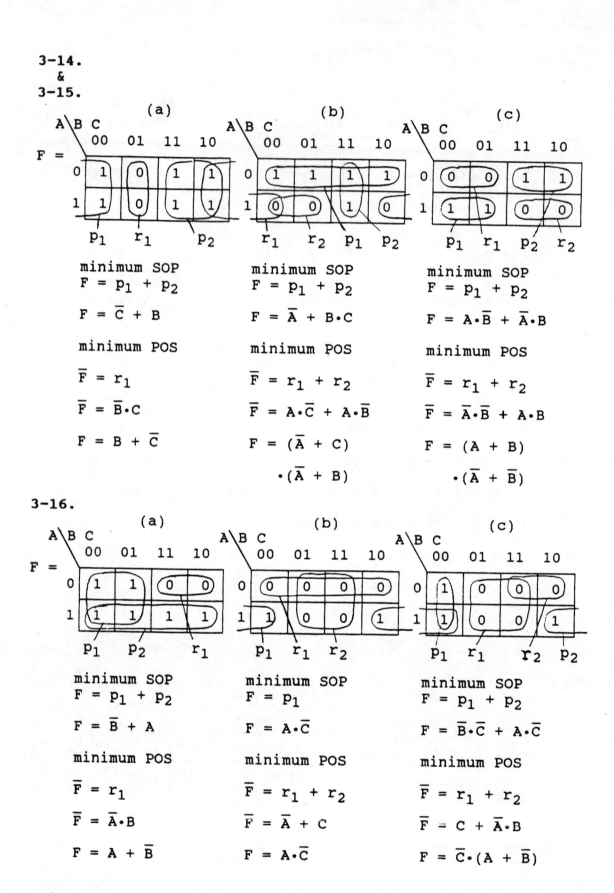

(a)

A\B C

F =

minimum SOP
$F = p_1 + p_2$

$F = \overline{C} + B$

minimum POS

$\overline{F} = r_1$

$\overline{F} = \overline{B} \cdot C$

$F = B + \overline{C}$

(b)

A\B C

minimum SOP
$F = p_1 + p_2$

$F = \overline{A} + B \cdot C$

minimum POS

$\overline{F} = r_1 + r_2$

$\overline{F} = A \cdot \overline{C} + A \cdot \overline{B}$

$F = (\overline{A} + C)$
$\quad \cdot (\overline{A} + B)$

(c)

A\B C

minimum SOP
$F = p_1 + p_2$

$F = A \cdot \overline{B} + \overline{A} \cdot B$

minimum POS

$\overline{F} = r_1 + r_2$

$\overline{F} = \overline{A} \cdot \overline{B} + A \cdot B$

$F = (A + B)$
$\quad \cdot (\overline{A} + \overline{B})$

3-16.

(a)

A\B C

F =

minimum SOP
$F = p_1 + p_2$

$F = \overline{B} + A$

minimum POS

$\overline{F} = r_1$

$\overline{F} = \overline{A} \cdot B$

$F = A + \overline{B}$

(b)

A\B C

minimum SOP
$F = p_1$

$F = A \cdot \overline{C}$

minimum POS

$\overline{F} = r_1 + r_2$

$\overline{F} = \overline{A} + C$

$F = A \cdot \overline{C}$

(c)

A\B C

minimum SOP
$F = p_1 + p_2$

$F = \overline{B} \cdot \overline{C} + A \cdot \overline{C}$

minimum POS

$\overline{F} = r_1 + r_2$

$\overline{F} = C + \overline{A} \cdot B$

$F = \overline{C} \cdot (A + \overline{B})$

3-17. If necessary, first convert the function or
& complement of the function into a SOP form.
3-18.

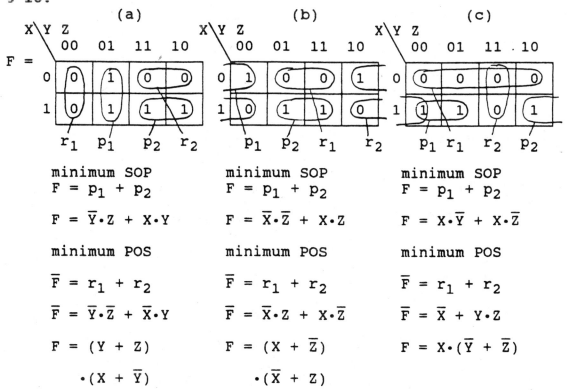

(a)	(b)	(c)
minimum SOP $F = p_1 + p_2$	minimum SOP $F = p_1 + p_2$	minimum SOP $F = p_1 + p_2$
$F = \overline{Y} \cdot Z + X \cdot Y$	$F = \overline{X} \cdot \overline{Z} + X \cdot Z$	$F = X \cdot \overline{Y} + X \cdot \overline{Z}$
minimum POS	minimum POS	minimum POS
$\overline{F} = r_1 + r_2$	$\overline{F} = r_1 + r_2$	$\overline{F} = r_1 + r_2$
$\overline{F} = \overline{Y} \cdot \overline{Z} + \overline{X} \cdot Y$	$\overline{F} = \overline{X} \cdot Z + X \cdot \overline{Z}$	$\overline{F} = \overline{X} + Y \cdot Z$
$F = (Y + Z)$	$F = (X + \overline{Z})$	$F = X \cdot (\overline{Y} + \overline{Z})$
$\cdot (X + \overline{Y})$	$\cdot (\overline{X} + Z)$	

3-19. If necessary, first convert the function or
& complement of the function into a SOP form.
3-20.

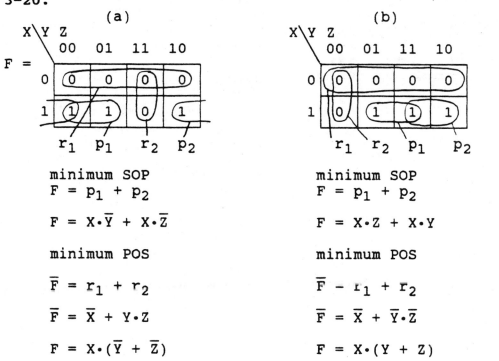

(a)	(b)
minimum SOP $F = p_1 + p_2$	minimum SOP $F = p_1 + p_2$
$F = X \cdot \overline{Y} + X \cdot \overline{Z}$	$F = X \cdot Z + X \cdot Y$
minimum POS	minimum POS
$\overline{F} = r_1 + r_2$	$\overline{F} = r_1 + r_2$
$\overline{F} = \overline{X} + Y \cdot Z$	$\overline{F} = \overline{X} + \overline{Y} \cdot \overline{Z}$
$F = X \cdot (\overline{Y} + \overline{Z})$	$F = X \cdot (Y + Z)$

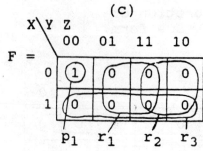

(c)

minimum SOP

$F = p_1$

$F = \overline{X} \cdot \overline{Y} \cdot \overline{Z}$

minimum POS

$\overline{F} = r_1 + r_2 + r_3$

$\overline{F} = Z + Y + X$

$F = \overline{Z} \cdot \overline{Y} \cdot \overline{X})$

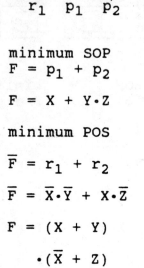

(d)

minimum SOP

$F = p_1 + p_2$

$F = X + Y \cdot Z$

minimum POS

$\overline{F} = r_1 + r_2$

$\overline{F} = \overline{X} \cdot \overline{Y} + X \cdot \overline{Z}$

$F = (X + Y)$

$\quad \cdot (\overline{X} + Z)$

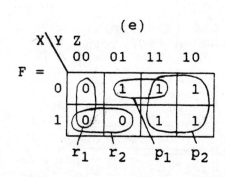

(e)

minimum SOP

$F = p_1 + p_2$

$F = \overline{X} \cdot Z + Y$

minimum POS

$\overline{F} = r_1 + r_2$

$\overline{F} = \overline{Y} \cdot \overline{Z} + X \cdot \overline{Y}$

$F = (Y + Z)$

$\quad \cdot (\overline{X} + Y)$

46

Section 3-4 Four-and Five-Variable Karnaugh Maps

3-21.

(a)

$F(X,Y,Z) = $

Z \ X Y	00	01	11	10
0	m_0	m_2	m_6	m_4
1	m_1	m_3	m_7	m_5

(b)

$F(X,Y,Z) = $

Y Z \ X	0	1
00	m_0	m_4
01	m_1	m_5
11	m_3	m_7
10	m_2	m_6

(c)

$F(W,X,Y,Z) = $

Y Z \ W X	00	01	11	10
00	m_0	m_4	m_{12}	m_8
01	m_1	m_5	m_{13}	m_9
11	m_3	m_7	m_{15}	m_{11}
10	m_2	m_6	m_{14}	m_{10}

3-22.
&
3-23.

(a) $F = F(W,X,Y,Z)$

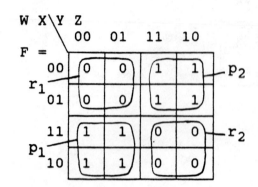

minimum SOP for the 1s
$$F = p_1 + p_2$$

$$F = W \cdot \overline{Y} + \overline{W} \cdot Y$$

minimum SOP for the 0s

$$\overline{F} = r_1 + r_2$$

$$\overline{F} = \overline{W} \cdot \overline{Y} + W \cdot Y$$

(b) $F = F(W,X,Y,Z)$

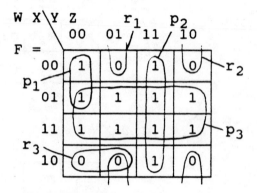

minimum SOP for the 1s
$$F = p_1 + p_2 + p_3$$

$$F = \overline{W} \cdot \overline{Y} \cdot \overline{Z} + Y \cdot Z + X$$

minimum SOP for the 0s

$$\overline{F} = r_1 + r_2 + r_3$$

$$\overline{F} = \overline{X} \cdot \overline{Y} \cdot Z + \overline{X} \cdot Y \cdot \overline{Z} + W \cdot \overline{X} \cdot \overline{Y}$$

(c) $F = F(W,X,Y,Z)$

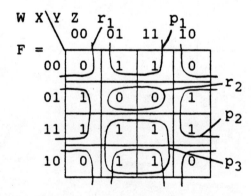

minimum SOP for the 1s
$$F = p_1 + p_2 + p_3$$

$$F = \overline{X} \cdot Z + X \cdot \overline{Z} + W \cdot Z$$

minimum SOP for the 0s

$$\overline{F} = r_1 + r_2$$

$$\overline{F} = \overline{X} \cdot \overline{Z} + \overline{W} \cdot X \cdot Z$$

3-24.

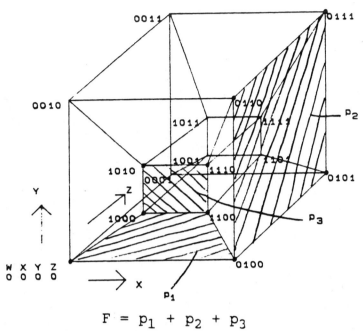

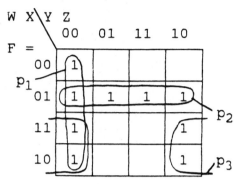

$$F = F(W,X,Y,Z)$$

$$F = p_1 + p_2 + p_3$$

$$F = \overline{Y} \cdot \overline{Z} + \overline{W} \cdot X + W \cdot \overline{Z}$$

$$F = p_1 + p_2 + p_3$$

$$F = \overline{Y} \cdot \overline{Z} + \overline{W} \cdot X + W \cdot \overline{Z}$$

3-25.
&
3-26.

(a) $F = F(W,X,Y,Z)$

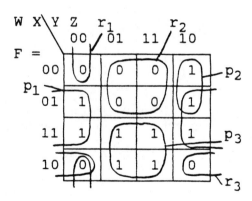

minimum SOP for F
$$F = p_1 + p_2 + p_3$$

$$F = X \cdot \overline{Z} + \overline{W} \cdot Y \cdot \overline{Z} + W \cdot Z$$

minimum SOP for $\overline{F}$

$$\overline{F} = r_1 + r_2 + r_3$$

$$\overline{F} = \overline{X} \cdot \overline{Y} \cdot \overline{Z} + \overline{W} \cdot Z + W \cdot \overline{X} \cdot \overline{Z}$$

(b) $F = F(W,X,Y,Z)$

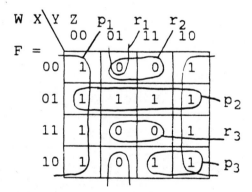

minimum SOP for F
$$F = p_1 + p_2 + p_3$$

$$F = \overline{Z} + \overline{W} \cdot X + W \cdot \overline{X} \cdot Y$$

minimum SOP for $\overline{F}$

$$\overline{F} = r_1 + r_2 + r_3$$

$$\overline{F} = \overline{X} \cdot \overline{Y} \cdot Z + \overline{W} \cdot \overline{X} \cdot Z + W \cdot X \cdot Z$$

49

(c) $F = F(W,X,Y,Z)$

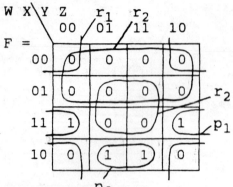

minimum SOP for F
$$F = p_1 + p_2$$

$$F = W \cdot X \cdot \overline{Z} + W \cdot \overline{X} \cdot Y$$

minimum SOP for $\overline{F}$

$$\overline{F} = r_1 + r_2 + r_3$$

$$\overline{F} = \overline{X} \cdot \overline{Z} + \overline{W} + X \cdot Z$$

3-27.

(a) $F = F(V,W,X,Y,Z)$

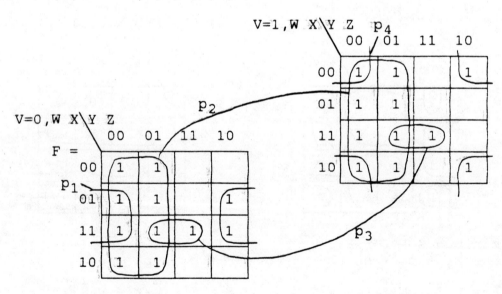

minimum SOP
$$F = p_1 + p_2 + p_3 + p_4$$

$$F = \overline{V} \cdot X \cdot \overline{Z} + \overline{Y} + W \cdot X \cdot Z + V \cdot \overline{X} \cdot \overline{Z}$$

(b) $F = F(V,W,X,Y,Z)$

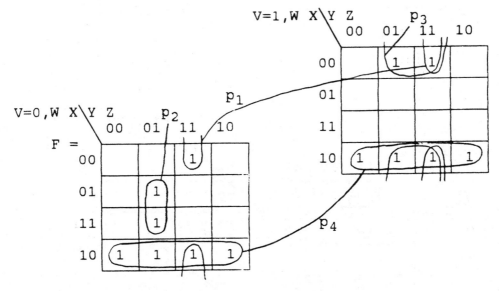

minimum SOP
$F = p_1 + p_2 + p_3 + p_4$

$F = \bar{X} \cdot Y \cdot Z + \bar{V} \cdot X \cdot \bar{Y} \cdot Z + V \cdot \bar{X} \cdot Z + W \cdot \bar{X}$

(c) $F = F(V,W,X,Y,Z)$

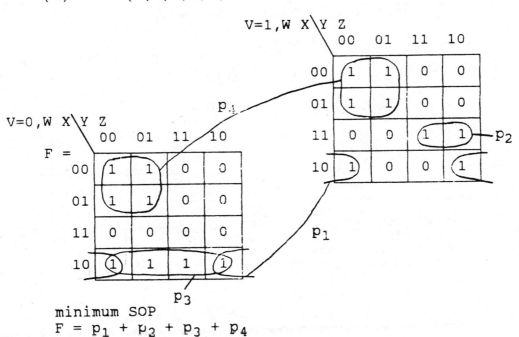

minimum SOP
$F = p_1 + p_2 + p_3 + p_4$

$F = W \cdot \bar{X} \cdot \bar{Z} + V \cdot W \cdot X \cdot Y + \bar{V} \cdot W \cdot \bar{X} + \bar{W} \cdot \bar{Y}$

3-28.

(a) $F = F(A,B,C,D,E) = \pi M(2,3,7,10,11,19,22,$
$23,27,30)$

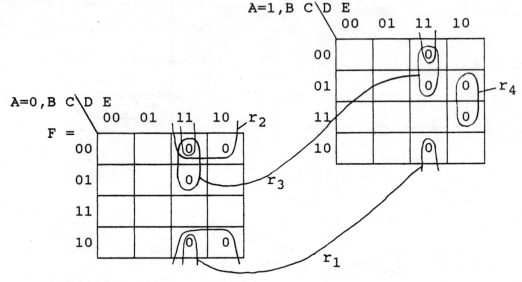

minimum SOP for the 0s

$\overline{F} = r_1 + r_2 + r_3 + r_4$

$\overline{F} = \overline{C}\cdot D\cdot\overline{E} + \overline{A}\cdot\overline{C}\cdot D + \overline{B}\cdot D\cdot E + A\cdot C\cdot D\cdot\overline{E}$

(b) $F = F(A,B,C,D,E) = \pi M(0,1,2,4,6,7,12,14,15,16,$
$18,20,21,22,23,28,29,30,$
$31)$

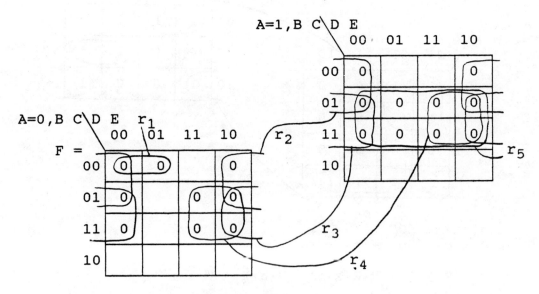

minimum SOP for the 0s

$\overline{F} = r_1 + r_2 + r_3 + r_4 + r_5$

52

$$\overline{F} = \overline{A} \cdot \overline{B} \cdot \overline{C} \cdot \overline{D} + \overline{B} \cdot \overline{E} + C \cdot \overline{E} + C \cdot D \mid A \cdot C$$

(c) $F = F(A,B,C,D,E)$

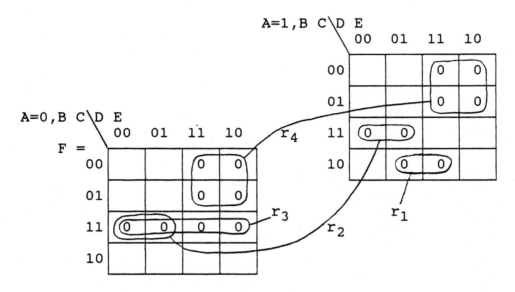

minimum SOP for the 0s

$$\overline{F} = r_1 + r_2 + r_3 + r_4$$

$$\overline{F} = A \cdot B \cdot \overline{C} \cdot D + B \cdot C \cdot \overline{D} + \overline{A} \cdot B \cdot C + \overline{B} \cdot E$$

3-29.

V W\X Y Z

F(V,W,X,Y,Z) =	000	001	011	010	110	111	101	100
00	m_0	m_1	m_3	m_2	m_6	m_7	m_5	m_4
01	m_8	m_9	m_{11}	m_{10}	m_{14}	m_{15}	m_{13}	m_{12}
11	m_{24}	m_{25}	m_{27}	m_{26}	m_{30}	m_{31}	m_{29}	m_{28}
10	m_{16}	m_{17}	m_{19}	m_{18}	m_{22}	m_{23}	m_{21}	m_{20}

Section 3-5 Multiple Function Minimization

3-30.

(a) Fi = Fi(X,Y,Z)   where i = 1, 2

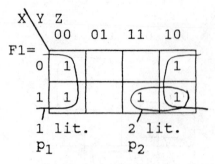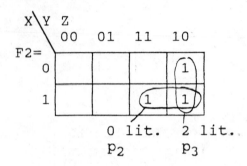

$$F1 = p_1 + p_2 = \overline{Z} + X \cdot Y$$

$$F2 = \underline{p_2} + p_3 = \underline{X \cdot Y} + Y \cdot \overline{Z} \qquad 5 \text{ lit. total}$$

(b) Fi = Fi(X,Y,Z)   where i = 1, 2

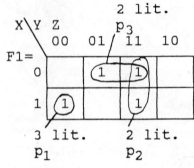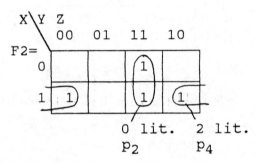

$$F1 = p_1 + p_2 + p_3 = X \cdot \overline{Y} \cdot \overline{Z} + Y \cdot Z + \overline{X} \cdot Z$$

$$F2 = \underline{p_2} + p_4 = \underline{Y \cdot Z} + X \cdot \overline{Z} \qquad 9 \text{ lit. total}$$

(c) Fi = Fi(A,B,C)   where i = 1, 2, 3

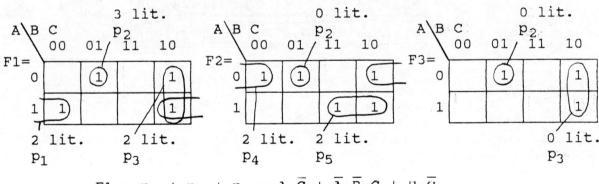

$$F1 = p_1 + p_2 + p_3 = A \cdot \overline{C} + \overline{A} \cdot \overline{B} \cdot C + B \cdot \overline{C}$$

$$F2 = \underline{p_2} + p_4 + p_5 = \underline{\overline{A} \cdot \overline{B} \cdot C} + \overline{A} \cdot \overline{C} + A \cdot B$$

$$F3 = \underline{p_2} + \underline{p_3} = \underline{\overline{A} \cdot \overline{B} \cdot C} + \underline{B \cdot \overline{C}} \qquad 11 \text{ lit. total}$$

3-31.

(a) $F_i = F_i(X,Y,Z)$ where $i = 1, 2$

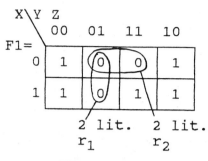
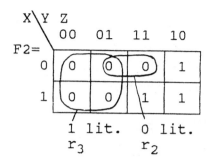

$$\overline{F1} = r_1 + r_2 = \overline{Y} \cdot Z + \overline{X} \cdot Z$$

$$\overline{F2} = \underline{r_2} + r_3 = \underline{\overline{X} \cdot Z} + \overline{Y} \qquad 5 \text{ lit. total}$$

(b) $F_i = F_i(X,Y,Z)$ where $i = 1, 2$

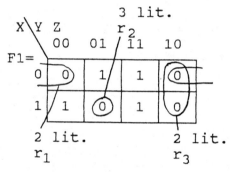
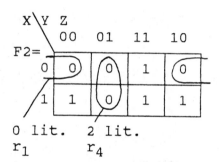

$$\overline{F1} = r_1 + r_2 + r_3 = \overline{X} \cdot \overline{Z} + X \cdot \overline{Y} \cdot Z + X \cdot \overline{Z}$$

$$\overline{F2} = \underline{r_1} + r_4 = \underline{\overline{X} \cdot \overline{Z}} + \overline{Y} \cdot Z \qquad 9 \text{ lit. total}$$

(c) $F_i = F_i(A,B,C)$ where $i = 1, 2, 3$

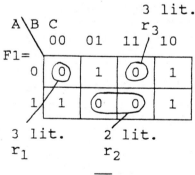
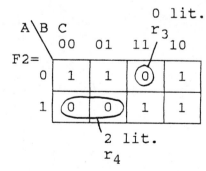
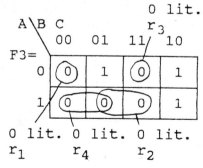

$$\overline{F1} = r_1 + r_2 + r_3 = \overline{A} \cdot \overline{B} \cdot \overline{C} + A \cdot C + \overline{A} \cdot B \cdot C$$

$$\overline{F2} = \underline{r_3} + r_4 = \underline{\overline{A} \cdot B \cdot C} + A \cdot \overline{B}$$

$$\overline{F3} = \underline{r_1} + \underline{r_2} + \underline{r_3} + \underline{r_4} = \underline{\overline{A} \cdot \overline{B} \cdot \overline{C}} + \underline{A \cdot C}$$

$$+ \underline{\overline{A} \cdot B \cdot C} + \underline{A \cdot \overline{B}} \qquad 10 \text{ lit. total}$$

3-32.

(a) Fi = Fi(A B C D) where i = 1, 2

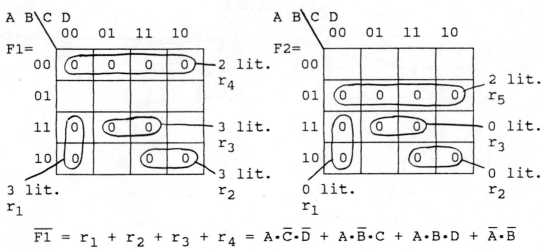

$$\overline{F1} = r_1 + r_2 + r_3 + r_4 = A\cdot\overline{C}\cdot\overline{D} + A\cdot\overline{B}\cdot C + A\cdot B\cdot D + \overline{A}\cdot\overline{B}$$

$$\overline{F2} = \underline{r_1} + \underline{r_2} + \underline{r_3} + r_5 = \underline{A\cdot\overline{C}\cdot\overline{D}} + \underline{A\cdot\overline{B}\cdot C} + \underline{A\cdot B\cdot D} + \overline{A}\cdot B$$

13 lit. total

(b) Fi = Fi(W X Y Z) where i = 1, 2

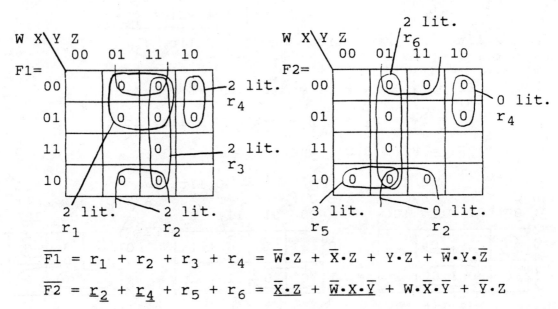

$$\overline{F1} = r_1 + r_2 + r_3 + r_4 = \overline{W}\cdot Z + \overline{X}\cdot Z + Y\cdot Z + \overline{W}\cdot Y\cdot\overline{Z}$$

$$\overline{F2} = \underline{r_2} + \underline{r_4} + r_5 + r_6 = \underline{\overline{X}\cdot Z} + \underline{\overline{W}\cdot X\cdot\overline{Y}} + W\cdot\overline{X}\cdot\overline{Y} + \overline{Y}\cdot Z$$

14 lit. total

3-33.

(a) Fi = Fi(A B C D) where i = 1, 2

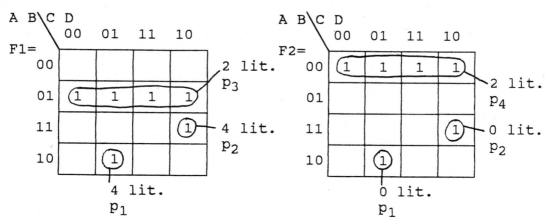

$$F1 = p_1 + p_2 + p_3 = A \cdot \overline{B} \cdot \overline{C} \cdot D + A \cdot B \cdot C \cdot \overline{D} + \overline{A} \cdot B$$

$$F2 = \underline{p_1} + \underline{p_2} + p_4 = \underline{A \cdot \overline{B} \cdot \overline{C} \cdot D} + \underline{A \cdot B \cdot C \cdot \overline{D}} + \overline{A} \cdot \overline{B}$$

12 lit. total

(b) Fi = Fi(W X Y Z) where i = 1, 2

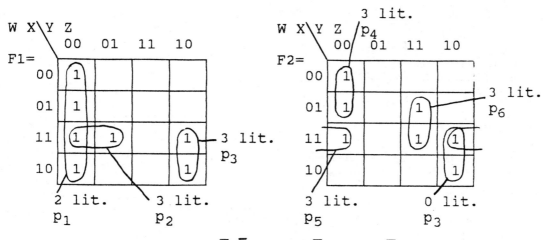

$$F1 = p_1 + p_2 + p_3 = \overline{Y} \cdot \overline{Z} + W \cdot X \cdot \overline{Y} + W \cdot Y \cdot \overline{Z}$$

$$F2 = \underline{p_3} + p_4 + p_5 + p_6 = \underline{W \cdot Y \cdot \overline{Z}} + \overline{W} \cdot \overline{Y} \cdot Z + W \cdot X \cdot \overline{Z} + X \cdot Y \cdot Z$$

17 lit. total

Section 3-6 Don't Care Output Conditions

3-34.

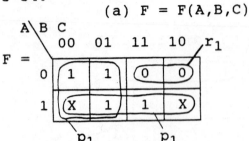

(a) $F = F(A,B,C)$

minimum SOP

$$F = p_1 + p_2 = \overline{B} + A$$

minimum POS

$$\overline{F} = r_1 = \overline{A} \cdot B$$

$$F = A + \overline{B}$$

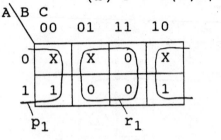

(b) $F = F(A,B,C)$

minimum SOP

$$F = p_1 = \overline{C}$$

minimum POS

$$\overline{F} = r_1 = C$$

$$F = \overline{C}$$

(c) $F = F(A,B,C)$

minimum SOP

$$F = p_1 + p_2 = \overline{B} \cdot \overline{C} + A \cdot B \cdot C$$

minimum POS

$$\overline{F} = r_1 + r_2 + r_3 = A \cdot \overline{C} + \overline{B} \cdot C + \overline{A} \cdot B$$

$$F = (\overline{A} + C) \cdot (B + \overline{C}) \cdot (A + \overline{B})$$

3-35. The minimum SOP expression of a function is algebraically identical to the minimum POS expression of that function when a don't care is used only as a 1 or only as a 0.

3-36. The minimum SOP expression of a function is <u>not</u> algebraically identical to the minimum POS expression of that function when a don't care is used both as a 1 and as a 0.

3-37.

(a) $F = F(W,X,Y,Z)$

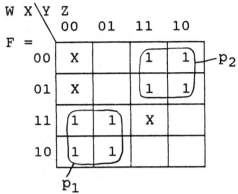

$F = p_1 + p_2$

$F = W \cdot \overline{Y} + \overline{W} \cdot Y$

(b) $F = F(W,X,Y,Z)$

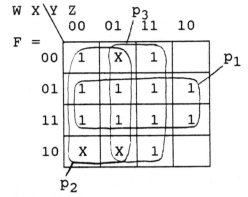

$F = p_1 + p_2 + p_3$

$F = X + \overline{Y} + Z$

(c) $F = F(W,X,Y,Z)$

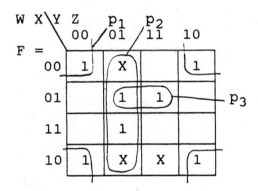

$F = p_1 + p_2 + p_3$

$F = \overline{X} \cdot \overline{Z} + \overline{Y} \cdot Z + \overline{W} \cdot X \cdot Z$

3-38. Starting with the POS expression

$F1 = (X + Y) \cdot (\overline{X} + \overline{Y})$

$= X \cdot \overline{X} + X \cdot \overline{Y} + Y \cdot \overline{X} + Y \cdot \overline{Y}$ by Postulate P4b
(AND is distributive over OR)

$= X \cdot \overline{Y} + \overline{X} \cdot Y = \overline{X} \cdot Y + X \cdot \overline{Y}$

Therefore $(X + Y) \cdot (\overline{X} + \overline{Y})$ and $\overline{X} \cdot Y + X \cdot \overline{Y}$ are algebraically identical since one expression can be obtained algebraically from the other.

Starting with the POS expression

$$F2 = (A + C) \cdot (B + \overline{C})$$

$$= A \cdot B + A \cdot \overline{C} + C \cdot B + C \cdot \overline{C} \qquad \text{by Postulate P4b}$$
(AND is distributive over OR)

$$= A \cdot B + A \cdot \overline{C} + B \cdot C$$

$$= A \cdot \overline{C} + B \cdot C \qquad \text{by Theorem T8a}$$
(Consensus Theorem)

Therefore $(A + C) \cdot (B + \overline{C})$ and $A \cdot \overline{C} + B \cdot C$ are algebraically identical since one expression can be obtained algebraically from the other.

3-39.

(a) $F = F(W,X,Y,Z)$ (b) $F = F(W,X,Y,Z)$

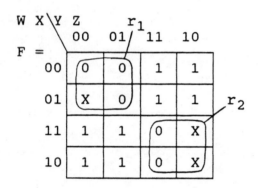 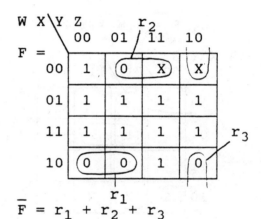

$\overline{F} = r_1 + r_2$ $\overline{F} = r_1 + r_2 + r_3$

$\overline{F} = \overline{W} \cdot \overline{Y} + W \cdot Y$ $\overline{F} = W \cdot \overline{X} \cdot \overline{Y} + \overline{W} \cdot \overline{X} \cdot Z + \overline{X} \cdot Y \cdot \overline{Z}$

$F = (W + Y) \cdot (\overline{W} + \overline{Y})$ $F = (\overline{W} + X + Y) \cdot (W + X +$

$\overline{Z}) \cdot (X + \overline{Y} + Z)$

(c) $F = F(W,X,Y,Z)$

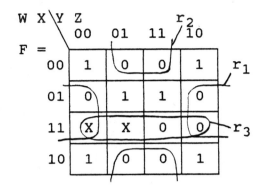

$\overline{F} = r_1 + r_2 + r_3$

$\overline{F} = X \cdot \overline{Z} + \overline{X} \cdot Z + W \cdot X$

$F = (\overline{X} + Z) \cdot (X + \overline{Z}) \cdot (\overline{W} + \overline{X})$

3-40.

(a) $F = F(W,X,Y,Z)$

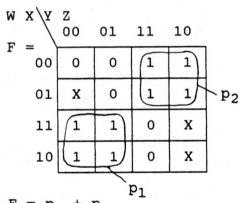

$F = p_1 + p_2$

$F = W \cdot \overline{Y} + \overline{W} \cdot Y$

(b) $F = F(W,X,Y,Z)$

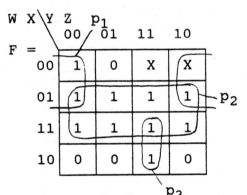

$F = p_1 + p_2 + p_3$

$F = \overline{W} \cdot \overline{Z} + X + W \cdot Y \cdot Z$

(c) $F = F(W,X,Y,Z)$

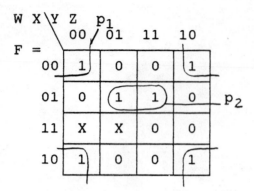

$F = p_1 + p_2$

$F = \overline{X} \cdot \overline{Z} + \overline{W} \cdot X \cdot Z$

3-41.

$F3 = F3(I3,I2,I1,I0)$ $F2 = F2(I3,I2,I1,I0)$

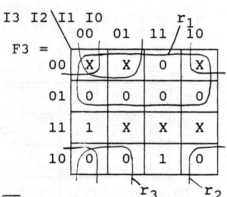 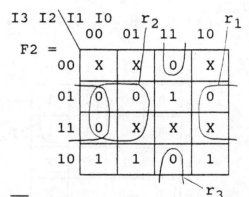

$\overline{F3} = r_1 + r_2 + r_3$ $\overline{F2} = r_1 + r_2 + r_3$

$\overline{F3} = \overline{I3} + \overline{I2} \cdot \overline{I0} + \overline{I2} \cdot \overline{I1}$ $\overline{F2} = I2 \cdot \overline{I0} + I2 \cdot \overline{I1} +$

$\phantom{\overline{F2} =} \overline{I2} \cdot I1 \cdot I0$

$F3 = I3 \cdot (I2 + I0) \cdot (I2 + I1)$ $F2 = (\overline{I2} + I0) \cdot (\overline{I2} + I1)$

$ \cdot (I2 + \overline{I1} + \overline{I0})$

$$F1 = F1(I3, I2, I2, I0)$$

$$F1 =$$

I3 I2 \ I1 I0	00	01	11	10
00	X	X	0	X
01	0	1	0	1
11	0	X	X	X
10	0	1	0	1

r_1 is above column 11, r_2 points to the 11 column.

$$\overline{F1} = r_1 + r_2$$

$$\overline{F1} = \overline{I1} \cdot \overline{I0} + I1 \cdot I0$$

$$F1 = (I1 + I0) \cdot (\overline{I1} + \overline{I0})$$

By inspecting the XS3 to BCD truth table in Example 3-7,

one can see that $F0 = \overline{I0}$.

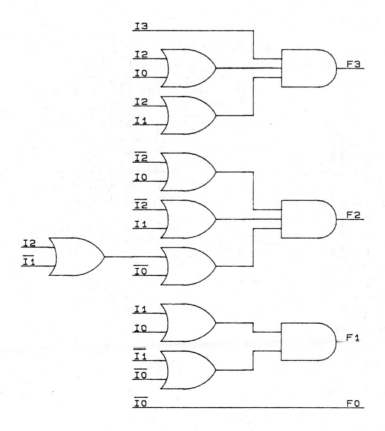

3-42.

(a) $F = F(A,B,C,D,E)$

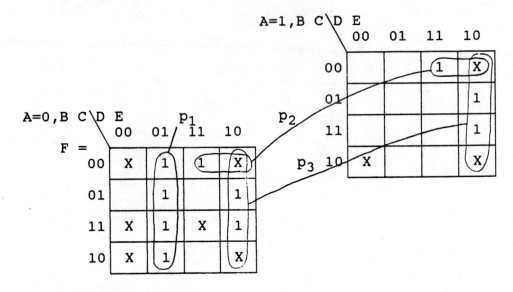

minimum SOP for F

$F = p_1 + p_2 + p_3$

$F = \overline{A} \cdot \overline{D} \cdot E + \overline{B} \cdot \overline{C} \cdot D + D \cdot \overline{E}$

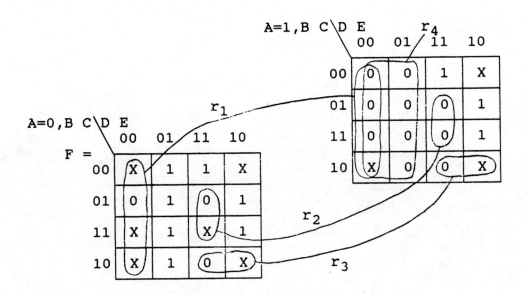

minimum POS for F

$\overline{F} = r_1 + r_2 + r_3 + r_4$

$\overline{F} = \overline{D} \cdot \overline{E} + C \cdot D \cdot E + B \cdot \overline{C} \cdot D + \overline{D}$

$F = (D + E) \cdot (\overline{C} + \overline{D} + \overline{E}) \cdot (\overline{B} + C + \overline{D}) \cdot D$

(b) F = F(A,B,C,D,E)

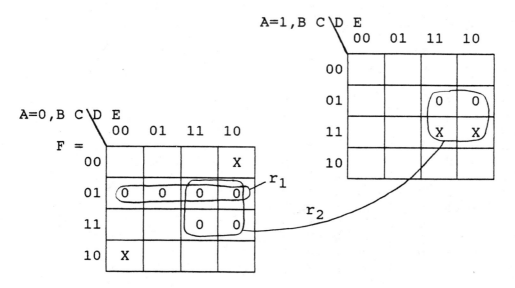

minimum POS for F

$\overline{F} = r_1 + r_2$

$\overline{F} = \overline{A} \cdot \overline{B} \cdot C + C \cdot D$

$F = (A + B + \overline{C}) \cdot (\overline{C} + \overline{D})$

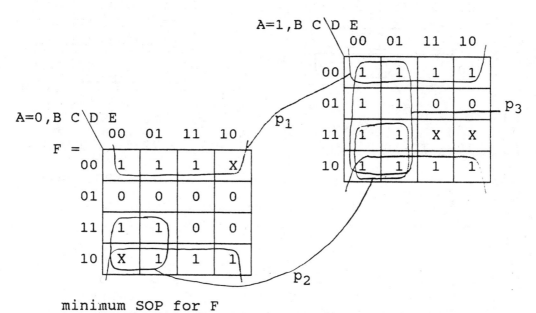

minimum SOP for F

$F = p_1 + p_2 + p_3$

$F = \overline{C} + B \cdot \overline{D} + A \cdot \overline{D}$

65

3-43.

	BCD				2421		
I3	I2	I1	I0	F3	F2	F1	F0
0	0	0	0	0	0	0	0
0	0	0	1	0	0	0	1
0	0	1	0	0	0	1	0
0	0	1	1	0	0	1	1
0	1	0	0	0	1	0	0
0	1	0	1	1	0	1	1
0	1	1	0	1	1	0	0
0	1	1	1	1	1	0	1
1	0	0	0	1	1	1	0
1	0	0	1	1	1	1	1
1	0	1	0	X	X	X	X
1	0	1	1	X	X	X	X
1	1	0	0	X	X	X	X
1	1	0	1	X	X	X	X
1	1	1	0	X	X	X	X
1	1	1	1	X	X	X	X

$$F3 = F3(I3, I2, I1, I0)$$

$$F2 = F2(I3, I2, I1, I0)$$

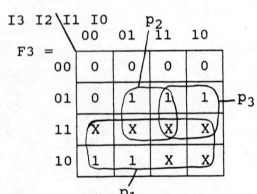

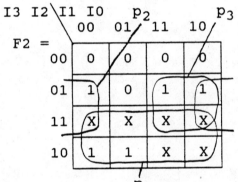

$$F3 = p_1 + p_2 + p_3$$

$$F3 = I3 + I2 \cdot I0 + I2 \cdot I1$$

$$F2 = p_1 + p_2 + p_3$$

$$F2 = I3 + I2 \cdot \overline{I0} + \underline{I2 \cdot I1}$$

(no cost)

$$F1 = F1(I3, I2, I2, I0)$$

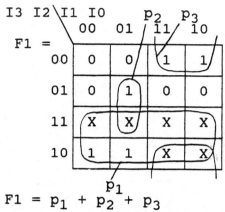

$$F1 = p_1 + p_2 + p_3$$

$$F1 = I3 + I2 \cdot \overline{I1} \cdot I0 + \overline{I2} \cdot I1$$

By inspecting the BCD to 2421 truth table illustrated above one can see that $F0 = I0$.

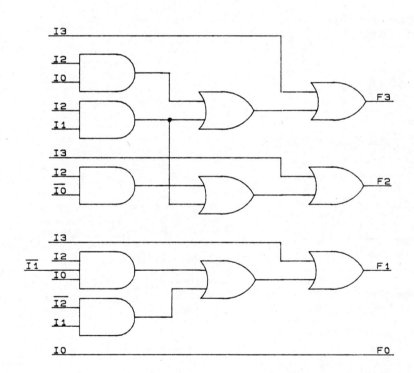

Section 3-7 Additional Functional Forms

3-44.

$$F = F(W,X,Y,Z)$$

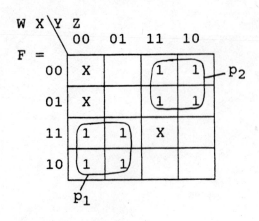

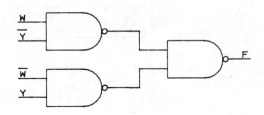

$$F = p_1 + p_2$$

$$F = W \cdot \overline{Y} + \overline{W} \cdot Y$$

$$F = \overline{\overline{W \cdot \overline{Y} + \overline{W} \cdot Y}} = \overline{(\overline{W \cdot \overline{Y}}) \cdot (\overline{\overline{W} \cdot Y})}$$

Recall that the NAND/NAND form is
a SOP derived form (see Fig. 3-20).

3-45.

$$F = F(W,X,Y,Z)$$

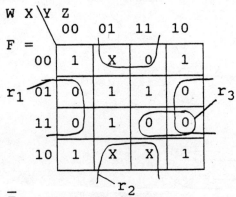

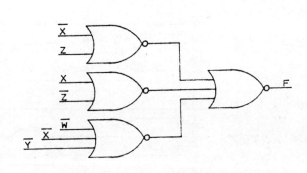

$$\overline{F} = r_1 + r_2 + r_3$$

$$\overline{F} = X \cdot \overline{Z} + \overline{X} \cdot Z + W \cdot X \cdot Y$$

$$F = (\overline{X} + Z) \cdot (X + \overline{Z}) \cdot (\overline{W} + \overline{X} + \overline{Y})$$

$$F = \overline{\overline{(\overline{X} + Z) \cdot (X + \overline{Z}) \cdot (\overline{W} + \overline{X} + \overline{Y})}}$$

$$F = \overline{(\overline{\overline{X} + Z}) + (\overline{X + \overline{Z}}) + (\overline{\overline{W} + \overline{X} + \overline{Y}})}$$

Recall that the NOR/NOR form is
a POS derived form (see Fig. 3-21).

3-46.

$$F = F(W,X,Y,Z)$$

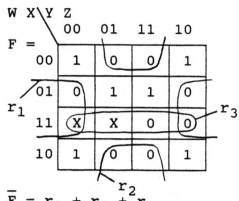

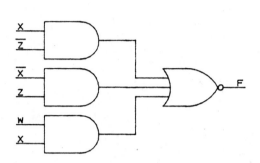

$$\overline{F} = r_1 + r_2 + r_3$$

$$\overline{F} = X \cdot \overline{Z} + \overline{X} \cdot Z + W \cdot X$$

$$F = \overline{X \cdot \overline{Z} + \overline{X} \cdot Z + W \cdot X}$$

Recall that the AND/NOR form is
a POS derived form (see Fig. 3-21).

3-47.

$$F = F(W,X,Y,Z)$$

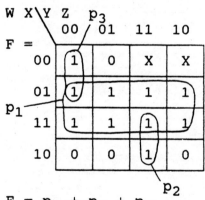

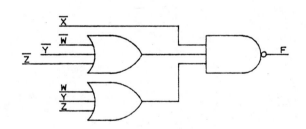

$$F = p_1 + p_2 + p_3$$

$$F = X + W \cdot Y \cdot Z + \overline{W} \cdot \overline{Y} \cdot \overline{Z}$$

$$F = \overline{\overline{X + W \cdot Y \cdot Z + \overline{W} \cdot \overline{Y} \cdot \overline{Z}}}$$

$$F = \overline{\overline{X} \cdot (\overline{W \cdot Y \cdot Z}) \cdot (\overline{\overline{W} \cdot \overline{Y} \cdot \overline{Z}})}$$

$$F = \overline{\overline{X} \cdot (\overline{W} + \overline{Y} + \overline{Z}) \cdot (W + Y + Z)}$$

Recall that the OR/NAND form is
a SOP derived form (see Fig. 3-20).

69

3-48.

(a) AND/OR -- SOP form

$$F1 = \overline{X} \cdot Y + X \cdot \overline{Y}$$

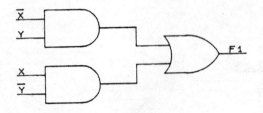

(b) NAND/NAND -- derived SOP form

$$F1 = \overline{X} \cdot Y + X \cdot \overline{Y}$$
$$F1 = \overline{\overline{\overline{X} \cdot Y + X \cdot \overline{Y}}}$$
$$F1 = \overline{(\overline{\overline{X} \cdot Y}) \cdot (\overline{X \cdot \overline{Y}})}$$

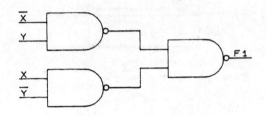

(c) NOR/NOR -- derived POS form

$$F1 = (X + Y) \cdot (\overline{X} + \overline{Y})$$
$$F1 = \overline{\overline{(X + Y) \cdot (\overline{X} + \overline{Y})}}$$
$$F1 = \overline{\overline{(X + Y)} + \overline{(\overline{X} + \overline{Y})}}$$

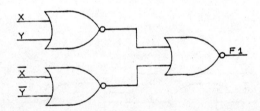

(d) NAND/AND -- derived POS form

$$F1 = (X + Y) \cdot (\overline{X} + \overline{Y})$$
$$F1 = \overline{\overline{(X + Y) \cdot (\overline{X} + \overline{Y})}}$$
$$F1 = \overline{\overline{(X + Y)} + \overline{(\overline{X} + \overline{Y})}}$$
$$F1 = \overline{(\overline{X} \cdot \overline{Y}) + (X \cdot Y)}$$
$$F1 = \overline{(\overline{X} \cdot \overline{Y})} \cdot \overline{(X \cdot Y)}$$

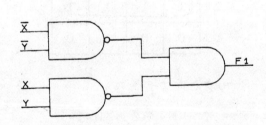

3.49.

$$F = (A + C) \cdot (B + \bar{C})$$

The NAND/NAND form is derived from the SOP form of the equation and not the POS form.

$$\bar{F} = \bar{A} \cdot \bar{C} + \bar{B} \cdot C \quad \text{-- Converted to the SOP form of } \bar{F} \text{ so the}$$
0s of the function can be easily plotted.

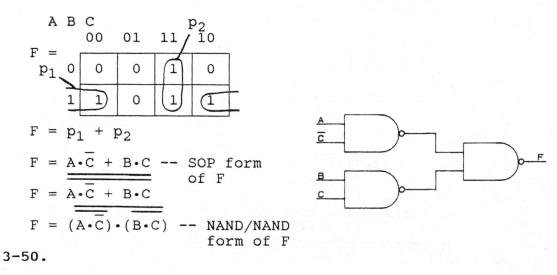

$$F = p_1 + p_2$$

$$F = A \cdot \bar{C} + B \cdot C \quad \text{-- SOP form}$$
$$\text{of F}$$

$$F = \overline{\overline{A \cdot \bar{C} + B \cdot C}}$$

$$F = \overline{\overline{(A \cdot \bar{C})} \cdot \overline{(B \cdot C)}} \quad \text{-- NAND/NAND}$$
$$\text{form of F}$$

3-50.

$$F = A \cdot \bar{C} + B \cdot C$$

The NOR/NOR form is derived from the POS form of the equation and not the SOP form. Using the SOP form of the function, the 1s can be easily plotted.

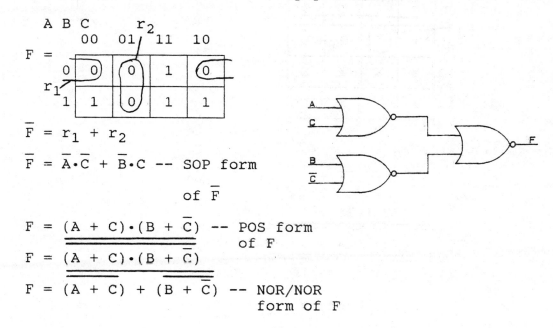

$$\bar{F} = r_1 + r_2$$

$$\bar{F} = \bar{A} \cdot \bar{C} + \bar{B} \cdot C \quad \text{-- SOP form}$$
$$\text{of } \bar{F}$$

$$F = (A + C) \cdot (B + \bar{C}) \quad \text{-- POS form}$$
$$\text{of F}$$

$$F = \overline{\overline{(A + C)} \cdot \overline{(B + \bar{C})}}$$

$$F = \overline{\overline{(A + C)} + \overline{(B + \bar{C})}} \quad \text{-- NOR/NOR}$$
$$\text{form of F}$$

3-51. The NOR/NOR form is a POS derived form.

$$F3 = F3(I3, I2, I1, I0)$$

F3 map (I3 I2 \ I1 I0):

I3 I2 \ I1 I0	00	01	11	10
00	0	0	0	0
01	0	1	1	1
11	X	X	X	X
10	1	1	X	X

$$\overline{F3} = r_1 + r_2$$

$$\overline{F3} = \overline{I3} \cdot \overline{I2} + \overline{I3} \cdot \overline{I1} \cdot \overline{I0}$$

$$F3 = (I3 + I2) \cdot (I3 + I1 + I0)$$

$$F3 = \overline{\overline{(I3 + I2) \cdot (I3 + I1 + I0)}}$$

$$F3 = \overline{\overline{(I3 + I2)} + \overline{(I3 + I1 + I0)}}$$

$$F2 = F2(I3, I2, I1, I0)$$

F2 map (I3 I2 \ I1 I0):

I3 I2 \ I1 I0	00	01	11	10
00	0	0	0	0
01	1	0	1	1
11	X	X	X	X
10	1	1	X	X

$$\overline{F2} = r_1 + r_2$$

$$\overline{F2} = \overline{I3} \cdot \overline{I2} + \overline{I3} \cdot \overline{I1} \cdot I0$$

$$F2 = (I3 + I2) \cdot (I3 + I1 + \overline{I0})$$

$$F2 = \overline{\overline{(I3 + I2) \cdot (I3 + I1 + \overline{I0})}}$$

$$F2 = \overline{\overline{(I3 + I2)} + \overline{(I3 + I1 + \overline{I0})}}$$

(no cost)

$$F1 = F1(I3, I2, I2, I0)$$

I3 I2\I1 I0

	00	01	11	10
00	0	0	1	1
01	0	1	0	0
11	X	X	X	X
10	1	1	X	X

F1 = (with groupings r_1, r_2, r_3)

$$\overline{F1} = r_1 + r_2 + r_3$$

$$\overline{F1} = \overline{I3} \cdot \overline{I2} \cdot \overline{I1} + I2 \cdot \overline{I0} + I2 \cdot I1$$

$$F1 = (I3 + I2 + I1) \cdot (\overline{I2} + I0) \cdot (\overline{I2} + \overline{I1})$$

$$F1 = \overline{\overline{(I3 + I2 + I1) \cdot (\overline{I2} + I0) \cdot (\overline{I2} + \overline{I1})}}$$

$$F1 = \overline{\overline{(I3 + I2 + I1)} + \overline{(\overline{I2} + I0)} + \overline{(\overline{I2} + \overline{I1})}}$$

By inspecting the truth table, F0 = I0.

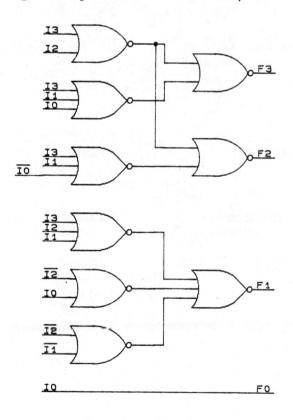

73

Section 4-3   Minimizing Two-Variable Functions

4-1.

(a) $F(X,Y) = \Sigma m(0,1)$

Step 1

Column 1 (0-cubes)	Column 2 (1-cubes)
0 00 √	0,1 0-
1 01 √	

$F = \overline{X}$ minimum SOP for 1s

(b) $F(X,Y) = \Sigma m(2,3)$

Step 1

Column 1 (0-cubes)	Column 2 (1-cubes)
2 10 √	2,3 1-
3 11 √	

$F = X$ minimum SOP for 1s

(c) $F(X,Y) = \Sigma m(0,2)$

Step 1

Column 1 (0-cubes)	Column 2 (1-cubes)
0 00 √	0,2 -0
2 10 √	

$F = \overline{Y}$ minimum SOP for 1s

(d) $F(X,Y) = \Sigma m(0,3)$

Step 1

Column 1 (0-cubes)	
0 00 √	group 0
3 11 √	group 2

No adjacent minterms.

$F = \overline{X}\cdot\overline{Y} + X\cdot Y$ minimum SOP for 1s

4-2.

(a) $F(X,Y) = \Sigma m(0,1)$
$F(X,Y) = \pi M(2,3)$

Step 1

Column 1 (0-cubes)	Column 2 (1-cubes)
2 10 √	2,3 1-
3 11 √	

$\overline{F} = X$ minimum SOP for 0s

(b) $F(X,Y) = \Sigma m(2,3)$
$F(X,Y) = \pi M(0,1)$

Step 1

Column 1 (0-cubes)	Column 2 (1-cubes)
0 00 √	0,1 0-
1 01 √	

$\overline{F} = \overline{X}$ minimum SOP for 0s

(c) $F(X,Y) = \Sigma m(0,2)$
   $F(X,Y) = \pi M(1,3)$

Step 1

Column 1 (0-cubes)	Column 2 (1-cubes)
1 01 √	1,3 -1
3 11 √	

$\overline{F} = Y$ minimum SOP for 0s

(d) $F(X,Y) = \Sigma m(0,3)$
   $F(X,Y) = \pi M(1,2)$

Step 1

Column 1
(0-cubes)

1 01	group 1
2 10	group 1

No adjacent minterms.

$\overline{F} = \overline{X} \cdot Y + X \cdot \overline{Y}$ minimum SOP for 0s

4-3.

(a) $F(X,Y) = \Sigma m(0,1,2)$

Step 1

Column 1 (0-cubes)	Column 2 (1-cubes)
0 00 √	0,1 0-
	0,2 -0
1 01 √	
2 10 √	

$F = \overline{X} + \overline{Y}$ minimum SOP for 1s

(b) $F(X,Y) = \Sigma m(0,2,3)$

Step 1

Column 1 (0-cubes)	Column 2 (1-cubes)
0 00 √	0,2 -0
2 10 √	2,3 1-
3 11 √	

$F = \overline{Y} + X$ minimum SOP for 1s

(c) $F(X,Y) = \Sigma m(0,1,3)$

Step 1

Column 1 (0-cubes)	Column 2 (1-cubes)
0 00 √	0,1 0-
1 01 √	1,3 -1
3 11 √	

$F = \overline{X} + Y$ minimum SOP for 1s

4-4.

(a) $F(X,Y) = \Sigma m(3)$
    $F(X,Y) = \pi M(0,1,2)$

   Step 1

Column 1 (0-cubes)	Column 2 (1-cubes)
0 00 √	0,1 0-
	0,2 -0
1 01 √	
2 10 √	

$\overline{F} = \overline{X} + \overline{Y}$
$F = X \cdot Y$ minimum POS for 0s

(b) $F(X,Y) = \Sigma m(1)$
    $F(X,Y) = \pi M(0,2,3)$

   Step 1

Column 1 (0-cubes)	Column 2 (1-cubes)
0 00 √	0,2 -0
2 10 √	2,3 1-
3 11 √	

$\overline{F} = \overline{Y} + X$

$F = Y \cdot \overline{X}$ minimum POS for 0s

(c) $F(X,Y) = \Sigma m(2)$
    $F(X,Y) = \pi M(0,1,3)$

   Step 1

Column 1 (0-cubes)	Column 2 (1-cubes)
0 00 √	0,1 0-
1 01 √	1,3 -1
3 11 √	

$\overline{F} = \overline{X} + Y$

$F = X \cdot \overline{Y}$ minimum POS for 0s

Section 4-4 Minimizing Three-and Four-Variable Functions

4-5.   (a) $F(X,Y,Z) = \Sigma m(0,3,4,6)$

   Step 1

Column 1 (0-cubes)	Column 2 (1-cubes)
0 000 √	0,4 -00
4 100 √	4,6 1-0
3 011	
6 110 √	

   Step 2

		0	3	4	6
*	3 011		√		
*	0,4 -00	√		√	
*	4,6 1-0			√	√

$F = \overline{X} \cdot Y \cdot Z + \overline{Y} \cdot \overline{Z} + X \cdot \overline{Z}$
minimum SOP for 1s

76

(b) $F(X,Y,Z) = \Sigma m(0,2,3,6,7)$

Step 1

Column 1 (0-cubes)	Column 2 (1-cubes)	Column 3 (2-cubes)
0 000 √	0,2 0-0	no combinations
2 010 √	2,3 01- √ 2,6 -10 √	2,3,6,7 -1- ~~2,6,3,7, 1-~~
3 011 √ 6 110 √	3,7 -11 √ 6,7 11- √	
7 111 √		

Step 2

```
                      0 2 3 6 7
  *       0,2 0-0  | √ √
  *   2,3,6,7 -1-  |   √ √ √ √
```

$F = \overline{X} \cdot \overline{Z} + Y$ minimum SOP for 1s

(c) $F(X,Y,Z) = \Sigma m(0,2,4,5,6,7)$

Step 1

Column 1 (0-cubes)	Column 2 (1-cubes)	Column 3 (2-cubes)
0 000 √	0,2 0-0 √ 0,4 -00 √	0,2,4,6 --0 ~~0,4,2,6 0~~
2 010 √ 4 100 √	2,6 -10 √ 4,5 10- √ 4,6 1-0 √	4,5,6,7 1-- ~~4,6,5,7 1~~
5 101 √ 6 110 √	5,7 1-1 √ 6,7 11- √	
7 111 √		

Step 2

```
                        0 2 4 5 6 7
  *   0,2,4,6 --0  | √ √ √   √
  *   4,5,6,7 1--  |     √ √ √ √
```

$F = \overline{Z} + X$ minimum SOP for 1s

77

4-6.   (a) $F(A,B,C) = \Sigma m(0,2,3,4,6,7) + d(5)$

Step 1

Column 1 (0-cubes)	Column 2 (1-cubes)	Column 3 (2-cubes)
0 000 √	0,2 0-0 √	0,2,4,6 --0
	0,4 -00 √	~~0,4,2,6 0~~
2 010 √		
4 100 √	2,3 01- √	2,3,6,7 -1-
	2,6 -10 √	~~2,6,3,7 1~~
3 011 √	4,5 10- √	4,5,6,7 1--
5 101 √	4,6 1-0 √	~~4,6,5,7 1~~
6 110 √		
	3,7 -11 √	
7 111 √	5,7 1-1 √	
	6,7 11- √	

Step 2

```
                 0 2 3 4 6 7
*  0,2,4,6 --0  | √ √   √ √
*  2,3,6,7 -1-  |   √ √   √ √
   4,5,6,7 1--  |       √ √ √
```

$F = \overline{C} + B$ minimum SOP for 1s

(b) $F(A,B,C) = \Sigma m(0,1,2,3,7) + d(6)$

Step 1

Column 1 (0-cubes)	Column 2 (1-cubes)	Column 3 (2-cubes)
0 000 √	0,1 00- √	0,1,2,3 0--
	0,2 0-0 √	~~0,2,1,3 0~~
1 001 √		
2 010 √	1,3 0-1 √	2,3,6,7 -1-
	2,3 01- √	~~2,6,3,7 1~~
3 011 √	2,6 -10 √	
6 110 √		
	3,7 -11 √	
7 111 √	6,7 11- √	

Step 2

```
                 0 1 2 3 7
*  0,1,2,3 0--  | √ √ √ √
*  2,3,6,7 -1-  |     √ √ √
```

$F = \overline{A} + B$ minimum SOP for 1s

(c) $F(A,B,C) = \pi M(0,7) \cdot d(1)$

$F(A,B,C) = \Sigma m(2,3,4,5,6) + d(1)$

Step 1

Column 1 (0-cubes)

```
1  001 √
2  010 √
4  100 √
─────────
3  011 √
5  101 √
6  110 √
```

Column 2 (1-cubes)

```
1,3  0-1
1,5  -01
2,3  01-
2,6  -10
4,5  10-
4,6  1-0
```

Step 2

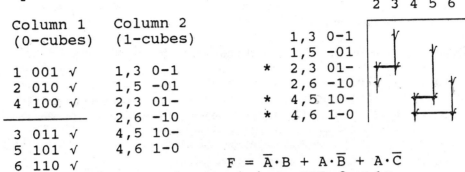

```
      1,3  0-1
      1,5  -01
  *   2,3  01-
      2,6  -10
  *   4,5  10-
  *   4,6  1-0
```

$F = \overline{A} \cdot B + A \cdot \overline{B} + A \cdot \overline{C}$
minimum SOP for 1s

Note that this is not the only minimum SOP form for the 1s of the function.

4-7. (a) $F(W,X,Y,Z) = \Sigma m(2,3,6,7,8,9,12,13)$

Step 1

Column 1 (0-cubes)

```
 2  0010 √
 8  1000 √
──────────
 3  0011 √
 6  0110 √
 9  1001 √
12  1100 √
──────────
 7  0111 √
13  1101 √
```

Column 2 (1-cubes)

```
 0,3   001- √
 2,6   0-10 √
 8,9   100- √
 8,12  1-00 √
───────────
 3,7   0-11 √
 6,7   011- √
 9,13  1-01 √
12,13  110- √
```

Column 3 (2-cubes)

```
2,3,6,7    0-1-
2,6,3,7    0-1-
8,9,12,13  1-0-
8,12,9,13  1-0-
```

Step 2

```
           2  3  6  7  8  9  12  13
  *   2,3,6,7    0-1-  │ √  √  √  √
  *   8,9,12,13  1-0-  │          √  √   √   √
```

$F = \overline{W} \cdot Y + W \cdot \overline{Y}$ minimum SOP for 1s

79

(b) $F(W,X,Y,Z) = \Sigma m(0,3,4,5,6,7,11,12,13,14,15)$

Step 1

Column 1 (0-cubes)	Column 2 (1-cubes)	Column 3 (2-cubes)	Column 4 (3-cubes)
0 0000 √	0,4 0-00	no combinations	no combinations
4 0100 √	4,5 010- √	4,5,6,7 01-- √	4,5,6,7,12,13,14,15 -1--
	4,6 01-0 √	4,5,12,13 -10- √	~~4,5,12,13,6,7,14,15 1~~
3 0011 √	4,12 -100 √	~~4,6,5,7 01~~	~~4,6,12,14,5,7,13,15 1~~
5 0101 √		4,6,12,14 -1-0 √	
6 0110 √	3,7 0-11 √	~~4,12,5,13 -10~~	
12 1100 √	3,11 -011 √	~~4,12,6,14 1-0~~	
	5,7 01-1 √		
7 0111 √	5,13 -101 √	3,7,11,15 --11	
11 1011 √	6,7 011- √	~~3,11,7,15 11~~	
13 1101 √	6,14 -110 √	5,7,13,15 -1-1 √	
14 1110 √	12,13 110- √	~~5,13,7,15 1-1~~	
	12,14 11-0 √	6,7,14,15 -11- √	
15 1111 √		~~6,14,7,15 11~~	
	7,15 -111 √	12,13,14,15 11-- √	
	11,15 1-11 √	~~12,14,13,15 11~~	
	13,15 11-1 √		
	14,15 111- √		

Step 2

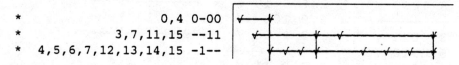

```
*                          0,4 0-00
*                  3,7,11,15 --11
*   4,5,6,7,12,13,14,15 -1--
```

$F = \overline{W} \cdot \overline{Y} \cdot \overline{Z} + Y \cdot Z + X$ minimum SOP for 1s

(c) $F(W,X,Y,Z) = \Sigma m(0,2,5,7,8,10) + d(13,15)$

Step 1

Column 1 (0-cubes)	Column 2 (1-cubes)	Column 3 (2-cubes)

```
Column 1              Column 2              Column 3
(0-cubes)             (1-cubes)             (2-cubes)

0  0000  √           0,2  00-0  √          0,2,8,10  -0-0  √
_____          0,8  -000  √          0̶,̶8̶,̶2̶,̶1̶0̶  ̶-̶0̶-̶0̶
2  0010  √           _____
8  1000  √           2,10  -010  √         no combinations
_____          8,10  10-0  √         _____
5  0101  √           _____           5,7,13,15  -1-1  √
10 1010  √           5,7  01-1  √          5̶,̶1̶3̶,̶7̶,̶1̶5̶  ̶-̶1̶-̶1̶
_____          5,13  -101  √
7  0111  √           _____
13 1101  √           7,15  -111  √
_____          13,15  11-1  √
15 1111  √
```

Step 2

```
                          0  2  5  7  8  10
 *  0,2,8,10  -0-0  │  √  √        √  √
 *  5,7,13,15 -1-1  │        √  √
```

$F = \overline{X}\cdot\overline{Z} + X\cdot Z$ minimum SOP for 1s

Section 4-5 The Abbreviated Decimal Method

4-8.   (a) $F(W,X,Y,Z) = \Sigma m(2,3,6,7,8,9,12,13) + d(0,4,15)$
         $F(W,X,Y,Z) = \pi M(1,5,10,11,14)\cdot d(0,4,15)$

Step 1

```
Column 1              Column 2              Column 3
(0-cubes)             (1-cubes)             (2-cubes)

0000   0  √          0,1  (1)  √           0,1,4,5  (1,4)  √
_____          0,4  (4)  √           0̶,̶4̶,̶1̶,̶5̶  ̶(̶4̶,̶1̶)̶
0001   1  √          _____
0100   4  √          1,5  (4)  √           no combinations
_____          4,5  (1)  √           _____
0101   5  √          _____           10,11,14,15  (1,4)  √
1010  10  √          10,11  (1)  √         1̶0̶,̶1̶4̶,̶1̶1̶,̶1̶5̶  ̶(̶4̶,̶1̶)̶
_____          10,14  (4)  √
1011  11  √          _____
1110  14  √          11,15  (4)  √
_____          14,15  (1)  √
1111  15  √
```

Step 2

	1	5	10	11	14

```
*      0,1,4,5 (1,4)     √—√
*    10,11,14,15 (1,4)          √  √   √
```

$$0,1,4,5\ (1,4) = 0\text{±}0\text{±} = \overline{W}\cdot\overline{Y}$$
$$10,11,14,15\ (1,4) = 1\text{±}1\text{±} = W\cdot Y$$

$$\overline{F} = \overline{W}\cdot\overline{Y} + W\cdot Y \quad \text{minimum SOP for 0s}$$

(b) $F(W,X,Y,Z) = \pi M(0,3,4,5,6,7,11,12,13,14,15)$
$$\cdot d(2,8,9)$$

Step 1

Column 1 (0-cubes)	Column 2 (1-cubes)	Column 3 (2-cubes)	Column 4 (3-cubes)
0000 0 √	0,2 (2) √	0,2,4,6,(2,4)	no combinations
———	0,4 (4) √	0̶,̶4̶,̶2̶,̶6̶ ̶(̶4̶,̶2̶)̶	
0010 2 √	0,8 (8) √	0,4,8,12 (4,8)	4,5,6,7,12,13,14,15 (1,2,8)
0100 4 √	———	0̶,̶8̶,̶4̶,̶1̶2̶ ̶(̶8̶,̶4̶)̶	4̶,̶5̶,̶1̶2̶,̶1̶3̶,̶6̶,̶7̶,̶1̶4̶,̶1̶5̶ ̶(̶1̶,̶8̶,̶2̶)̶
1000 8 √	2,3 (1) √	———	4̶,̶6̶,̶1̶2̶,̶1̶4̶,̶5̶,̶7̶,̶1̶3̶,̶1̶5̶ ̶(̶2̶,̶8̶,̶1̶)̶
———	2,6 (4) √	2,3,6,7 (1,4)	
0011 3 √	4,5 (1) √	2̶,̶6̶,̶3̶,̶7̶ ̶(̶4̶,̶1̶)̶	
0101 5 √	4,6 (2) √	4,5,6,7 (1,2) √	
0110 6 √	4,12 (8) √	4,5,12,13 (1,8) √	
1001 9 √	8,9 (1) √	4̶,̶6̶,̶5̶,̶7̶ ̶(̶2̶,̶1̶)̶	
1100 12 √	8,12 (4) √	4,6,12,14 (2,8) √	
———		4̶,̶1̶2̶,̶5̶,̶1̶3̶ ̶(̶8̶,̶1̶)̶	
0111 7 √	3,7 (4) √	4̶,̶1̶2̶,̶6̶,̶1̶4̶ ̶(̶8̶,̶2̶)̶	
1011 11 √	3,11 (8) √	8,9,12,13 (1,4)	
1101 13 √	5,7 (2) √	8̶,̶1̶2̶,̶9̶,̶1̶3̶ ̶(̶4̶,̶1̶)̶	
1110 14 √	5,13 (8) √	———	
———	6,7 (1) √	3,7,11,15 (4,8)	
1111 15 √	6,14 (8) √	3̶,̶1̶1̶,̶7̶,̶1̶5̶ ̶(̶8̶,̶4̶)̶	
	9,11 (2) √	5,7,13,15 (2,8) √	
	9,13 (4) √	5̶,̶1̶3̶,̶7̶,̶1̶5̶ ̶(̶8̶,̶2̶)̶	
	12,13 (1) √	6,7,14,15 (1,8) √	
	12,14 (2) √	6̶,̶1̶4̶,̶7̶,̶1̶5̶ ̶(̶8̶,̶1̶)̶	
	———	9,11,13,15 (2,4)	
	7,15 (8) √	9̶,̶1̶3̶,̶1̶1̶,̶1̶5̶ ̶(̶4̶,̶2̶)̶	
	11,15 (4) √	12,13,14,15 (1,2) √	
	13,15 (2) √	1̶2̶,̶1̶4̶,̶1̶3̶,̶1̶5̶ ̶(̶2̶,̶1̶)̶	
	14,15 (1) √		

82

Step 2

```
                                    0  3  4  5  6  7 11 12 13 14 15
*            0,2,4,6  (2,4)
             0,4,8,12 (4,8)
             2,3,6,7  (1,4)
             8,9,12,13 (1,4)
*            3,7,11,15 (4,8)
             9,11,13,15 (2,4)
* 4,5,6,7,12,13,14,15 (1,2,8)
```

$$0,2,4,6 \ (2,4) = 0\text{\textpm}\text{\textpm}0 = \overline{W}\cdot\overline{Z}$$
$$3,7,11,15 \ (4,8) = \text{\textpm}\text{\textpm}11 = Y\cdot Z$$
$$4,5,6,7,12,13,14,15 \ (1,2,8) = \text{\textpm}1\text{\textpm}\text{\textpm} = X$$

$$\overline{F} = \overline{W}\cdot\overline{Z} + Y\cdot Z + X \quad \text{minimum SOP for 0s}$$

(c) $F(W,X,Y,Z) = \Sigma m(0,1,2,3,4,5,6,7,9,11,13,15)$
$$+ \ d(10,14)$$
$$F(W,X,Y,Z) = \pi M(8,12)\cdot d(10,14)$$

Step 1

Column 1 (0-cubes)	Column 2 (1-cubes)	Column 3 (2-cubes)
1000 8 √	8,10 (2) √	8,10,12,14 (2,4)
	8,12 (4) √	~~8,12,10,14 (4,2)~~
1010 10 √		
1100 12 √	10,14 (4) √	
	12,14 (2) √	
1110 14 √		

$$8,10,12,14 \ (2,4) = 1\text{\textpm}\text{\textpm}0 = W\cdot\overline{Z}$$

$$\overline{F} = W\cdot\overline{Z} \quad \text{minimum SOP for 0s}$$

4-9.   (a) $F(V,W,X,Y,Z) = \pi M(0,1,4,5,6,8,9,12,13,14,15,16,$
$17,18,20,21,24,25,26,28,29,31)$
$F(V,W,X,Y,Z) = \Sigma m(2,3,7,10,11,19,22,23,27,30)$

Step 1

Column 1 (0-cubes)	Column 2 (1-cubes)	Column 3 (2-cubes)

```
Column 1              Column 2              Column 3
(0-cubes)             (1-cubes)             (2-cubes)

00010   2 √           2,3   (1)  √          2,3,10,11  (1,8)
─────────             2,10  (8)  √          2̶,̶1̶0̶,̶3̶,̶1̶1̶ ̶(̶8̶,̶1̶)̶
00011   3 √           ───────────           ──────────────────
01010  10 √           3,7   (4)  √          3,7,19,23  (4,16)
─────────             3,11  (8)  √          3,11,19,27 (8,16)
00111   7 √           3,19  (16) √          3̶,̶1̶9̶,̶7̶,̶2̶3̶ ̶(̶1̶6̶,̶4̶)̶
01011  11 √           10,11 (1)  √          3̶,̶1̶9̶,̶1̶1̶,̶2̶7̶ ̶(̶1̶6̶,̶8̶)̶
10011  19 √           ───────────
10110  22 √           7,23  (16) √
─────────             11,27 (16)
10111  23 √           19,23 (4)  √
11011  27 √           19,27 (8)  √
11110  30 √           22,23 (1)
                      22,30 (8)
```

Step 2

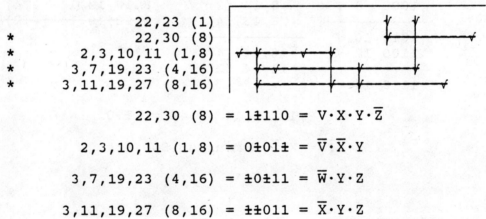

```
                                  2  3  7 10 11 19 22 23 27 30

             22,23 (1)
*            22,30 (8)
*      2,3,10,11  (1,8)
*      3,7,19,23  (4,16)
*      3,11,19,27 (8,16)
```

$$22,30 \ (8) = 1\text{±}110 = V\cdot X\cdot Y\cdot \overline{Z}$$

$$2,3,10,11 \ (1,8) = 0\text{±}01\text{±} = \overline{V}\cdot \overline{X}\cdot Y$$

$$3,7,19,23 \ (4,16) = \text{±}0\text{±}11 = \overline{W}\cdot Y\cdot Z$$

$$3,11,19,27 \ (8,16) = \text{±}\text{±}011 = \overline{X}\cdot Y\cdot Z$$

$$F = V\cdot X\cdot Y\cdot \overline{Z} + \overline{V}\cdot \overline{X}\cdot Y + \overline{W}\cdot Y\cdot Z + \overline{X}\cdot Y\cdot Z \quad \text{minimum SOP for 1s}$$

(b) $F(V,W,X,Y,Z) = \Sigma m(3,5,8,9,11,13,19,25,26)$
$+ d(10,17,24,27)$

Step 1

Column 1 (0-cubes)	Column 2 (1-cubes)	Column 3 (2-cubes)	Column 3 (2-cubes)
00100 8 √	8,9 (1) √	8,9,10,11 (1,2) √	8,9,10,11,24,25,26,27 (1,2,16)
———————	8,10 (2) √	8,9,24,25 (1,16) √	~~8,9,24,25,10,11,26,27 (1,16,2)~~
00011 3 √	8,24 (16) √	~~8,10,9,11 (2,1)~~	~~8,10,24,26,9,11,25,27 (2,16,1)~~
00101 5 √	———————	8,10,24,26 (2,16) √	
01001 9 √	3,11 (8) √	~~8,24,9,25 (16,1)~~	
01010 10 √	3,19 (16) √	~~8,24,10,26 (16,2)~~	
10001 17 √	5,13 (8)	———————	
11000 24 √	9,11 (2) √	3,11,19,27 (8,16)	
———————	9,13 (4)	~~3,19,11,27 (16,8)~~	
01011 11 √	9,25 (16) √	9,11,25,27 (2,16) √	
01101 13 √	10,11 (1) √	~~9,25,11,27 (16,2)~~	
10011 19 √	10,26 (16) √	10,11,26,27 (1,16) √	
11001 25 √	17,19 (2) √	~~10,26,11,27 (16,1)~~	
11010 26 √	17,25 (8) √	17,19,25,27 (2,8)	
———————	24,25 (1) √	~~17,25,19,27 (8,2)~~	
11011 27 √	24,26 (2) √	24,25,26,27 (1,2) √	
	———————	~~24,26,25,27 (2,1)~~	
	11,27 (16) √		
	19,27 (8) √		
	25,27 (2) √		
	26,27 (1) √		

Step 2

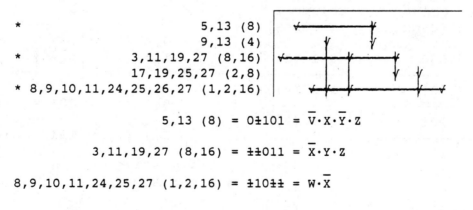

		3	5	8	9	11	13	19	25	26

*   5,13 (8)

                 9,13 (4)

*   3,11,19,27 (8,16)

       17,19,25,27 (2,8)

* 8,9,10,11,24,25,26,27 (1,2,16)

$$5,13\ (8) = 0\text{-}101 = \overline{V}\cdot X\cdot\overline{Y}\cdot Z$$

$$3,11,19,27\ (8,16) = \text{-}\text{-}011 = \overline{X}\cdot Y\cdot Z$$

$$8,9,10,11,24,25,27\ (1,2,16) = \text{-}10\text{-}\text{-} = W\cdot\overline{X}$$

$$F = \overline{V}\cdot X\cdot\overline{Y}\cdot Z + \overline{X}\cdot Y\cdot Z + W\cdot\overline{Y} \quad \text{minimum SOP for 1s}$$

4-10. (a) $F(X,Y,Z) = \overline{X} \cdot \overline{Y} \cdot Z + X \cdot \overline{Y} \cdot Z + X \cdot Y$
Expand the function so it can be written in minterm compact form.

$F(X,Y,Z) = \overline{X} \cdot \overline{Y} \cdot Z + X \cdot \overline{Y} \cdot Z + X \cdot Y \cdot (Z + \overline{Z})$

$F(X,Y,Z) = \Sigma m(1,5,7,6)$

Step 1   Step 2

Column 1 (0-cubes)	Column 2 (1-cubes)
001 1 √	1,5 (4)
101 5 √	5,7 (2)
110 6 √	6,7 (1)
111 7 √	

```
                    1 5 7 6
          * 1,5 (4)  | ✓ ✓
            5,7 (2)  |  ✓ ✓
          * 6,7 (1)  |    ✓ ✓
```

$1,5 \ (4) = \text{±}01 = \overline{Y} \cdot Z$

$6,7 \ (1) = 11\text{±} = X \cdot Y$

$F = \overline{Y} \cdot Z + X \cdot Y$
minimum SOP for F

(b) $F(A,B,C) = \overline{A} \cdot (\overline{B} \cdot \overline{C} + \overline{B} \cdot C) + A \cdot (\overline{B} \cdot C + B \cdot C)$

$F(A,B,C) = \overline{A} \cdot \overline{B} \cdot \overline{C} + \overline{A} \cdot \overline{B} \cdot C + A \cdot \overline{B} \cdot C + A \cdot B \cdot C$

$F(A,B,C) = \Sigma m(0,1,5,7)$

Step 1   Step 2

Column 1 (0-cubes)	Column 2 (1-cubes)
000 0 √	0,1 (1)
001 1 √	1,5 (4)
101 5 √	5,7 (2)
111 7 √	

```
                    0 1 5 7
          * 0,1 (1)  | ✓ ✓
            1,5 (4)  |  ✓ ✓
          * 5,7 (2)  |    ✓ ✓
```

$0,1 \ (1) = 00\text{±} = \overline{A} \cdot \overline{B}$

$5,7 \ (2) = 1\text{±}1 = A \cdot C$

$F = \overline{A} \cdot \overline{B} + A \cdot C$
minimum SOP for F

(c) $\overline{F}(X,Y,Z) = \overline{X} + \overline{X}\cdot Z + \overline{X}\cdot Y + X\cdot Y\cdot Z$

$\overline{F}(X,Y,Z) = \overline{X}\cdot(Y + \overline{Y})\cdot(Z + \overline{Z}) + \overline{X}\cdot(Y + \overline{Y})\cdot Z$

$\qquad\qquad\quad + \overline{X}\cdot Y\cdot(Z + \overline{Z}) + X\cdot Y\cdot Z$

$\overline{F}(X,Y,Z) = \overline{X}\cdot Y\cdot Z + \overline{X}\cdot Y\cdot\overline{Z} + \overline{X}\cdot\overline{Y}\cdot Z + \overline{X}\cdot\overline{Y}\cdot\overline{Z} + \overline{X}\cdot Y\cdot Z$

$\qquad\qquad\quad + \overline{X}\cdot\overline{Y}\cdot Z + \overline{X}\cdot Y\cdot Z + \overline{X}\cdot Y\cdot\overline{Z} + X\cdot Y\cdot Z$

$\overline{F}(X,Y,Z) = \Sigma m(3,2,1,0,7)$

$F(X,Y,Z) = \Sigma m(4,5,6)$

Step 1   Step 2

Column 1 (0-cubes)	Column 2 (1-cubes)

$\qquad$ 4 5 6

100 4 √   4,5 (1)   * 4,5 (1)
_____   4,6 (2)   * 4,6 (2)
101 5 √
110 6 √   4,5 (1) = 10$\overline{\text{1}}$ = X·$\overline{Y}$

$\qquad\qquad\qquad\qquad\qquad$ 4,6 (2) = 1$\overline{\text{1}}$0 = X·$\overline{Z}$

$\qquad\qquad\qquad$ F = X·$\overline{Y}$ + X·$\overline{Z}$
$\qquad\qquad\qquad$ minimum SOP for F

Section 4-6 Karnaugh Map Descriptions with Map-Entered Variables

4-11. (a) $F(X,Y,Z) = \Sigma m(0, 1\cdot Z, 2)$
$\qquad\qquad$ where $m = m(X,Y)$

Step 1$\qquad$Step 2

$p_1 = \overline{X}\cdot Z \qquad p_2 = \overline{Y}$

$F = p_1 + p_2$

$\qquad = \overline{X}\cdot Z + \overline{Y}$

87

(b) $F(X,Y,Z) = \Sigma m(1, 2\cdot Z, 3\cdot \overline{Z}) + d(0)$
      where $m = m(X,Y)$

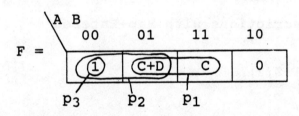

F =

$\begin{array}{c|cccc}X\ Y & 00 & 01 & 11 & 10 \\\hline & - & 1 & Z & Z \end{array}$

p_3 ... p_2 ... p_1

Step 1   Step 2

$p_1 = Y\cdot \overline{Z}$ $p_3 = \overline{X}$

$p_2 = \overline{Y}\cdot Z$

$F = p_1 + p_2 + p_3$

$\quad = Y\cdot \overline{Z} + \overline{Y}\cdot Z + \overline{X}$

(c) $F(X,Y,Z) = \Sigma m(0\cdot Z, 3) + d(1)$
      where $m = m(X,Y)$

F =

$\begin{array}{c|cccc}X\ Y & 00 & 01 & 11 & 10 \\\hline & Z & - & 1 & 0 \end{array}$

p_1 ... p_2

Step 1   Step 2

$p_1 = \overline{X}\cdot Z$ $p_2 = Y$

$F = p_1 + p_2$

$\quad = \overline{X}\cdot Z + Y$

4-12.   (a) $F(A,B,C,D) = \Sigma m(0, 1\cdot(C + D), 3\cdot C)$
            where $m = m(A,B)$

F =

$\begin{array}{c|cccc}A\ B & 00 & 01 & 11 & 10 \\\hline & 1 & C+D & C & 0 \end{array}$

p_3 ... p_2 ... p_1

Step 1   Step 2

$p_1 = B\cdot C$ $p_3 = \overline{A}\cdot \overline{B}$

$p_2 = \overline{A}\cdot D$

$F = p_1 + p_2 + p_3$

$\quad = B\cdot C + \overline{A}\cdot D + \overline{A}\cdot \overline{B}$

(b) $F(A,B,C,D) = \Sigma m(0\cdot C, 2\cdot C\cdot D, 3)$
      where $m = m(A,B)$

F =

$\begin{array}{c|cccc}A\ B & 00 & 01 & 11 & 10 \\\hline & C & 0 & 1 & C\cdot D \end{array}$

p_1 ... p_3 ... p_2

Step 1   Step 2

$p_1 = \overline{A}\cdot \overline{B}\cdot C$ $p_3 = A\cdot B$
$p_2 = A\cdot C\cdot D$

$F = p_1 + p_2 + p_3$

$\quad = \overline{A}\cdot \overline{B}\cdot C + A\cdot C\cdot D + A\cdot B$

(c) $F(W,X,Y,Z) = \Sigma m(0 \cdot Y \cdot \overline{Z}, 1, 2 \cdot \overline{Z}) + d(3)$
         where $m = m(W,X)$

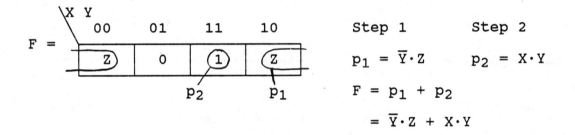

Step 1   Step 2

$p_1 = \overline{W} \cdot Y \cdot \overline{Z}$ $p_3 = X$

$p_2 = W \cdot \overline{Z}$

$F = p_1 + p_2 + p_3$

$\quad = \overline{W} \cdot Y \cdot \overline{Z} + W \cdot \overline{Z} + X$

4-13. (a) $F(X,Y,Z) = \overline{X} \cdot \overline{Y} \cdot Z + X \cdot \overline{Y} \cdot Z + X \cdot Y$
         where $m = m(X,Y)$

$F(X,Y,Z) = m_0 \cdot Z + m_2 \cdot Z + m_3$

$F =$

\X Y	00	01	11	10
	Z	0	1	Z

Step 1   Step 2

$p_1 = \overline{Y} \cdot Z$ $p_2 = X \cdot Y$

$F = p_1 + p_2$

$\quad = \overline{Y} \cdot Z + X \cdot Y$

(b) $F(X,Y,Z) = \overline{X} \cdot \overline{Y} \cdot \overline{Z} + \overline{X} \cdot Y + X \cdot Y \cdot Z$
         where $m = m(X,Y)$

$F(X,Y,Z) = m_0 \cdot \overline{Z} + m_1 + m_3 \cdot Z$

$F =$

\X Y	00	01	11	10
	$\overline{Z}$	1	Z	0

Step 1   Step 2

$p_1 = \overline{X} \cdot \overline{Z}$ double covered
                      1 therefore
$p_2 = Y \cdot Z$ redundant

$F = p_1 + p_2$

$\quad = \overline{X} \cdot \overline{Z} + Y \cdot Z$

(c) $F(X,Y,Z) = \overline{X} + X \cdot Y \cdot Z$
         where $m = m(X,Y)$

$F(X,Y,Z) = \overline{X} \cdot (Y + \overline{Y}) + X \cdot Y \cdot Z$

$\quad\quad = \overline{X} \cdot Y + \overline{X} \cdot \overline{Y} + X \cdot Y \cdot Z$
$\quad\quad = m_1 + m_0 + m_3 \cdot Z$

89

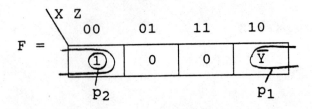

F =

X Y	00	01	11	10
	1	1	Z	0

p_2 p_1

Step 1   Step 2

$p_1 = Y \cdot Z$ $p_2 = \overline{X}$

$F = p_1 + p_2$

$\quad = Y \cdot Z + \overline{X}$

(d) $F(X,Z,Y) = \overline{X} \cdot \overline{Z} + X \cdot \overline{Z} \cdot \overline{Y}$
         where $m = m(X,Z)$

$\quad F(X,Z,Y) = m_0 + m_2 \cdot \overline{Y}$

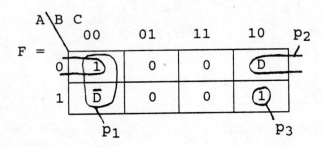

F =

X Z	00	01	11	10
	1	0	0	$\overline{Y}$

p_2 p_1

Step 1   Step 2

$p_1 = \overline{Z} \cdot \overline{Y}$ $p_2 = \overline{X} \cdot \overline{Z}$

$F = p_1 + p_2$

$\quad = \overline{Z} \cdot \overline{Y} + \overline{X} \cdot \overline{Z}$

4-14.   (a) $F(A,B,C,D) = \Sigma m(0, 2 \cdot D, 4 \cdot \overline{D}, 6)$
            where $m = m(A,B,C)$

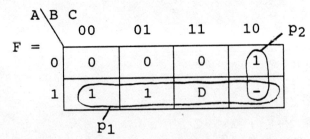

F =

A\B C	00	01	11	10	p_2
0	1	0	0	D	
1	$\overline{D}$	0	0	1	

p_1 p_3

Step 1   Step 2

$p_1 = \overline{B} \cdot \overline{C} \cdot \overline{D}$ 1 at m_0 is
                     double

$p_2 = \overline{A} \cdot \overline{C} \cdot D$ covered

                     $p_3 = A \cdot B \cdot \overline{C}$

$F = p_1 + p_2 + p_3$

$\quad = \overline{B} \cdot \overline{C} \cdot \overline{D} + \overline{A} \cdot \overline{C} \cdot D + A \cdot B \cdot \overline{C}$
this function is cyclic
(two minimum coverings
are possible)

(b) $F(A,B,C,D) = \Sigma m(2, 4, 5, 7 \cdot D) + d(6)$
         where $m = m(A,B,C)$

F =

A\B C	00	01	11	10	p_2
0	0	0	0	1	
1	1	1	D	-	

p_1

Step 1   Step 2

$p_1 = A \cdot D$ $p_2 = B \cdot \overline{C}$

$F = p_1 + p_2$

$F = A \cdot D + B \cdot \overline{C}$

(c) $F(A,B,D,C) = \Sigma m(1, 2 \cdot C, 5, 6 \cdot C) + d(7)$
        where $m = m(A,B,D)$

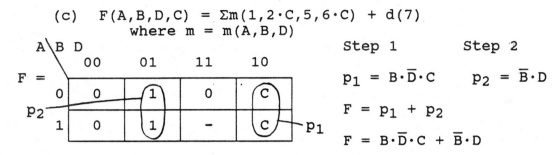

Step 1

$p_1 = B \cdot \overline{D} \cdot C$

Step 2

$p_2 = \overline{B} \cdot D$

$F = p_1 + p_2$

$F = B \cdot \overline{D} \cdot C + \overline{B} \cdot D$

Section 4-7 Adding Variables to a Design

4-15. (a) $F(U,V,W,X,Y,Z) = \Sigma m(0 \cdot X, 1, 3 \cdot Y, 6 \cdot \overline{Z})$
            where $m = m(U,V,W)$

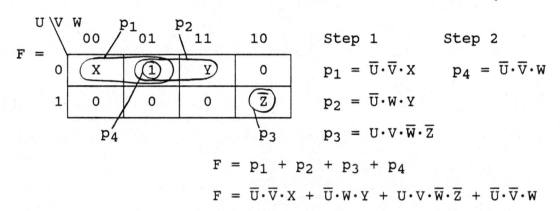

Step 1

$p_1 = \overline{U} \cdot \overline{V} \cdot X$

$p_2 = \overline{U} \cdot W \cdot Y$

$p_3 = U \cdot V \cdot \overline{W} \cdot \overline{Z}$

Step 2

$p_4 = \overline{U} \cdot \overline{V} \cdot W$

$F = p_1 + p_2 + p_3 + p_4$

$F = \overline{U} \cdot \overline{V} \cdot X + \overline{U} \cdot W \cdot Y + U \cdot V \cdot \overline{W} \cdot \overline{Z} + \overline{U} \cdot \overline{V} \cdot W$

(b) $F(A,B,C,D,E) = \Sigma m(2, 3 \cdot D \cdot E, 7 \cdot E) + d(4,6)$
        where $m = m(A,B,C)$

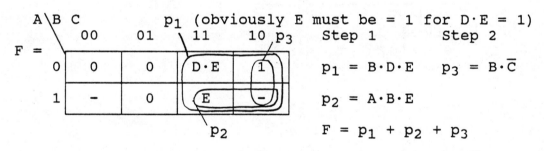

p_1 (obviously E must be $= 1$ for $D \cdot E = 1$)

Step 1

$p_1 = B \cdot D \cdot E$

$p_2 = A \cdot B \cdot E$

Step 2

$p_3 = B \cdot \overline{C}$

$F = p_1 + p_2 + p_3$

$F = B \cdot D \cdot E + A \cdot B \cdot E + B \cdot \overline{C}$

(c) $F(T,U,V,W,X,Y,Z) = \Sigma m(0 \cdot W, 1 \cdot X, 3 \cdot Y, 4, 6 \cdot \overline{Z}) + d(5,7)$
        where $m = m(T,U,V)$

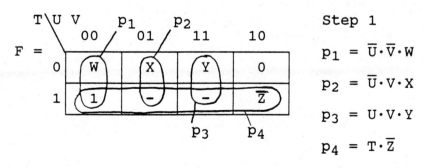

Step 1

$p_1 = \overline{U} \cdot \overline{V} \cdot W$

$p_2 = \overline{U} \cdot V \cdot X$

$p_3 = U \cdot V \cdot Y$

$p_4 = T \cdot \overline{Z}$

$$F = p_1 + p_2 + p_3 + p_4$$

$$F = \overline{U}\cdot\overline{V}\cdot W + \overline{U}\cdot V\cdot X + U\cdot V\cdot Y + T\cdot\overline{Z}$$

4-16. (a) $F(V,W,X,Y,Z) = \Sigma m(1,6\cdot Z,7\cdot Z,9,10\cdot\overline{Z},11\cdot\overline{Z},15) + d(14)$
where $m = m(V,W,X,Y)$

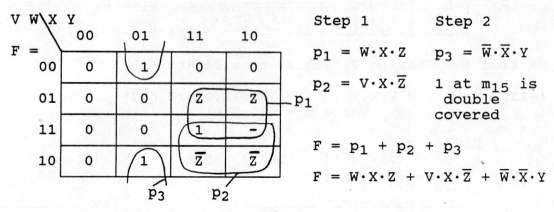

Step 1

$p_1 = W\cdot X\cdot Z$

$p_2 = V\cdot X\cdot\overline{Z}$

Step 2

$p_3 = \overline{W}\cdot\overline{X}\cdot Y$

1 at m_{15} is
double
covered

$F = p_1 + p_2 + p_3$

$F = W\cdot X\cdot Z + V\cdot X\cdot\overline{Z} + \overline{W}\cdot\overline{X}\cdot Y$

(b) $F(A,B,C,D,E,G) = \Sigma m(2\cdot(E + G),4,5,9\cdot\overline{G},10\cdot(E + G),12,13) + d(6,14)$
where $m = m(A,B,C,D)$

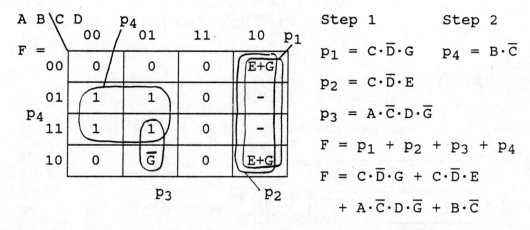

Step 1

$p_1 = C\cdot\overline{D}\cdot G$

$p_2 = C\cdot\overline{D}\cdot E$

$p_3 = A\cdot\overline{C}\cdot D\cdot\overline{G}$

$F = p_1 + p_2 + p_3 + p_4$

$F = C\cdot\overline{D}\cdot G + C\cdot\overline{D}\cdot E$

$+ A\cdot\overline{C}\cdot D\cdot\overline{G} + B\cdot\overline{C}$

Step 2

$p_4 = B\cdot\overline{C}$

4-17.

A\B C	00	01	11	10
F2 = 0	0	$\overline{D}$	1	0
1	–	–	1	0

$\quad\quad\quad\quad p_1 \quad p_2$

Step 1

$p_1 = C\cdot\overline{D}$

$F2 = p_1 + p_2$

$F2 = C\cdot\overline{D} + B\cdot C$

Step 2

$p_2 = B\cdot C$

92

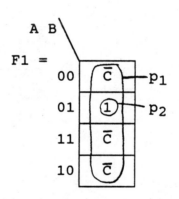

A\B C
$$F3 =$$

	00	01	11	10
0	0	1	$\overline{D}\cdot\overline{E}$	0
1	–	–	1	0

p_2 p_1

p_3

Step 1   Step 2

$p_1 = C\cdot\overline{D}\cdot\overline{E}$ $p_2 = \overline{B}\cdot C$

$p_3 = A\cdot C$

$F3 = p_1 + p_2 + p_3$

$F3 = C\cdot\overline{D}\cdot\overline{E} + \overline{B}\cdot C + A\cdot C$

A\B C
$$F4 =$$

	00	01	11	10
0	0	1	1	0
1	–	–	$D+\overline{E}$	0

p_3

p_1 p_2

Step 1   Step 2

$p_1 = C\cdot D$ $p_3 = \overline{A}\cdot C$

$p_2 = C\cdot\overline{E}$

$F4 = p_1 + p_2 + p_3$

$F4 = C\cdot D + C\cdot\overline{E} + \overline{A}\cdot C$

Section 4-8   Reducing the Map Size

4-18.   (a)

A	B	C	F1	F1
0	0	0	1	
0	0	1	0	$\overline{C}$
0	1	0	1	
0	1	1	1	1
1	0	0	1	
1	0	1	0	$\overline{C}$
1	1	0	1	
1	1	1	0	$\overline{C}$

A B
$$F1 =$$

00	$\overline{C}$
01	①
11	$\overline{C}$
10	$\overline{C}$

p_1

p_2

Step 1   Step2

$p_1 = \overline{C}$ $p_2 = \overline{A}\cdot B$

$F1 = \overline{C} + \overline{A}\cdot B$

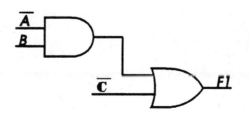

(b)

A	B	C	F2	F2
0	0	0	1	
0	0	1	1	1
0	1	0	0	
0	1	1	1	C
1	0	0	1	
1	0	1	0	$\overline{C}$
1	1	0	0	
1	1	1	0	0

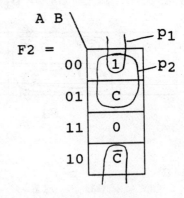

Step 1

$p_1 = \overline{B} \cdot \overline{C}$

$p_2 = \overline{A} \cdot C$

$F2 = p_1 + p_2$

$F2 = \overline{B} \cdot \overline{C} + \overline{A} \cdot C$

Step 2

1 at m_0 is double covered

(c)

A	B	C	F3	F3
0	0	0	1	
0	0	1	0	$\overline{C}$
0	1	0	1	
0	1	1	0	$\overline{C}$
1	0	0	1	
1	0	1	0	$\overline{C}$
1	1	0	0	
1	1	1	–	0,C

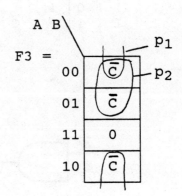

Step 1

$p_1 = \overline{B} \cdot \overline{C}$

$p_2 = \overline{A} \cdot \overline{C}$

94

$$F3 = p_1 + p_2$$

$$F3 = \overline{B} \cdot \overline{C} + \overline{A} \cdot \overline{C}$$

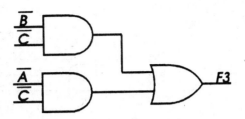

4-19. (a)

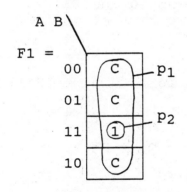

$$F1 = p_1 + p_2$$

$$F1 = C + A \cdot B$$

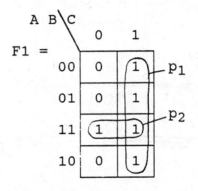

$$F1 = p_1 + p_2$$

$$F1 = C + A \cdot B$$

(b)

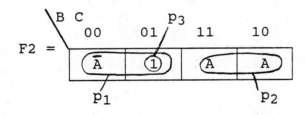

$$F2 = p_1 + p_2 + p_3$$

$$F2 = \overline{B} \cdot \overline{A} + B \cdot A + \overline{B} \cdot C$$

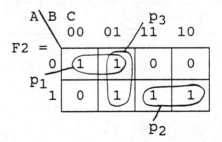

$$F2 = p_1 + p_2 + p_3$$

$$F2 = \overline{A} \cdot \overline{B} + \overline{B} \cdot C + A \cdot B$$

(c)

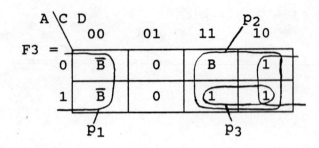

$F3 = p_1 + p_2 + p_3$

$F3 = \overline{D} \cdot \overline{B} + C \cdot B + A \cdot C$

$F3 = p_1 + p_2 + p_3$

$F3 = \overline{B} \cdot \overline{D} + B \cdot C + A \cdot C$

4-20. The following maps were drawn first and then filled in by inspection by comparing columns I0 and F3, I0 and F2, I0 and F1, and I0 and F0 two rows at a time.

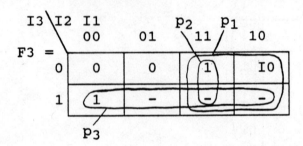

Step 1

$p_1 = I2 \cdot I0$

Step 2

$p_2 = I2 \cdot I1$

$p_3 = I3$

$F3 = p_1 + p_2 + p_3$

$F3 = I3 \cdot I0 + I2 \cdot I1 + I3$

Step 1

$p_1 = I2 \cdot \overline{I1} \cdot \overline{I0}$

$p_2 = \overline{I2} \cdot I0$

Step 2

$p_3 = \overline{I2} \cdot I1$

$F2 = p_1 + p_2 + p_3$

$F2 = I2 \cdot \overline{I1} \cdot \overline{I0} + \overline{I2} \cdot I0 + \overline{I2} \cdot I1$

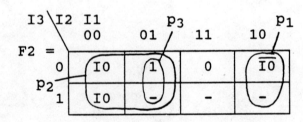

Step 1

$p_1 = \overline{I1} \cdot \overline{I0}$

$p_2 = I1 \cdot I0$

$F1 = p_1 + p_2$

$F1 = \overline{I1} \cdot \overline{I0} + I1 \cdot I0$

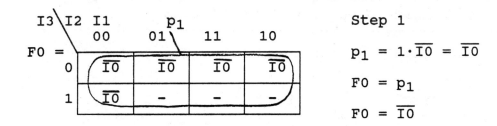

Step 1

$$p_1 = 1 \cdot \overline{I0} = \overline{I0}$$

$$F0 = p_1$$

$$F0 = \overline{I0}$$

97

Section 5-2   Gate-Level Combinational Logic Design

5-1.

(a) $F = A + \tilde{} B \cdot C$

(b) $F = \overline{\tilde{} A} + \tilde{} B \cdot C$

(c) $F = A + \tilde{} B \cdot \overline{\tilde{} C}$

(d) $\overline{\tilde{} F} = \overline{\tilde{} A} + \overline{B} \cdot C$

5-2. Available signals: F, W, ~X, ~Y, Z

$$F = Y \cdot Z + X \cdot Y + W \cdot \overline{X} \cdot \overline{Y}$$

5-3.

(a) $\overline{F} = \tilde{} X \cdot \overline{Y} + Y \cdot \overline{Z} + \overline{\tilde{} W} \cdot \overline{Y} \cdot Z$

(b) $\tilde{} F = \overline{X} \cdot \tilde{} Y + \overline{\tilde{} Y} \cdot \overline{Z} + W \cdot \tilde{} Y \cdot Z$

(c) $\overline{F} = \tilde{} X \cdot \overline{Y} + Y \cdot \overline{Z} + W \cdot \overline{Y} \cdot Z$

(d) $\tilde{} F = \overline{X} \cdot \tilde{} Y + \overline{\tilde{} Y} \cdot \tilde{} Z + W \cdot \tilde{} Y \cdot \overline{\tilde{} Z}$

5-4.

(a) Signal list: F,A,~B
(b) Signal list: F[NL],A,~B
(c) Signal list: F,A[NL],~B
(d) Signal list: F,A[NL],~B[NL]
(e) Signal list: F[NL],A,~B[NL]

5-5.

(a) Signal list: F1,~F2[NL],X[NL],~Y[NL],~Z
(b) Signal list: F1[NL],~F2,X,~Y,~Z[NL]
(c) Signal list: F1,~F2[NL],X,~Y,~Z
(d) Signal list: F1,~F2,X[NL],~Y[NL],~Z
(e) Signal list: F1[NL],~F2,X[NL],~Y[NL],~Z[NL]
(f) Signal list: F1[NL],~F2,X,~Y,~Z

Section 5-3   Logic Conventions and Polarity Indication

5-6. 74ALS11 device
$V_{NH} = V_{OH}(MIN) - V_{IH}(MIN) = 2.5 - 2.0 = .5 \text{ V}$
$V_{NL} = V_{IL}(MAX) - V_{OL}(MAX) = 0.8 - 0.4 = .4 \text{ V}$

74AS11 device.
$V_{NH} = V_{OH}(MIN) - V_{IH}(MIN) = 2.5 - 2.0 = .5 \text{ V}$
$V_{NL} = V_{IL}(MAX) - V_{OL}(MAX) = 0.8 - 0.5 = .3 \text{ V}$

The 74ALS11 has a slightly better noise margin than the 74AS11.

5-7. 74ACT1011 (CMOS) device
$$V_{NH} = V_{OH}(MIN) - V_{IH}(MIN) = 3.8 - 2.0 = 1.8 \text{ V}$$
$$V_{NL} = V_{IL}(MAX) - V_{OL}(MAX) = 0.8 - 0.44 = .36 \text{ V}$$

74AC1011 device
$$V_{NH} = V_{OH}(MIN) - V_{IH}(MIN) = 3.8 - 3.15 = .65 \text{ V}$$
$$V_{NL} = V_{IL}(MAX) - V_{OL}(MAX) = 1.35 - 0.44 = .91 \text{ V}$$

The 74ACT1011 has a better high level noises margin while the 74AC1011 has a better low level noises margin.

5-8.

(a)   (b)

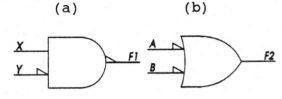

X	Y	F1
L	L	H
L	H	H
H	L	L
H	H	H

A	B	F2
L	L	H
L	H	H
H	L	H
H	H	L

5-9.

W	X	~Y	OUT1
L	L	L	L
L	L	H	L
L	H	L	H
L	H	H	L
H	L	L	L
H	L	H	L
H	H	L	L
H	H	H	L

X	Z	OUT2
L	L	L
L	H	H
H	L	L
H	H	L

OUT1	OUT2	F
L	L	L
L	H	H
H	L	H
H	H	H

5-10.

A	~B	OUT1
0	0	0
0	1	0
1	0	1
1	1	0

A	C[NL]	OUT2
0	0	1
0	1	0
1	0	0
1	1	0

OUT1	OUT2	F
0	0	0
0	1	1
1	0	1
1	1	1

A	~B	OUT1
L	L	L
L	H	L
H	L	H
H	H	L

A	C[NL]	OUT2
L	H	H
L	L	L
H	H	L
H	L	L

OUT1	OUT2	F
L	L	L
L	H	H
H	L	H
H	H	H

5-11.

X	Y	Z	F1	F2	F3
0	0	0	1	1	1
0	0	1	1	1	0
0	1	0	1	0	1
0	1	1	1	0	0
1	0	0	0	1	1
1	0	1	0	1	0
1	1	0	0	0	1
1	1	1	0	0	0

$F1 = \overline{X}$ $F2 = \overline{Y}$ $F3 = \overline{Z}$

Signal list: F1,F2,F3,X,Y,Z

LLP: $F1 = X$ $F2 = Y$ $F3 = Z$
NIP: 1   0   1   0   1   0

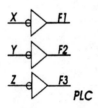

PLC

5-12.

X	Y	Z	F
0	0	0	0
0	0	1	0
0	1	0	0
0	1	1	1
1	0	0	0
1	0	1	1
1	1	0	1
1	1	1	1

$F = $

X \ Y Z	00	01	11	10
0	0	0	(1) p_1	0
1	0	(1) p_2	(1)	1 p_3

$F = p_1 + p_2 + p_3$
$\quad = Y \cdot Z + X \cdot Z + X \cdot Y$

Signal list: F,X,Y,Z

LLP: $F = Y \cdot Z + X \cdot Z + X \cdot Y$
NIP: 1   1 1   1 1   1 1

100

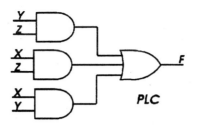

PLC

5-13.

X	Y	Z	F
0	0	0	1
0	0	1	1
0	1	0	1
0	1	1	0
1	0	0	1
1	0	1	0
1	1	0	0
1	1	1	0

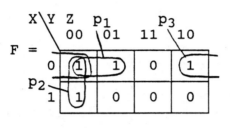

$$F = p_1 + p_2 + p_3$$

$$= \overline{X} \cdot \overline{Y} + \overline{Y} \cdot \overline{Z} + \overline{X} \cdot \overline{Z}$$

Signal list: F,X,Y,Z

LLP: $F = X \cdot Y + Y \cdot Z + X \cdot Z$
PIP: H   L L   L L   L L

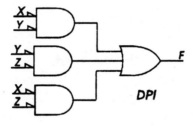

DPI

5-14.

A1	A0	B1	B0	C0	C2	S1	S0
0	0	0	0	0	0	0	0
0	0	0	0	1	0	0	1
0	0	0	1	0	0	0	1
0	0	0	1	1	0	1	0
0	0	1	0	0	0	1	0
0	0	1	0	1	0	1	1
0	0	1	1	0	0	1	1
0	0	1	1	1	1	0	0
0	1	0	0	0	0	0	1
0	1	0	0	1	0	1	0
0	1	0	1	0	0	1	0
0	1	0	1	1	0	1	1
0	1	1	0	0	0	1	1
0	1	1	0	1	1	0	0
0	1	1	1	0	1	0	0
0	1	1	1	1	1	0	1
1	0	0	0	0	0	1	0
1	0	0	0	1	0	1	1
1	0	0	1	0	0	1	1
1	0	0	1	1	1	0	0
1	0	1	0	0	1	0	0
1	0	1	0	1	1	0	1
1	0	1	1	0	1	0	1
1	0	1	1	1	1	1	0
1	1	0	0	0	0	1	1
1	1	0	0	1	1	0	0
1	1	0	1	0	1	0	0
1	1	0	1	1	1	0	1
1	1	1	0	0	1	0	1
1	1	1	0	1	1	1	0
1	1	1	1	0	1	1	0
1	1	1	1	1	1	1	1

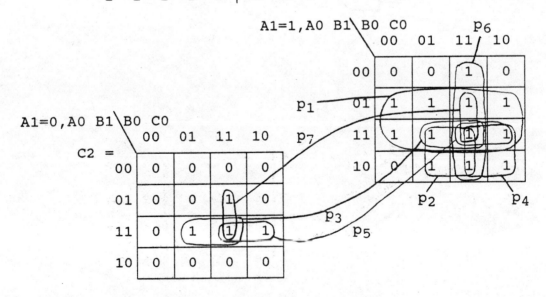

$$C2 = p_1 + p_2 + p_3 + p_4 + p_5 + p_6 + p_7$$

$$C2 = A1 \cdot B1 + A1 \cdot A0 \cdot C0 + A0 \cdot B1 \cdot C0 + A1 \cdot A0 \cdot B0 + A0 \cdot B1 \cdot B0$$
$$+ A1 \cdot B0 \cdot C0 + B1 \cdot B0 \cdot C0$$

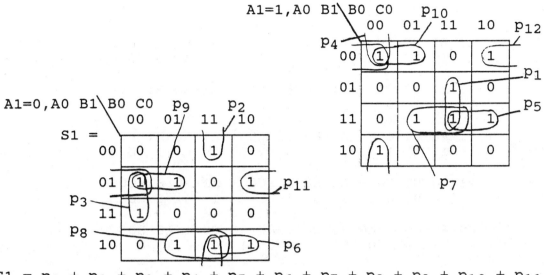

$$S1 = p_1 + p_2 + p_3 + p_4 + p_5 + p_6 + p_7 + p_8 + p_9 + p_{10} + p_{11} + p_{12}$$

$$S1 = A1 \cdot B1 \cdot B0 \cdot C0 + \overline{A1} \cdot \overline{B1} \cdot B0 \cdot C0 + \overline{A1} \cdot B1 \cdot \overline{B0} \cdot \overline{C0}$$

$$+ A1 \cdot \overline{B1} \cdot \overline{B0} \cdot \overline{C0} + A1 \cdot A0 \cdot B1 \cdot B0 + \overline{A1} \cdot A0 \cdot \overline{B1} \cdot B0$$

$$+ A1 \cdot A0 \cdot B1 \cdot C0 + \overline{A1} \cdot A0 \cdot \overline{B1} \cdot C0 + \overline{A1} \cdot \overline{A0} \cdot B1 \cdot \overline{B0}$$

$$+ A1 \cdot \overline{A0} \cdot \overline{B1} \cdot \overline{B0} + \overline{A1} \cdot \overline{A0} \cdot B1 \cdot \overline{C0} + A1 \cdot \overline{A0} \cdot \overline{B1} \cdot \overline{C0}$$

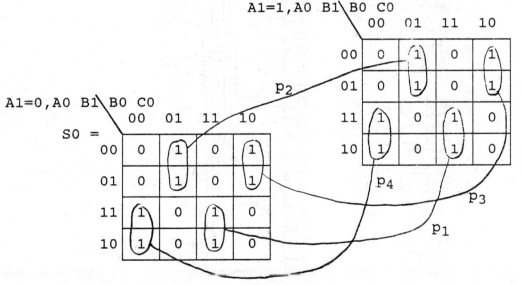

$$S0 = p_1 + p_2 + p_3 + p_4$$

$$S0 = A0 \cdot B0 \cdot C0 + \overline{A0} \cdot \overline{B0} \cdot C0 + \overline{A0} \cdot B0 \cdot \overline{C0} + A0 \cdot \overline{B0} \cdot \overline{C0}$$

Signal list: C2,S1,S0,A1,A0,B1,B0,C0

```
LLP: C2 = A1·B1 + A1·A0·C0 + A0·B1·C0 + A1·A0·B0 + A0·B1·B0
NIP: 1    1 1    1 1 1    1 1 1    1 1 1    1 1 1

        + A1·B0·C0 + B1·B0·C0
          1 1 1    1 1 1


LLP: S1 = A1·B1·B0·C0 + A1·B1·B0·C0 + A1·B1·B0·C0
NIP: 1    1 1 1 1    0 0 1 1    0 1 0 0

        + A1·B1·B0·C0 + A1·A0·B1·B0 + A1·A0·B1·B0
          1 0 0 0    1 1 1 1    0 1 0 1

        + A1·A0·B1·C0 + A1·A0·B1·C0 + A1·A0·B1·B0
          1 1 1 1    0 1 0 1    0 0 1 0

        + A1·A0·B1·B0 + A1·A0·B1·C0 + A1·A0·B1·C0
          1 0 0 0    0 0 1 0    1 0 0 0


LLP: S0 = A0·B0·C0 + A0·B0·C0 + A0·B0·C0 + A0·B0·C0
NIP: 1    1 1 1    0 0 1    0 1 0    1 0 0
```

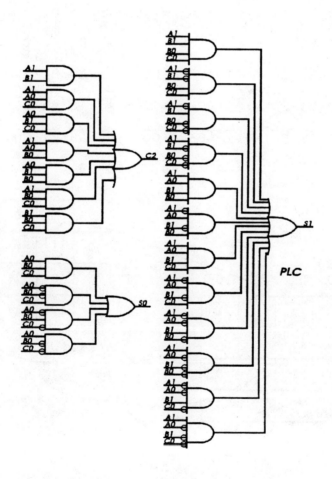

PLC

5-15.

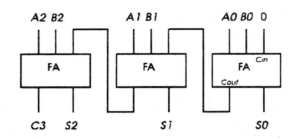

5-16.

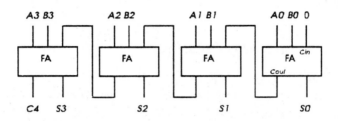

5-17.

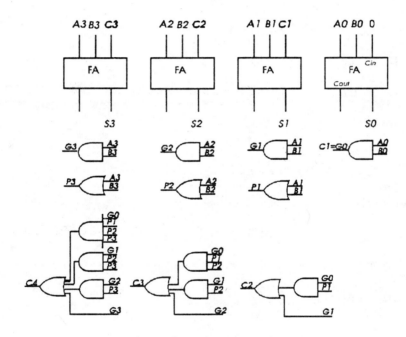

Section 5-5   Applying the Top-Down design Process to Other Realistic Problems

5-18.

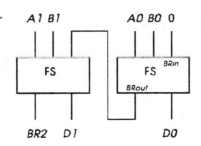

5-19. Use full subtractors for the difference bits. The borrow bits are then obtained as follows:

$$BR1 = \overline{A0} \cdot B0$$
$$= G0 \qquad \text{where G0 is the borrow generate term for stage 0.}$$

$$BR_{i+1} = \overline{A_i} \cdot B_i + BR_i \cdot \overline{A_i} \cdot + BR_i \cdot B_i$$

Note that the only difference in this form and the form of the carry term for the adder is BR_{i+1} for C_i and $\overline{A_i}$ for A_i. Therefore

$$BR2 = \overline{A1} \cdot B1 + BR1 \cdot \overline{A1} + BR1 \cdot B1$$

$$= G1 + BR1 \cdot (\overline{A1} + B1)$$
$$= G1 + G0 \cdot P1 \qquad \text{where G1 is the borrow generate}$$
term for stage 1 and P1 is the borrow propagate term for stage 1.

$$BR3 = \overline{A2} \cdot B2 + BR2 \cdot (\overline{A2} + B2)$$

$$= G2 + (G1 + G0 \cdot P1) \cdot P2$$

$$= G2 + G1 \cdot P2 + G0 \cdot P1 \cdot P2 \qquad \text{where G2 is the borrow}$$
generate term for stage 2 and P2 is the borrow propagate term for stage 2.

$$G0 = \overline{A0} \cdot B0 \qquad G1 = \overline{A1} \cdot B1 \qquad G2 = \overline{A2} \cdot B2$$

$$P1 = \overline{A1} + B1 \qquad P2 = \overline{A2} + B2$$

$$BR2 = G1 + G0 \cdot P1 \qquad BR3 = G2 + G1 \cdot P2 + G0 \cdot P1 \cdot P2$$

Signal list: G0, G1, G2, P1, P2, A1, B1, A0, B0

LLP:	G0 = A0·B0		G1 = A1·B1		G2 = A2·B2	
NIP:	1	0 1	1	0 1	1	0 1

LLP:	P1 = A1 + B1		P2 = A2 + B2	
NIP:	1	0 1	1	0 1

LLP:	BR2 = G1 +	G0·P1	BR3 = G2 +	G1·P2	+ G0·P1·P2
NIP:	1	1 1 1	1	1 1 1	1 1 1

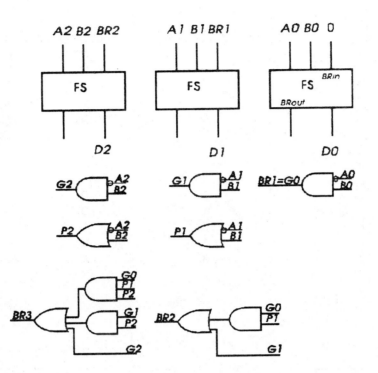

5-20.

		A0<B0	A0>B0			A0=B0
A0	B0	L1	G1	L1	G1	E1
0	0	0	0	0	0	1
0	1	1	0	0	1	0
1	0	0	1	1	0	0
1	1	0	0	1	1	X

$L1 = \overline{A0} \cdot B0$

$G1 = A0 \cdot \overline{B0}$

$E1 = \overline{L1} \cdot \overline{G1}$

Signal list: L1, G1, E1, A0, B0

LLP: L1 = A0·B0   G1 = A0·B0   E1 = L1·G1
NIP: 1   0   1   1   1   0   1   0   0

5-21.

A0<B0	A0>B0			A1 A0<B1 B0	A1 A0>B1 B0
L1	G1	A1	B1	L2	G2
0	0	0	0	0	0
0	0	0	1	1	0
0	0	1	0	0	1
0	0	1	1	0	0
0	1	0	0	0	1
0	1	0	1	1	0
0	1	1	0	0	1
0	1	1	1	0	1
1	0	0	0	1	0
1	0	0	1	1	0
1	0	1	0	0	1
1	0	1	1	1	0
1	1	0	0	X	X
1	1	0	1	X	X
1	1	1	0	X	X
1	1	1	1	X	X

L2	G2	E2
0	0	1
0	1	0
1	0	0
1	1	X

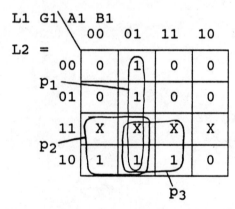

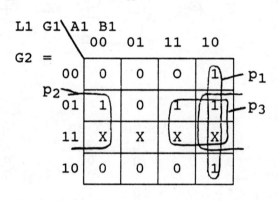

$$L2 = p_1 + p_2 + p_3 \qquad G2 = p_1 + p_2 + p_3 \qquad E2 = \overline{L2 \cdot G2}$$

$$= \overline{A1} \cdot B1 + L1 \cdot \overline{A1} + L1 \cdot B1 \qquad = A1 \cdot \overline{B1} + G1 \cdot \overline{B1} + G1 \cdot A1$$

Signal list: L2, G2, E2, L1, G1, A1, B1

LLP: L2=A1·B1 + L1·A1 + L1·B1   G2=A1·B1 + G1·B1 + G1·A1   E2=L2·G2
NIP: 1 0 1   1 0   1 1   1 1 0   1 0   1 1   1 0 0

108

5-22.

				OLES	OEQU	OGRE	
A1	A0	B1	B0	A<B	A=B	A>B	
0	0	0	0	0	1	0	where $A = A1\ A0$
0	0	0	1	1	0	0	and $B = B1\ B0$
0	0	1	0	1	0	0	
0	0	1	1	1	0	0	
0	1	0	0	0	0	1	
0	1	0	1	0	1	0	
0	1	1	0	1	0	0	
0	1	1	1	1	0	0	
1	0	0	0	0	0	1	
1	0	0	1	0	0	1	
1	0	1	0	0	1	0	
1	0	1	1	1	0	0	
1	1	0	0	0	0	1	
1	1	0	1	0	0	1	
1	1	1	0	0	0	1	
1	1	1	1	0	1	0	

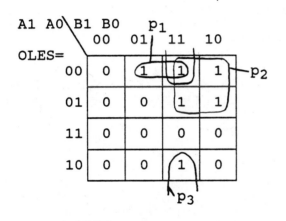

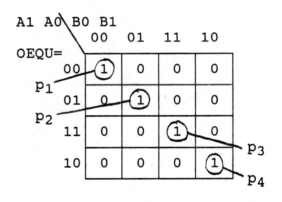

$OLES = p_1 + p_2 + p_3$

$$= \overline{A1}\cdot\overline{A0}\cdot B0 + \overline{A1}\cdot B1 + \overline{A0}\cdot B1\cdot B0$$

Signal list: OLES, A1, A0, B1, B0

$OEQU = p_1 + p_2 + p_3 + p_4$

$$= \overline{A1}\cdot\overline{A0}\cdot\overline{B1}\cdot\overline{B0} + \overline{A1}\cdot A0\cdot\overline{B1}\cdot B0 + A1\cdot A0\cdot B1\cdot B0 + A1\cdot\overline{A0}\cdot B1\cdot\overline{B0}$$

Signal list: OEQU, A1, A0, B1, B0

109

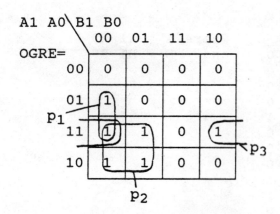

A1 A0 \ B1 B0

$$OGRE = p_1 + p_2 + p_3$$

$$= A0 \cdot \overline{B1} \cdot \overline{B0} + A1 \cdot \overline{B1} + A1 \cdot A0 \cdot \overline{B0}$$

Signal list: OGRE, A1, A0, B1, B0

```
LLP:   OLES = A1·A0·B0 + A1·B1 + A0·B1·B0
NIP:   1       0  0  1    0  1    0  1  1

LLP:   OEQU=A1·A0·B1·B0 + A1·A0·B1·B0 + A1·A0·B1·B0 + A1·A0·B1·B0
NIP:   1      0  0  0  0    0  1  0  1    1  1  1  1    1  0  1  0

LLP:=  OGRE = A0·B1·B0 + A1·B1 + A1·A0·B0
NIP:   1       1  0  0    1  0    1  1  0
```

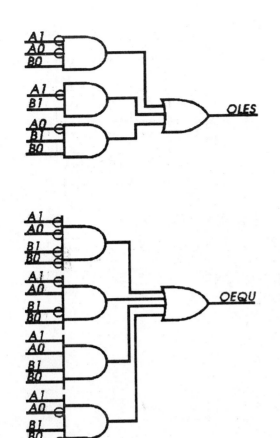

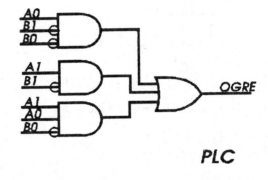

PLC

5-23.

	BCD				XS3			
I3	I2	I1	I0		F3	F2	F1	F0
0	0	0	0		0	0	1	1
0	0	0	1		0	1	0	0
0	0	1	0		0	1	0	1
0	0	1	1		0	1	1	0
0	1	0	0		0	1	1	1
0	1	0	1		1	0	0	0
0	1	1	0		1	0	0	1
0	1	1	1		1	0	1	0
1	0	0	0		1	0	1	1
1	0	0	1		1	1	0	0
1	0	1	0		X	X	X	X
1	0	1	1		X	X	X	X
1	1	0	0		X	X	X	X
1	1	0	1		X	X	X	X
1	1	1	0		X	X	X	X
1	1	1	1		X	X	X	X

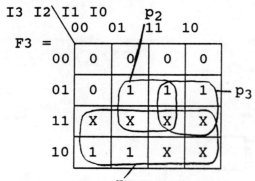

F3 = p₁ + p₂ + p₃

$$F3 = p_1 + p_2 + p_3$$

$$F3 = I3 + I2 \cdot I0 + I2 \cdot I1$$

Signal list: F3,I3,I2,I1,I0

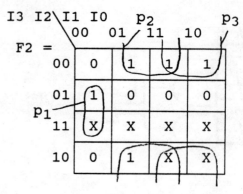

$$F2 = p_1 + p_2 + p_3$$

$$F2 = I2 \cdot \overline{I1} \cdot \overline{I0} + \overline{I2} \cdot I0 + \overline{I2} \cdot I1$$

Signal list: F2,I2,I1,I0

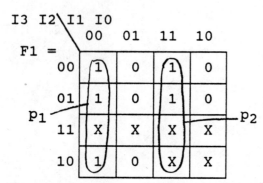

$$F1 = p_1 + p_2$$

$$F1 = \overline{I1} \cdot \overline{I0} + I1 \cdot I0$$

Signal list: F1, I1, I0

By inspecting the BCD to XS3 truth table one can see that $F0 = \overline{I0}$.

Signal list: F0, I0

```
LLP: F3 = I3 + I2·I0 + I2·I1
NIP: 1    1      1 1    1 1

LLP: F2 = I2·I1·I0 + I2·I0 + I2·I1
NIP: 1    1 0 0      0 1     0 1

LLP: F1 = I1·I0 + I1·I0
NIP: 1    0 0     1 1

LLP: F0 = I0
NIP: 1    0
```

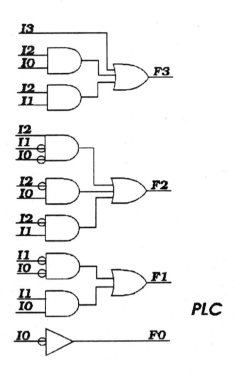

PLC

5-24.

4-BIT BINARY				BCD			
I3	I2	I1	I0	F3	F2	F1	F0
0	0	0	0	0	0	0	0
0	0	0	1	0	0	0	1
0	0	1	0	0	0	1	0
0	0	1	1	0	0	1	1
0	1	0	0	0	1	0	0
0	1	0	1	0	1	0	1
0	1	1	0	0	1	1	0
0	1	1	1	0	1	1	1
1	0	0	0	1	0	0	0
1	0	0	1	1	0	0	1
1	0	1	0	X	X	X	X
1	0	1	1	X	X	X	X
1	1	0	0	X	X	X	X
1	1	0	1	X	X	X	X
1	1	1	0	X	X	X	X
1	1	1	1	X	X	X	X

For this problem we are assuming only 4-bit BCD is required
at the output. Since the binary input bits 1010 through
1111 result in don't care outputs, each output F3 through F0
can simply be equated to I3 through I0 respectively thus
providing a simple four line connection.

5-25.

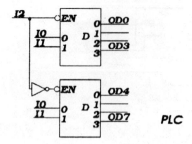

PLC

5-26.

Input Switches	BCD F3	F2	F1	F0
S0	0	0	0	0
S1	0	0	0	1
S2	0	0	1	0
S3	0	0	1	1
S4	0	1	0	0
S5	0	1	0	1
S6	0	1	1	0
S7	0	1	1	1
S8	1	0	0	0
S9	1	0	0	1

Assuming each input switch represents a 0 when pressed or actuated. One input is always actuated and all other input combinations provide don't care outputs.

$$F3 = \overline{S8} + \overline{S9}$$

$$F2 = \overline{S4} + \overline{S5} + \overline{S6} + \overline{S7}$$

$$F1 = \overline{S2} + \overline{S3} + \overline{S6} + \overline{S7}$$

$$F0 = \overline{S1} + \overline{S3} + \overline{S5} + \overline{S7} + \overline{S9}$$

Signal list: F3,F2,F1,F0,S1,S2,S3,S4,S5,S6,S7,S8,S9

LLP: F3 = S8 + S9
NIP: 1   0   0

LLP: F2 = S4 + S5 + S6 + S7
NIP: 1   0   0   0   0

LLP: F1 = S2 + S3 + S6 + S7
NIP: 1   0   0   0   0

LLP: F0 = S1 + S3 + S5 + S7 + S9
NIP: 1   0   0   0   0   0

114

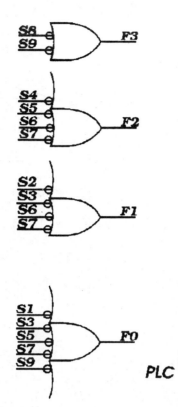

PLC

5-27.

EN	I3	I2	I1	I0	F2	F1	F0
0	X	X	X	X	0	0	0
1	1	X	X	X	0	0	1
1	0	1	X	X	0	1	0
1	0	0	1	X	0	1	1
1	0	0	0	1	1	0	0
1	0	0	0	0	0	0	0

Normally this would require a 5-variable map. By using map-
entered variables the map size can be reduce to 4-variables.
We could have assumed a 0 instead of a 1 is required to
enable the function. An overbar can still be added to each
EN in the final equations if is desired to allow a 0 to
enable the function rather than a 1.

I3	I2	I1	I0	F2	F1	F0
X	X	X	X	0	0	0
1	X	X	X	0	0	EN
0	1	X	X	0	EN	0
0	0	1	X	0	EN	EN
0	0	0	1	EN	0	0
0	0	0	0	0	0	0

I3 I2\I1 I0
         00   01   11   10

F2 =

00 | 0 | (EN) | 0 | 0   ← p_1

01 | 0 | 0 | 0 | 0

11 | 0 | 0 | 0 | 0

10 | 0 | 0 | 0 | 0

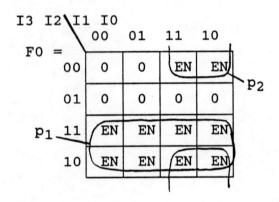

$$F2 = p_1$$

$$F2 = \overline{I3} \cdot \overline{I2} \cdot \overline{I1} \cdot I0 \cdot EN$$

Signal list: F2,I3,I2,I1,I0,EN

I3 I2\I1 I0
         00   01   11   10

F1 =

00 | 0 | 0 | EN | EN   ← p_2

01 | EN | EN | EN | EN   ← p_1

11 | 0 | 0 | 0 | 0

10 | 0 | 0 | 0 | 0

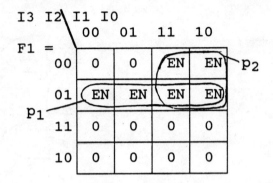

$$F1 = p_1 + p_2$$

$$F1 = \overline{I3} \cdot I2 \cdot EN + \overline{I3} \cdot I1 \cdot EN$$

Signal list: F1,I3,I2,I1,EN

I3 I2\I1 I0
         00   01   11   10

F0 =

00 | 0 | 0 | EN | EN   ← p_2

01 | 0 | 0 | 0 | 0

p_1 11 | EN | EN | EN | EN

10 | EN | EN | EN | EN

$$F0 = p_1 + p_2$$

$$F0 = I3 \cdot EN + \overline{I2} \cdot I1 \cdot EN$$

Signal list: F0,I3,I2,I1,EN

```
LLP: F2 = I3·I2·I1·I0·EN
NIP: 1    0  0  0  1  1

LLP: F1 = I3·I2·EN + I3·I1·EN
NIP: 1    0  1  1    0  1  1

LLP: F0 = I3·EN + I2·I1·EN
NIP: 1    1  1    0  1  1
```

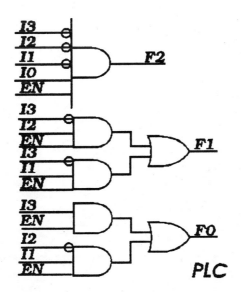

PLC

5-28.

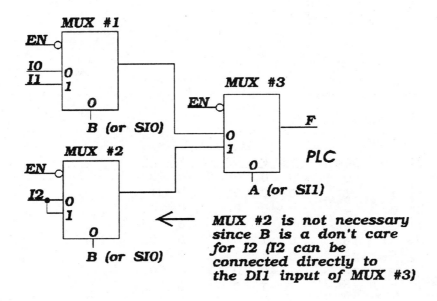

PLC

← MUX #2 is not necessary since B is a don't care for I2 (I2 can be connected directly to the DI1 input of MUX #3)

5-29.

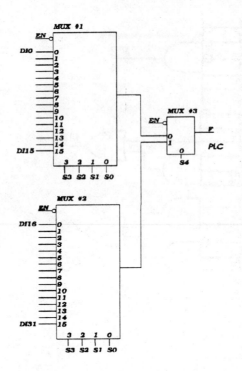

5-30.

X	Y	Z	Dec	Dec^2	Binary					
					F5	F4	F3	F2	F1	F0
0	0	0	0	0	0	0	0	0	0	0
0	0	1	1	1	0	0	0	0	0	1
0	1	0	2	4	0	0	0	1	0	0
0	1	1	3	9	0	0	1	0	0	1
1	0	0	4	16	0	1	0	0	0	0
1	0	1	5	25	0	1	1	0	0	1
1	1	0	6	36	1	0	0	1	0	0
1	1	1	7	49	1	1	0	0	0	1

By inspecting the truth table we can write

$$F5 = m_6 + m_7 = X \cdot Y$$
$$F4 = m_4 + m_5 + m_7$$
$$= m_4 + m_5 + m_5 + m_7$$

$$= X \cdot \overline{Y} + X \cdot Z$$
$$F3 = m_3 + m_5$$

$$= \overline{X} \cdot Y \cdot Z + X \cdot \overline{Y} \cdot Z$$
$$F2 = m_2 + m_6$$

$$= Y \cdot \overline{Z}$$
$$F1 = 0$$
$$F0 = Z$$

Signal list: F5,F4,F3,F2,F1,F0,X,Y,Z

```
LLP: F5 = X·Y
NIP: 1    1 1

LLP: F4 = X·Y + X·Z
NIP: 1    1 0    1 1

LLP: F3 = X·Y·Z + X·Y·Z
NIP: 1    0 1 1    1 0 1

LLP: F2 = Y·Z
NIP: 1    1 0

LLP: F1 = 0
NIP: Not applicable (connect F1 to logic 0)

LLP: F0 = Z
NIP: 1    1        for a Buffer (or just connect
                   F0 to Z directly)
```

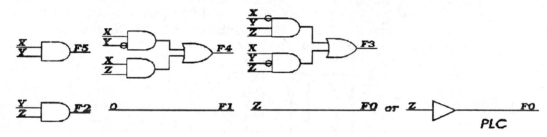

5-31.

X Y Z	F2	F1	F0
0 0 0	0	0	0
0 0 1	1	1	1
0 1 0	1	1	0
0 1 1	1	0	1
1 0 0	1	0	0
1 0 1	0	1	1
1 1 0	0	1	0
1 1 1	0	0	1

From the truth table F0 = Z

Signal list: F0, Z

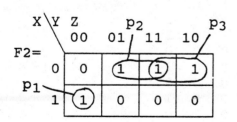

$F2 = p_1 + p_2 + p_3$

$F1 = p_1 + p_2$

119

$F2 = X \cdot \overline{Y} \cdot \overline{Z} + \overline{X} \cdot Z + \overline{X} \cdot Y$
Signal list: F2, X, Y, Z

$F1 = \overline{Y} \cdot Z + Y \cdot \overline{Z}$
$\quad = Y \oplus Z$
Signal list: F2, X, Y, Z

LLP: $F2 = X \cdot Y \cdot Z + X \cdot Z + X \cdot Y$
NIP: 1   1 0 0   0 1   0 1

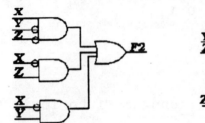

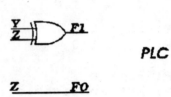

PLC

5-32. Since, 2's complement of N = (1's complement of N) + 1_{LSB}

then 1'C of N =   B2 B1 B0   3-bit number
                  + 0   0   1   add 1 to LSB (A2 A1 A0 = 001)
                    S2 S1 S0   2's complement of N

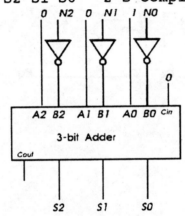

5-33. 4-BIT BINARY   2 groups of BCD bits required

I3	I2	I1	I0	F7	F6	F5	F4	F3	F2	F1	F0
0	0	0	0	0	0	0	0	0	0	0	0
0	0	0	1	0	0	0	0	0	0	0	1
0	0	1	0	0	0	0	0	0	0	1	0
0	0	1	1	0	0	0	0	0	0	1	1
0	1	0	0	0	0	0	0	0	1	0	0
0	1	0	1	0	0	0	0	0	1	0	1
0	1	1	0	0	0	0	0	0	1	1	0
0	1	1	1	0	0	0	0	0	1	1	1
1	0	0	0	0	0	0	0	1	0	0	0
1	0	0	1	0	0	0	0	1	0	0	1
1	0	1	0	0	0	0	1	0	0	0	0
1	0	1	1	0	0	0	1	0	0	0	1
1	1	0	0	0	0	0	1	0	0	1	0
1	1	0	1	0	0	0	1	0	0	1	1
1	1	1	0	0	0	0	1	0	1	0	0
1	1	1	1	0	0	0	1	0	1	0	1

However, F7 = F6 = F5 = 0, F0 = I0, and F3 = m_8 + m_9

= I3$\cdot\overline{I2}\cdot\overline{I1}$ by observing the truth table.

Signal list: F3, F0, I3, I2, I1, I0

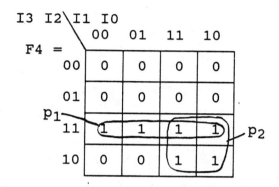

F4 = p_1 + p_2

F4 = I3$\cdot$I2 + I3$\cdot$I1

Signal list: F4, I3, I2, I1

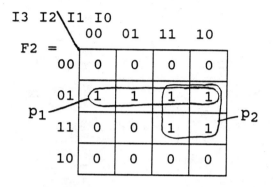

F2 = p_1 + p_2

F2 = $\overline{I3}\cdot$I2 + I2$\cdot$I1

Signal list: F2, I3, I2, I1

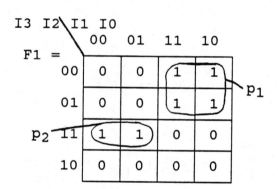

F1 = p_1 + p_2

F1 = $\overline{I3}\cdot$I1 + I3$\cdot$I2$\cdot\overline{I1}$

Signal list: F1, I3, I2, I1

LLP: F4 = I3$\cdot$I2 + I3$\cdot$I1
NIP: 1   1 1   1 1

LLP: F3 = I3$\cdot$I2$\cdot$I1
NIP: 1   1 0 0

LLP: F2 = I3$\cdot$I2 + I2$\cdot$I1
NIP: 1   0 1   1 1

```
LLP: F1 = I3·I1 + I3·I2·I1
NIP: 1    0  1    1  1  0

LLP: F0 = I0
NIP: 1     1
```

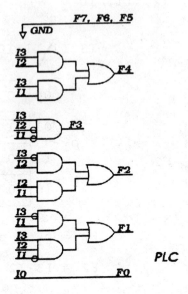

```
5-34.        2421 code      2-out-of-5 code
     m    I3 I2 I1 I0    F4 F3 F2 F1 F0
     0    0  0  0  0     0  0  0  1  1
     1    0  0  0  1     0  0  1  0  1
     2    0  0  1  0     0  0  1  1  0
     3    0  0  1  1     0  1  0  0  1
     4    0  1  0  0     0  1  0  1  0
    11    1  0  1  1     0  1  1  0  0
    12    1  1  0  0     1  0  0  0  1
    13    1  1  0  1     1  0  0  1  0
    14    1  1  1  0     1  0  1  0  0
    15    1  1  1  1     1  1  0  0  0
     5    0  1  0  1     X  X  X  X  X
     6    0  1  1  0     X  X  X  X  X
     7    0  1  1  1     X  X  X  X  X
     8    1  0  0  0     X  X  X  X  X
     9    1  0  0  1     X  X  X  X  X
    10    1  0  1  0     X  X  X  X  X
```

122

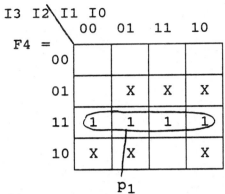

F4 = p_1

F4 = $I3 \cdot I2$

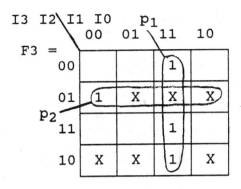

F3 = $p_1 + p_2$

F3 = $I1 \cdot I0 + \overline{I3} \cdot I2$

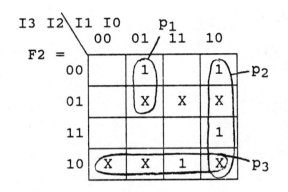

F2 = $p_1 + p_2 + p_3$

F2 = $\overline{I3} \cdot \overline{I1} \cdot I0 + I1 \cdot \overline{I0} + I3 \cdot \overline{I2}$

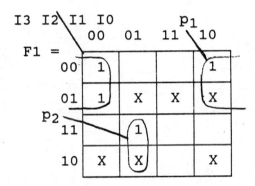

F1 = $p_1 + p_2$

F1 = $\overline{I3} \cdot \overline{I0} + I3 \cdot \overline{I1} \cdot I0$

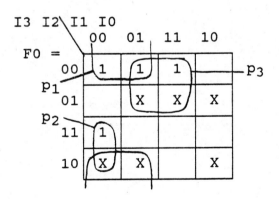

F0 = $p_1 + p_2 + p_3$

F0 = $\overline{I2} \cdot \overline{I1} + I3 \cdot \overline{I1} \cdot \overline{I0} + \overline{I3} \cdot I0$

Signal list: F4, F3, F2, F1, F0, I3, I2, I1, I0

LLP: F4 = $I3 \cdot I2$
NIP: 1   1 1

123

```
LLP: F3 = I1·I0 + I3·I2
NIP: 1     1 1     0 1

LLP: F2 = I3·I1·I0 + I1·I0 + I3·I2
NIP: 1     0  0  1     1  0     1  0

LLP: F1 = I3·I0 + I3·I1·I0
NIP: 1     0  0     1  0  1

LLP: F0 = I2·I1 + I3·I1·I0 + I3·I0
NIP: 1     0  0     1  0  0     0  1
```

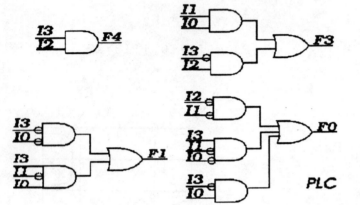
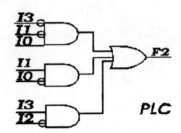

5-35.

A	B	C	F
0	0	0	1
0	0	1	0
0	1	0	0
0	1	1	1
1	0	0	0
1	0	1	1
1	1	0	1
1	1	1	1

$$F = \begin{array}{c|c|c|c|c} & 00 & 01 & 11 & 10 \\ \hline 0 & 1 & \boxed{0}\,r_1 & 1 & \boxed{0}\,r_2 \\ \hline 1 & \boxed{0}\,r_3 & 1 & 1 & 1 \end{array}$$

$$\overline{F} = r_1 + r_2 + r_3$$

$$= \overline{A}\cdot\overline{B}\cdot C + \overline{A}\cdot B\cdot\overline{C} + A\cdot\overline{B}\cdot\overline{C}$$

Signal list: F, A, B, C

```
LLP:  F = A·B·C + A·B·C + A·B·C
NIP:  0     0 0 1     0 1 0     1 0 0
```

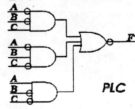

124

5-36.

A	B	C	D	F
0	0	0	0	0
0	0	0	1	1
0	0	1	0	1
0	0	1	1	0
0	1	0	0	1
0	1	0	1	0
0	1	1	0	0
0	1	1	1	1
1	0	0	0	1
1	0	0	1	0
1	0	1	0	0
1	0	1	1	1
1	1	0	0	0
1	1	0	1	1
1	1	1	0	1
1	1	1	1	0

A B\C D

$F_0 =$	00	01	11	10
00		1		1
01	1		1	
11		1		1
10	1		1	

$$F = m_1 + m_2 + m_4 + m_7 + m_8 + m_{11} + m_{13} + m_{14}$$

$$= \overline{A} \cdot \overline{B} \cdot \overline{C} \cdot D + \overline{A} \cdot \overline{B} \cdot C \cdot \overline{D} + \overline{A} \cdot B \cdot \overline{C} \cdot \overline{D} + \overline{A} \cdot B \cdot C \cdot D$$

$$+ A \cdot \overline{B} \cdot \overline{C} \cdot \overline{D} + A \cdot \overline{B} \cdot C \cdot D + A \cdot B \cdot \overline{C} \cdot D + A \cdot B \cdot C \cdot \overline{D}$$

$$= \overline{A} \cdot \overline{B} \cdot (C \oplus D) + \overline{A} \cdot B \cdot (\overline{C \oplus D})$$

$$+ A \cdot \overline{B} \cdot (\overline{C \oplus D}) + A \cdot B \cdot (C \oplus D)$$

$$= (C \oplus D) \cdot (\overline{A \oplus B}) + (\overline{C \oplus D}) \cdot (A \oplus B)$$
$$= (C \oplus D) \oplus (A \oplus B)$$

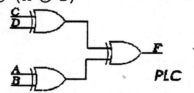

PLC

5-37.

Ai	Bi	Fi
0	0	1
0	1	0
1	0	0
1	1	1

A3 A2 A1 A0 = B3 B2 B1 B0   or F = 1

if   A3 = B3   or $F_3 = A3 \cdot B3 + \overline{A3} \cdot \overline{B3}$ = A3 $\odot$ B3

AND   A2 = B2   or $F_2 = A2 \cdot B2 + \overline{A2} \cdot \overline{B2}$ = A2 $\odot$ B2

AND   A1 = B1   or $F_1 = A1 \cdot B1 + \overline{A1} \cdot \overline{B1}$ = A1 $\odot$ B1

AND   A0 = B0   or $F_0 = A0 \cdot B0 + \overline{A0} \cdot \overline{B0}$ = A0 $\odot$ B0 otherwise F = 0

Therefore

$$F = F3 \cdot F2 \cdot F1 \cdot F0$$
$$F = (A3 \odot B3) \cdot (A2 \odot B2) \cdot (A1 \odot B1) \cdot (A0 \odot B0)$$

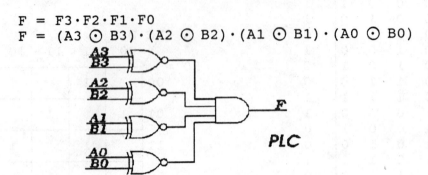

The circuit is not restricted to just BCD numbers since it will also detect the equivalent of any two 4-bit binary numbers.

5-38.

Ai	Bi	Ai=Bi	Ai>Bi
0	0	1	0
0	1	0	0
1	0	0	1
1	1	1	0

$$F = A3 \cdot \overline{B3} + (A3 \odot B3) \cdot A2 \cdot \overline{B2}$$

$$+ (A3 \odot B3) \cdot (A2 \odot B2) \cdot A1 \cdot \overline{B1}$$

$$+ (A3 \odot B3) \cdot (A2 \odot B2) \cdot (A1 \odot B1) \cdot A0 \cdot \overline{B0}$$

Signal list: F, A3, A2, A1, A0, B3, B2, B1, B0

```
LLP: F = A3·B3 + (A3 ⊙ B3)·A2·B2 + (A3 ⊙ B3)·(A2 ⊙ B2)·A1·B1
NIP: 1   1  0              1  0                          1  0
     + (A3 ⊙ B3)·(A2 ⊙ B2)·(A1 ⊙ B1)·A0·B0
                                      1  0
```

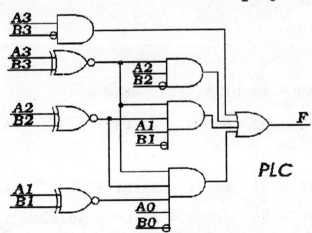

126

5-39.

Ai	Bi	Ai=Bi	Ai<Bi
0	0	1	0
0	1	0	1
1	0	0	0
1	1	1	0

$$F = \overline{A3} \cdot B3 + (A3 \odot B3) \cdot \overline{A2} \cdot B2$$

$$+ (A3 \odot B3) \cdot (A2 \odot B2) \cdot \overline{A1} \cdot B1$$

$$+ (A3 \odot B3) \cdot (A2 \odot B2) \cdot (A1 \odot B1) \cdot \overline{A0} \cdot B0$$

Signal list: F, A3, A2, A1, A0, B3, B2, B1, B0

LLP: F = A3·B3 + (A3 ⊙ B3)·A2·B2 + (A3 ⊙ B3)·(A2 ⊙ B2)·A1·B1
NIP: 1   0 1   0   1   0   1

$$+ (A3 \odot B3) \cdot (A2 \odot B2) \cdot (A1 \odot B1) \cdot A0 \cdot B0$$
                                                              0   1

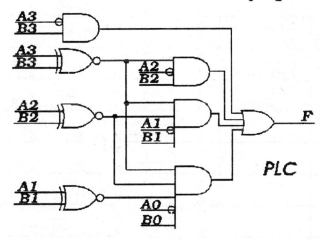

127

5-40.

A1	A0	B2	B1	B0	AxB = R	R3	R2	R1	R0	R3	R2	R1	R0
0	0	0	0	0	0x0 = 0	0	0	0	0				
0	0	0	0	1	0x1 = 0	0	0	0	0	0	0	0	0
0	0	0	1	0	0x2 = 0	0	0	0	0				
0	0	0	1	1	0x3 = 0	0	0	0	0	0	0	0	0
0	0	1	0	0	0x4 = 0	0	0	0	0				
0	0	1	0	1	0x5 = X	X	X	X	X	0,B0	0,B0	0,B0	0,B0
0	0	1	1	0	0x6 = X	X	X	X	X				
0	0	1	1	1	0x7 = X	X	X	X	X	X	X	X	X
0	1	0	0	0	1x0 = 0	0	0	0	0				
0	1	0	0	1	1x1 = 1	0	0	0	1	0	0	0	B0
0	1	0	1	0	1x2 = 2	0	0	1	0				
0	1	0	1	1	1x3 = 3	0	0	1	1	0	0	1	B0
0	1	1	0	0	1x4 = 4	0	1	0	0				
0	1	1	0	1	1x5 = X	X	X	X	X	0,B0	$\overline{B0}$,1	0,B0	0,B0
0	1	1	1	0	1x6 = X	X	X	X	X				
0	1	1	1	1	1x7 = X	X	X	X	X	X	X	X	X
1	0	0	0	0	2x0 = 0	0	0	0	0				
1	0	0	0	1	2x1 = 2	0	0	1	0	0	0	B0	0
1	0	0	1	0	2x2 = 4	0	1	0	0				
1	0	0	1	1	2x3 = 6	0	1	1	0	0	1	B0	0
1	0	1	0	0	2x4 = 8	1	0	0	0				
1	0	1	0	1	2x5 = X	X	X	X	X	$\overline{B0}$,1	0,B0	0,B0	0,B0
1	0	1	1	0	·	X	X	X	X				
1	0	1	1	1	·	X	X	X	X	X	X	X	X
1	1	0	0	0	·	X	X	X	X				
1	1	0	0	1		X	X	X	X	X	X	X	X
1	1	0	1	0		X	X	X	X				
1	1	0	1	1		X	X	X	X	X	X	X	X
1	1	1	0	0		X	X	X	X				
1	1	1	0	1		X	X	X	X	X	X	X	X
1	1	1	1	0		X	X	X	X				
1	1	1	1	1		X	X	X	X	X	X	X	X

R3 = :

A1 A0 \ B2 B1	00	01	11	10
00	0	0	–	0,B0
01	0	0	–	0,B0
11	–	–	–	–
10	0	0	–	$\overline{B0}$,1

p_1

$$R3 = p_1 = A1 \cdot B2$$

R2 = :

A1 A0 \ B2 B1	00	01	11	10
00	0	0	–	0,B0
01	0	0	–	$\overline{\overline{B0}}$,1
11	–	–	–	–
10	0	1	–	0,B0

p_2

p_1

$$R2 = p_1 + p_2$$
$$= A1 \cdot B1 + A0 \cdot B2$$

A1 A0\B2 B1

R1 =

A1 A0 \ B2 B1	00	01 (p_2)	11	10
00	0	0	–	0,B0
01	0	1	–	0,B0
11	–	–	–	–
10	B0	B0	–	0,B0

p_1

R0 =

A1 A0 \ B2 B1	00	01	11	10
00	0	0	–	0,B0
01	B0	B0	–	0,B0
11	–	–	–	–
10	0	0	–	0,B0

p_1

$$R1 = p_1 + p_2$$
$$= A1 \cdot B0 + A0 \cdot B1$$

$$R0 = p_1$$
$$= A0 \cdot B0$$

Signal list: R3, R2, R1, R0, A1, A0, B2, B1, B0

LLP: R3 = A1·B2
NIP: 1   1 1

LLP: R2 = A1·B1 + A0·B2
NIP: 1   1 1   1 1

LLP: R1 = A1·B0 + A0·B1
NIP: 1   1 1   1 1

LLP: R0 = A0·B0
NIP: 1   1 1

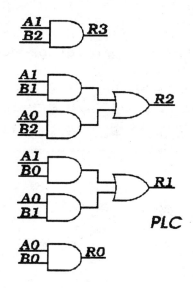

PLC

	B3	B2	B1	B0		
	x		A2	A1	A0	
		P03	P02	P01	P00	
	P13	P12	P11	P10		
C15	C14	C13	C12	0		
S15	S14	S13	S12	S11	S10	
P23	P22	P21	P20			
C26	C25	C24	C23	0		
S26	S25	S24	S23	S22	S21	S20

where, P00 = A0^B0   P12 = A1^B2
       P01 = A0^B1   P13 = A1^B3
       P02 = A0^B2   P20 = A2^B0
       P03 = A0^B3   P21 = A2^B1
       P10 = A1^B0   P22 = A2^B2
       P11 = A1^B1   P23 = A2^B3

Signal list: S26, S25, S24, S23, S21, S20, B3, B2, B1, B0, A2, A1, A0

Sample design documentation

LLP: P00 = A0^B0
NIP: 1   1 1

The rest of the equations have the same form and NIP.

For m (4) multiplicand bits and n (3) multiplier bits there are m + n (4 + 3 = 7) product bits S26 through S20. There are n - 1 (3 - 1 = 2) binary adders and each adder contains m (4) bits. m x n (4 x 3 = 12) partial products are required using two input AND gates.

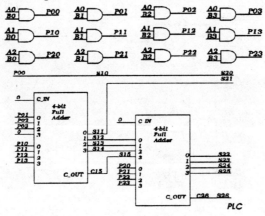

5-42.

```
                B1    B0
              x A1    A0
              A0xB1  A0xB0
      A1xB1   A1xB0
 C3   C2
 R3   R2      R1      R0
```

R0 = A0ˆB0   1 AND (ˆ represents multiplication)

R1 = A0ˆB1 + A1ˆB0   2 ANDs, 1 HA (+represents additions not ORing)

R2 = A1ˆB1 + C2   1 AND, 1 HA (+ represents additions not ORing)

R3 = C3

5-43.

```
   _0 Q          _0 Q          _0 Q          _0 Q
00√00         01√00         10√00         11√00
   _0            _0            _0            _0
00 R          00 R          00 R          00 R

   _0 Q          _1 Q          _0 Q          _0 Q
00√01         01√01         10√01         11√01
   _0            01            _0            _0
01 R          00 R          01 R          01 R

   _0 Q          _1 Q          _1 Q          _0 Q
00√10         01√10         10√10         11√10
   _0            01            10            _0
10 R          01 R          00 R          10 R
```

$$\begin{array}{llll}
\underline{0}\ Q & \underline{1}\ Q & \underline{1}\ Q & \underline{1}\ Q \\
00\sqrt{11} & 01\sqrt{11} & 10\sqrt{11} & 11\sqrt{11} \\
\underline{0} & \underline{01} & \underline{10} & \underline{11} \\
11\ R & 10\ R & 01\ R & 00\ R
\end{array}$$

A1	A0	B1	B0	Q0	R1	R0
0	0	0	0	0	0	0
0	0	0	1	0	0	1
0	0	1	0	0	1	0
0	0	1	1	0	1	1
0	1	0	0	0	0	0
0	1	0	1	1	0	0
0	1	1	0	1	0	1
0	1	1	1	1	1	0
1	0	0	0	0	0	0
1	0	0	1	0	0	1
1	0	1	0	1	0	0
1	0	1	1	1	0	1
1	1	0	0	0	0	0
1	1	0	1	0	0	1
1	1	1	0	0	1	0
1	1	1	1	1	0	0

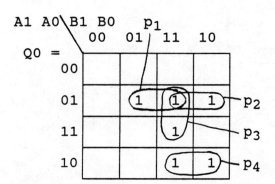

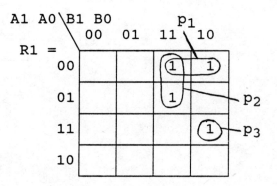

$Q0 = p_1 + p_2 + p_3 + p_4$

$ = \overline{A1}\cdot A0\cdot B0 + \overline{A1}\cdot A0\cdot B1$

$ + A0\cdot B1\cdot B0 + A1\cdot \overline{A0}\cdot B1$

Signal list: Q0,A1,A0,B1,B0

$R1 = p_1 + p_2 + p_3$

$ = \overline{A1}\cdot \overline{A0}\cdot B1 + \overline{A1}\cdot B1\cdot B0$

$ + A1\cdot A0\cdot B1\cdot \overline{B0}$

Signal list: R1,A1,A0,B1,B0

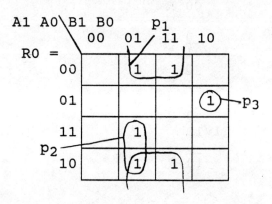

$$RO = p_1 + p_2 + p_3$$

$$= \overline{A0} \cdot B0 + A1 \cdot \overline{B1} \cdot B0 + \overline{A1} \cdot A0 \cdot B1 \cdot \overline{B0}$$

Signal list: R0,A1,A0,B1,B0

LLP: Q0 = A1·A0·B0 + A1·A0·B1 + A0·B1·B0 + A1·A0·B1
NIP: 1   0 1 1   0 1 1   1 1 1   1 0 1

LLP: R1 = A1·A0·B1 + A1·B1·B0 + A1·A0·B1·B0
NIP: 1   0 0 1   0 1 1   1 1 1 0

LLP: R0 = A0·B0 + A1·B1·B0 + A1·A0·B1·B0
NIP: 1   0 1   1 0 1   0 1 1 0

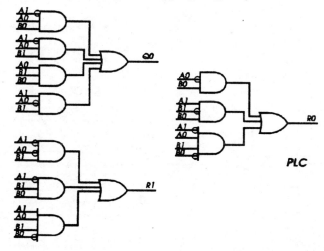

Section 5-6   Reviewing the Top-Down Design Process

5-44.

D	C	B	A	OA	OB	OC	OD	OE	OF	OG
0	0	0	0	1	1	1	1	1	1	0
0	0	0	1	0	1	1	0	0	0	0
0	0	1	0	1	1	0	1	1	0	1
0	0	1	1	1	1	1	1	0	0	1
0	1	0	0	0	1	1	0	0	1	1
0	1	0	1	1	0	1	1	0	1	1
0	1	1	0	1	0	1	1	1	1	1
0	1	1	1	1	1	1	0	0	0	0
1	0	0	0	1	1	1	1	1	1	1
1	0	0	1	1	1	1	1	0	1	1
1	0	1	0	X	X	X	X	X	X	X
1	0	1	1	X	X	X	X	X	X	X
1	1	0	0	X	X	X	X	X	X	X
1	1	0	1	X	X	X	X	X	X	X
1	1	1	0	X	X	X	X	X	X	X
1	1	1	1	X	X	X	X	X	X	X

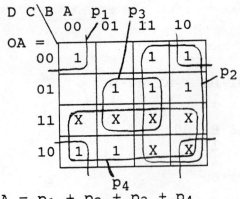

$$OA = p_1 + p_2 + p_3 + p_4$$

$$= \overline{C} \cdot \overline{A} + B + C \cdot A + D$$

$$= (C \odot B) + B + D$$

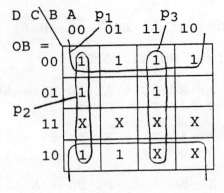

$$OB = p_1 + p_2 + p_3$$

$$= \overline{C} + \overline{B} \cdot \overline{A} + B \cdot A$$

$$= \overline{C} + (B \odot A)$$

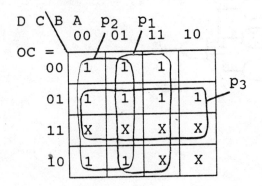

$$OC = p_1 + p_2 + p_3$$

$$= A + \overline{B} + C$$

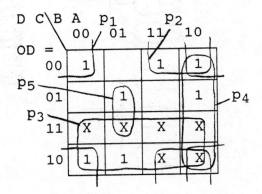

$$OD = p_1 + p_2 + p_3 + p_4 + p_5$$

$$= \overline{C} \cdot \overline{A} + \overline{C} \cdot B + D + B \cdot \overline{A} + C \cdot \overline{B} \cdot A$$

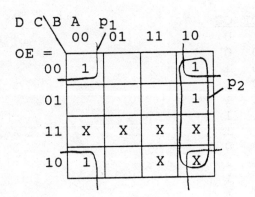

$$OE = p_1 + p_2$$

$$= \overline{C} \cdot \overline{A} + B \cdot \overline{A}$$

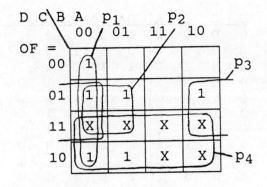

$$OF = p_1 + p_2 + p_3 + p_4$$

$$= \overline{B} \cdot \overline{A} + C \cdot \overline{B} + C \cdot \overline{A} + D$$

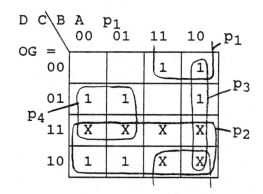

$$OG = p_1 + p_2 + p_3 + p_4$$

$$= \overline{C} \cdot B + D + B \cdot \overline{A} + C \cdot \overline{B}$$

$$= (C \oplus B) + D + B \cdot \overline{A}$$

For a common cathode display the signal logic convention for the destination of OA thorough OG would be positive logic.

Signal list: OA, OB, OC, OD, OE, OF, OG, D, C, B, A

```
LLP:  OA = (C ⊙ B) + B + D
NIP:  1                1   1

LLP:  OB = C + (B ⊙ A)
NIP:  1     0

LLP:  OC = A + B + C
NIP:  1   1   0   1

LLP:  OD = C·A + C·B + D + B·A + C·B·A
NIP:  1    0 0   0 1   1   1 0   1 0 1

LLP:  OE = C·A  +  B·A
NIP:  1    0 0      1 0

LLP:  OF = B·A  +  C·B  +  C·A + D
NIP:  1    0 0      1 0      1 0   1

LLP:  OG = (C ⊕ B) + D + B·A
NIP:  1              1   1 0
```

135

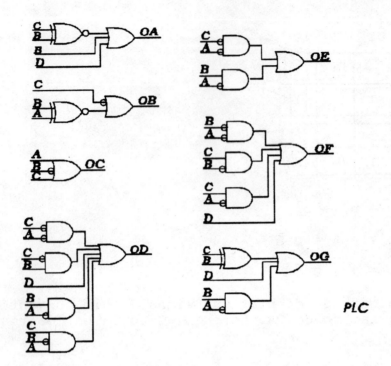

PLC

5-45.

A	B	C	D	AK6	AK5	AK4	AK3	AK2	AK1	AK0
0	0	0	0	0	1	1	0	0	0	0
0	0	0	1	0	1	1	0	0	0	1
0	0	1	0	0	1	1	0	0	1	0
0	0	1	1	0	1	1	0	0	1	1
0	1	0	0	0	1	1	0	1	0	0
0	1	0	1	0	1	1	0	1	0	1
0	1	1	0	0	1	1	0	1	1	0
0	1	1	1	0	1	1	0	1	1	1
1	0	0	0	0	1	1	1	0	0	0
1	0	0	1	0	1	1	1	0	0	1
1	0	1	0	1	0	0	0	0	0	1
1	0	1	1	1	0	0	0	0	1	0
1	1	0	0	1	0	0	0	0	1	1
1	1	0	1	1	0	0	0	1	0	0
1	1	1	0	1	0	0	0	1	0	1
1	1	1	1	1	0	0	0	1	1	0

From the truth table we can write by inspection

$AK5 = AK4 = \overline{AK6}$
$AK3 = m_8 + m_9 = A \cdot \overline{B} \cdot \overline{C}$

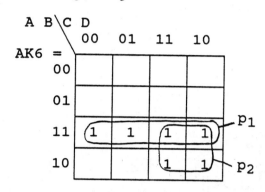

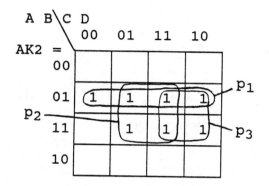

$AK6 = p_1 + p_2$

$\quad\quad = A \cdot B + A \cdot C$

$AK2 = p_1 + p_2 + p_3$

$\quad\quad = \overline{A} \cdot B + B \cdot D + B \cdot C$

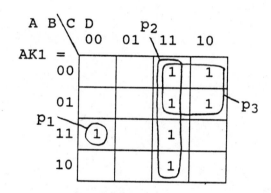

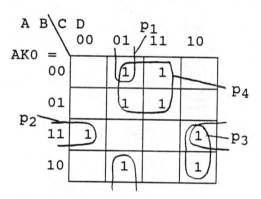

$AK1 = p_1 + p_2 + p_3$

$\quad\quad = A \cdot B \cdot \overline{C} \cdot \overline{D} + C \cdot D + \overline{A} \cdot C$

$AK0 = p_1 + p_2 + p_3 + p_4$

$\quad\quad = \overline{B} \cdot \overline{C} \cdot D + A \cdot B \cdot \overline{D} + A \cdot C \cdot \overline{D} + \overline{A} \cdot D$

Signal list: AK6, AK5, AK4, AK3, AK2, AK1, AK0, A, B, C, D

LLP: AK6 = A·B + A·C
NIP: 1   1 1   1 1

LLP: AK5 = AK4 = AK6
NIP: 1   1   0

LLP: AK3 = A·B·C
NIP: 1   1 0 0

LLP: AK2 = A·B + B·D + B·C
NIP: 1   0 1   1 1   1 1

LLP: AK1 = A·B·C·D + C·D + A·C
NIP: 1   1 1 0 0   1 1   0 1

LLP: AK0 = B·C·D + A·B·D + A·C·D + A·D
NIP: 1   001   110   110   01

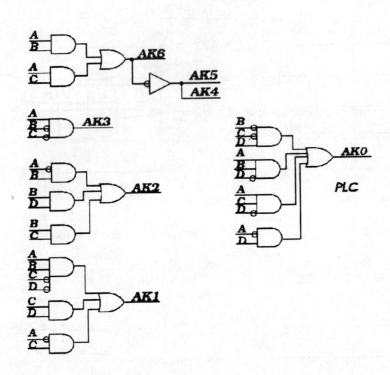

138

Section 6-2   Obtaining Realizable IC Circuits

6-1.

A B	F
0 0	0
0 1	1
1 0	1
1 1	1

$\overline{F} = \overline{A} \cdot \overline{B}$
Signal list: F, A, B

LLP: F = A•B
NIP: 0   0 0

$F = A + B$
Signal list: F, A, B

LLP: F = A + B
NIP: 1   1   1

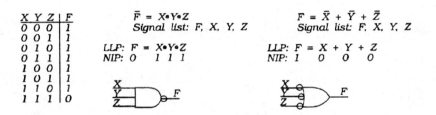

6-2.

X Y Z	F
0 0 0	1
0 0 1	1
0 1 0	1
0 1 1	1
1 0 0	1
1 0 1	1
1 1 0	1
1 1 1	0

$\overline{F} = X \cdot Y \cdot Z$
Signal list: F, X, Y, Z

LLP: F = X•Y•Z
NIP: 0   1 1 1

$F = \overline{X} + \overline{Y} + \overline{Z}$
Signal list: F, X, Y, Z

LLP: F = X + Y + Z
NIP: 1   0   0   0

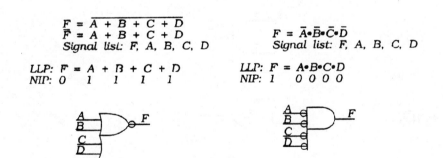

6-3.

$F = \overline{A + B + C + D}$
$\overline{F} = A + B + C + D$
Signal list: F, A, B, C, D

LLP: F = A + B + C + D
NIP: 0   1   1   1   1

$F = \overline{A} \cdot \overline{B} \cdot \overline{C} \cdot \overline{D}$
Signal list: F, A, B, C, D

LLP: F = A•B•C•D
NIP: 1   0 0 0 0

6-4.

$$F = X \cdot Y$$
Signal list: F, X, Y

$$\bar{F} = \bar{X} + \bar{Y}$$
Signal list: F, X, Y

LLP: $F = X \cdot Y$
NIP: 1   1 1

LLP: $F = X + Y$
NIP: 0   0   0

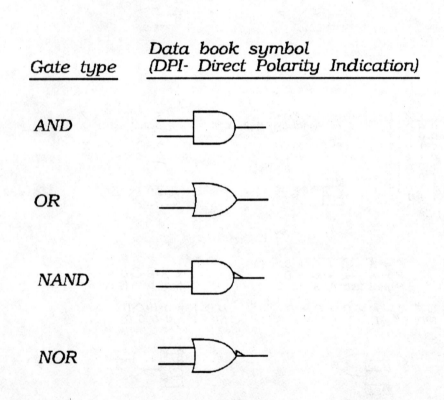

Data book symbol

DeMorgan equivalent symbol

6-5.

Gate type	Data book symbol (DPI- Direct Polarity Indication)
AND	
OR	
NAND	
NOR	

140

6-6.

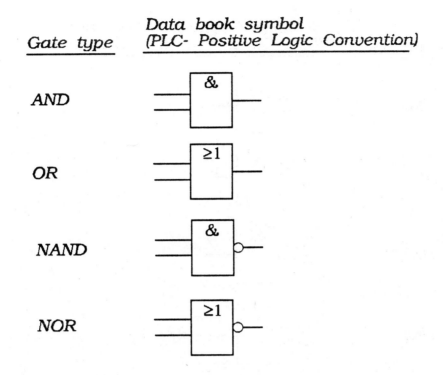

Gate type	Data book symbol (PLC- Positive Logic Convention)
AND	&
OR	≥1
NAND	& (with bubble)
NOR	≥1 (with bubble)

6-7.

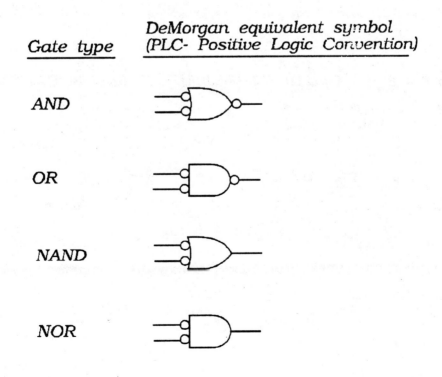

Gate type	DeMorgan equivalent symbol (PLC- Positive Logic Convention)
AND	
OR	
NAND	
NOR	

141

6-8.

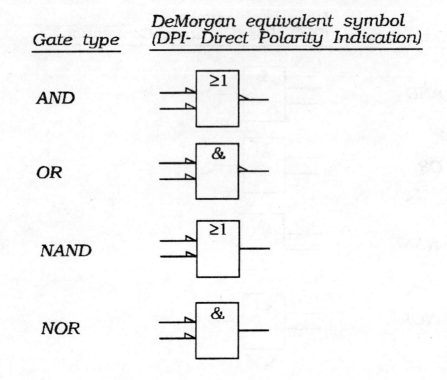

Gate type	DeMorgan equivalent symbol (DPI- Direct Polarity Indication)
AND	≥ 1
OR	$\&$
NAND	≥ 1
NOR	$\&$

6-9.

$F = A + B$
Signal list: F, A, B

LLP: $F = A + B$
NIP: 1 1 1

$F = A + B$
Signal list: F[NL],A[NL],B[NL]

LLP: $F = A + B$
NIP: 0 0 0

$\bar{F} = \bar{A}\bullet\bar{B}$
Signal list: F[NL],A[NL],B[NL]

LLP: $F = A\bullet B$
NIP: 1 1 1

PLC = NLC = NLC

6-10.

$\bar{F} = A \bullet B$
Signal list: F, A, B

$\bar{F} = A \bullet B$
Signal list: F[NL],A[NL],B[NL]

$F = \bar{A} + \bar{B}$
Signal list: F[NL],A[NL],B[NL]

LLP: $F = A \bullet B$
NIP: 0   1 1

LLP: $F = A \bullet B$
NIP: 1   0 0

LLP: $F = A + B$
NIP: 0   1   1

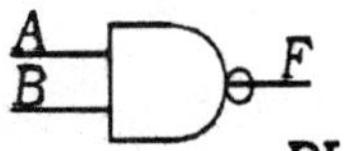

PLC 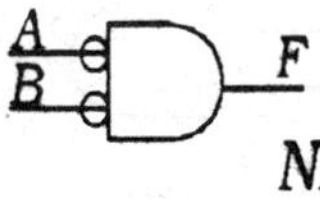   =   NLC 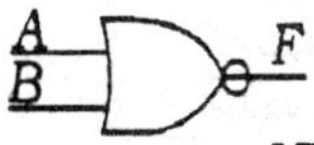   =   NLC

6-11.

$\bar{F} = A + B$
Signal list: F, A, B

$\bar{F} = A + B$
Signal list: F[NL],A[NL],B[NL]

$F = \bar{A} \bullet \bar{B}$
Signal list: F[NL],A[NL],B[NL]

LLP: $F = A + B$
NIP: 0   1   1

LLP: $F = A + B$
NIP: 1   0   0

LLP: $F = A \bullet B$
NIP: 0   1 1

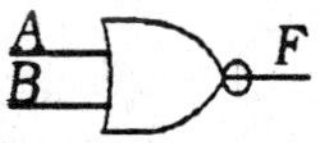

PLC 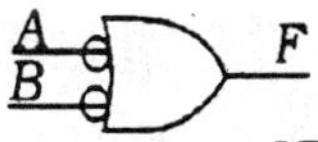   =   NLC 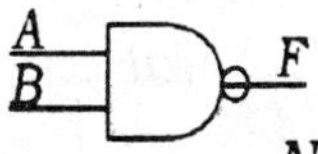   =   NLC

6-12.

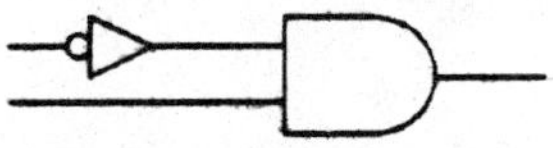

(a)

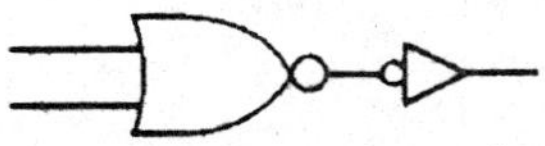

(b)

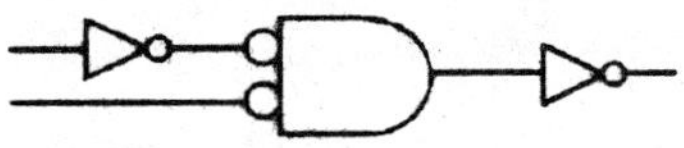

(c)

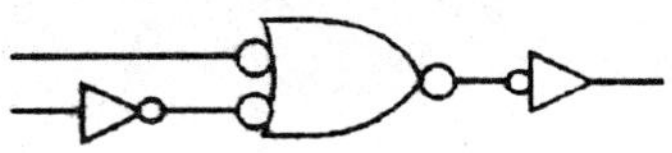

(d)

143

6-13.

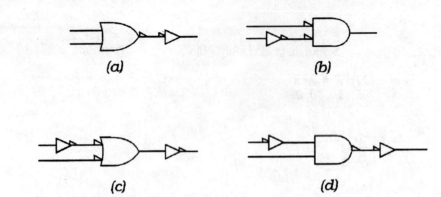

(a)

(b)

(c)

(d)

6-14.

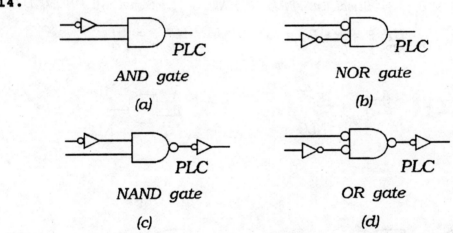

PLC

AND gate

(a)

PLC

NOR gate

(b)

PLC

NAND gate

(c)

PLC

OR gate

(d)

6-15.

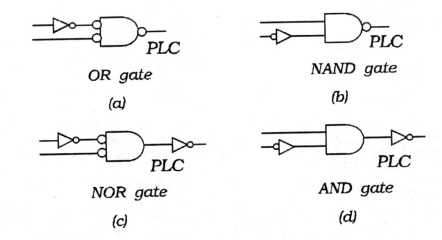

OR gate

(a)

NAND gate

(b)

NOR gate

(c)

AND gate

(d)

6-16.

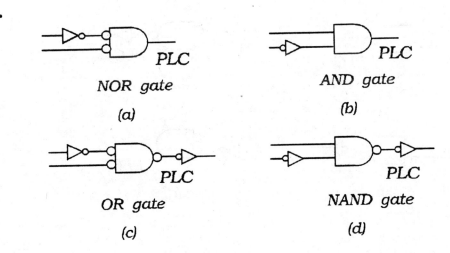

NOR gate

(a)

AND gate

(b)

OR gate

(c)

NAND gate

(d)

145

6-17.

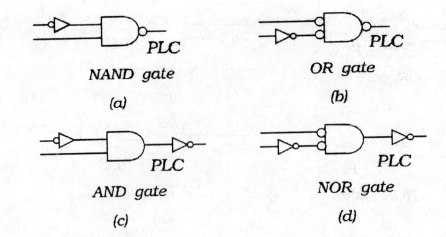

NAND gate

(a)

OR gate

(b)

AND gate

(c)

NOR gate

(d)

6-18.

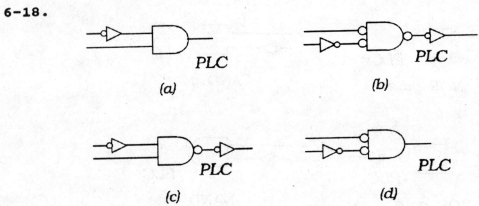

(a)

(b)

(c)

(d)

146

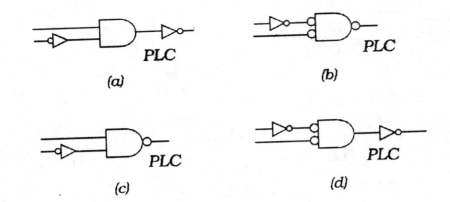

(a)

(b)

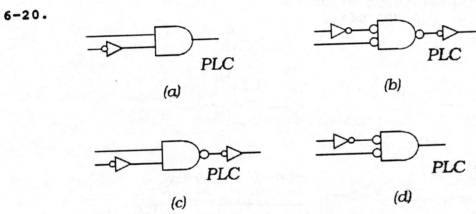

(c)

(d)

6-20.

(a)

(b)

(c)

(d)

6-21.

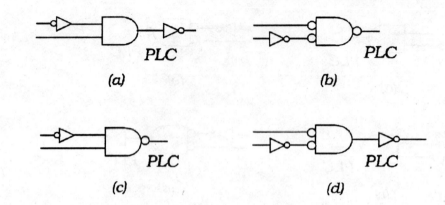

(a)　PLC

(b)　PLC

(c)　PLC

(d)　PLC

Section 6-3 Implementing Circuits with NAND Gates and NOR Gates

6-22.

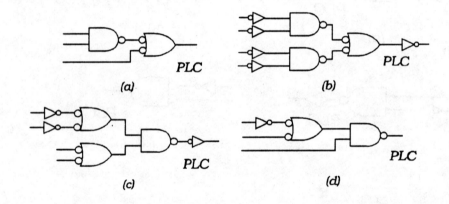

(a)　PLC

(b)　PLC

(c)　PLC

(d)　PLC

6-23.

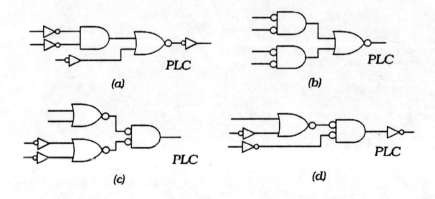

(a)　PLC

(b)　PLC

(c)　PLC

(d)　PLC

6-24.

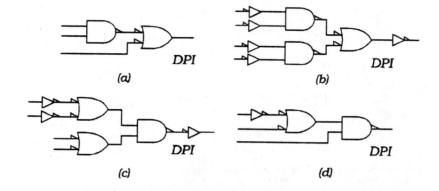

(a)   DPI

(b)   DPI

(c)   DPI

(d)   DPI

6-25.

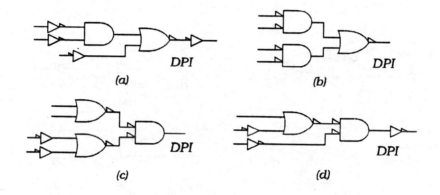

(a)   DPI

(b)   DPI

(c)   DPI

(d)   DPI

6-26.

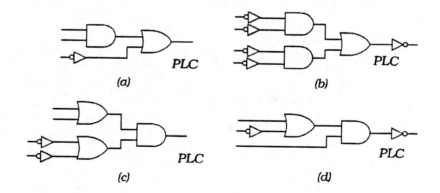

(a)   PLC

(b)   PLC

(c)   PLC

(d)   PLC

6-27.

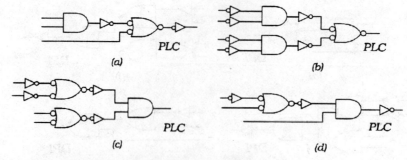

(a)

(b)

(c)

(d)

Inverters must be used on all interial signal
lines increasing the complexity of the circuit
compared to using only NAND gates or only NOR
gates which do not require these Inverters.

6-28.

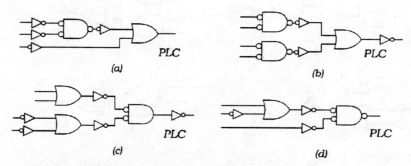

(a)

(b)

(c)

(d)

Inverters must be used on all interial signal
lines increasing the complexity of the circuit
compared to using only NAND gates or only NOR
gates which do not require these Inverters.

6-29.

X	Y	Z	F
0	0	0	0
0	0	1	1
0	1	0	1
0	1	1	0
1	0	0	1
1	0	1	0
1	1	0	0
1	1	1	0

X\Y Z

F =	00	01	11	10
0		1		1
1	1			

$F = m_1 + m_2 + m_4$

$F = \bar{X} \cdot \bar{Y} \cdot Z + \bar{X} \cdot Y \cdot \bar{Z} + X \cdot \bar{Y} \cdot \bar{Z}$
Signal list: F, X, Y, Z

LLP: $F = X \cdot Y \cdot Z + X \cdot Y \cdot Z + X \cdot Y \cdot Z$
NIP: 1   0 0 1   0 1 0   1 0 0

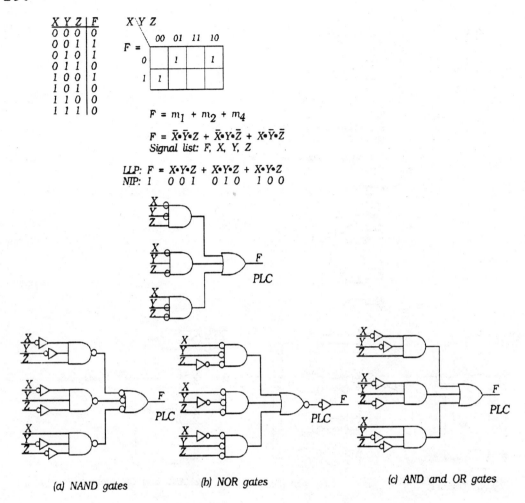

(a) NAND gates   (b) NOR gates   (c) AND and OR gates

6-30.

X	Y	Z	F
0	0	0	0
0	0	1	1
0	1	0	1
0	1	1	0
1	0	0	1
1	0	1	0
1	1	0	0
1	1	1	0

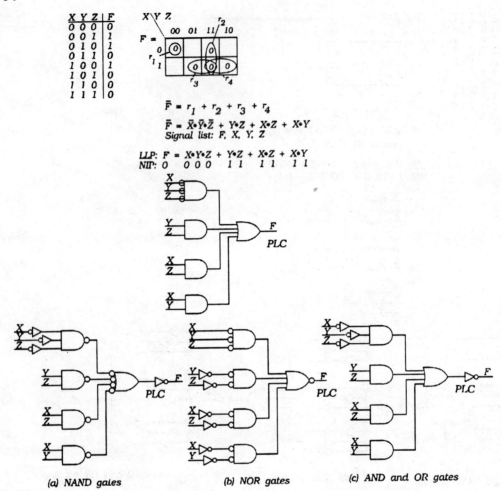

$$\bar{F} = r_1 + r_2 + r_3 + r_4$$

$$\bar{F} = \bar{X} \cdot \bar{Y} \cdot \bar{Z} + Y \cdot Z + X \cdot Z + X \cdot Y$$

Signal list: F, X, Y, Z

LLP: F = X·Y·Z + Y·Z + X·Z + X·Y

NIT: 0   0 0 0   1 1   1 1   1 1

PLC

(a) NAND gates   (b) NOR gates   (c) AND and OR gates

152

6-31.

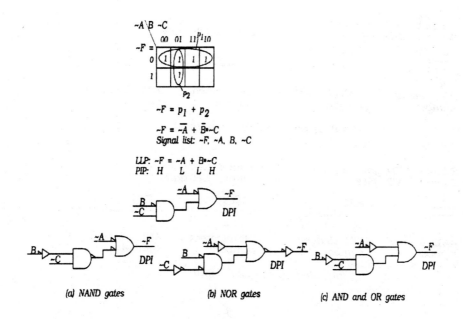

$$\sim F = p_1 + p_2$$

$$\sim F = \sim A + B \bullet \sim C$$
Signal list: $\sim F$, $\sim A$, B, $\sim C$

LLP: $\sim F = \sim A + B \bullet \sim C$
PIP: H   L   L   H

(a) NAND gates   (b) NOR gates   (c) AND and OR gates

6-32.

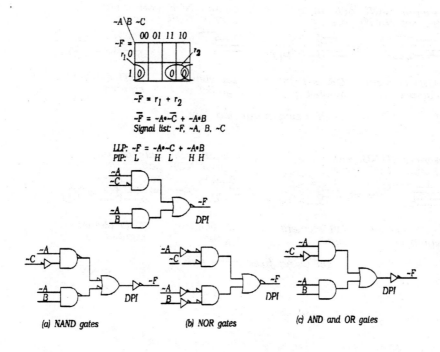

$$\overline{\sim F} = r_1 + r_2$$

$$\overline{\sim F} = \sim A \bullet \sim C + \sim A \bullet B$$
Signal list: $\sim F$, $\sim A$, B, $\sim C$

LLP: $\sim F = \sim A \bullet \sim C + \sim A \bullet B$
PIP: L   H L   H H

(a) NAND gates   (b) NOR gates   (c) AND and OR gates

6-33.

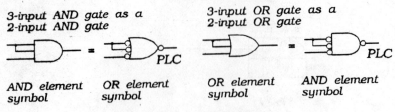

3-input AND gate as a
2-input AND gate

3-input OR gate as a
2-input OR gate

AND element   OR element   OR element   AND element
symbol   symbol   symbol   symbol

(a) using paralleled inputs

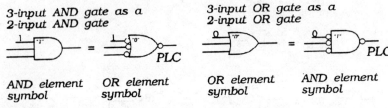

3-input AND gate as a
2-input AND gate

3-input OR gate as a
2-input OR gate

AND element   OR element   OR element   AND element
symbol   symbol   symbol   symbol

(b) using fixed-mode inputs

6-34.

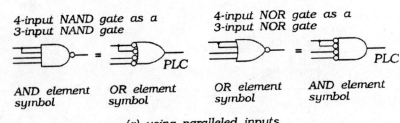

4-input NAND gate as a
3-input NAND gate

4-input NOR gate as a
3-input NOR gate

AND element   OR element   OR element   AND element
symbol   symbol   symbol   symbol

(a) using paralleled inputs

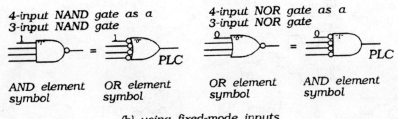

4-input NAND gate as a
3-input NAND gate

4-input NOR gate as a
3-input NOR gate

AND element   OR element   OR element   AND element
symbol   symbol   symbol   symbol

(b) using fixed-mode inputs

6-35.

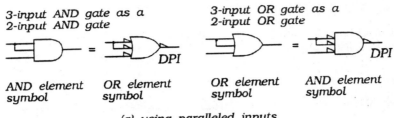

3-input AND gate as a
2-input AND gate

3-input OR gate as a
2-input OR gate

AND element
symbol

OR element
symbol

OR element
symbol

AND element
symbol

(a) using paralleled inputs

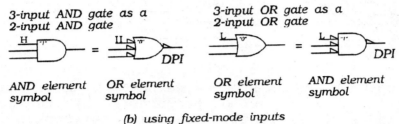

3-input AND gate as a
2-input AND gate

3-input OR gate as a
2-input OR gate

AND element
symbol

OR element
symbol

OR element
symbol

AND element
symbol

(b) using fixed-mode inputs

6-36.

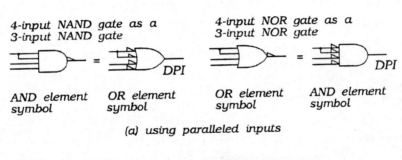

4-input NAND gate as a
3-input NAND gate

4-input NOR gate as a
3-input NOR gate

AND element
symbol

OR element
symbol

OR element
symbol

AND element
symbol

(a) using paralleled inputs

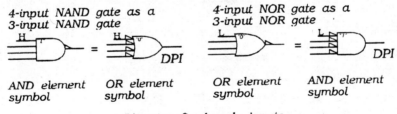

4-input NAND gate as a
3-input NAND gate

4-input NOR gate as a
3-input NOR gate

AND element
symbol

OR element
symbol

OR element
symbol

AND element
symbol

(b) using fixed-mode inputs

6-37.

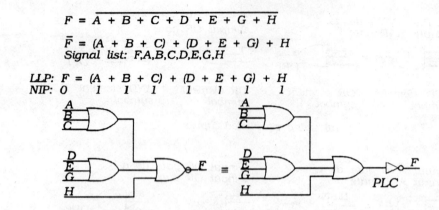

$$F = \overline{A + B + C + D + E + G + H}$$

$$\bar{F} = (A + B + C) + (D + E + G) + H$$
Signal list: F,A,B,C,D,E,G,H

LLP: F = (A + B + C) + (D + E + G) + H
NIP: 0   1   1   1   1   1   1   1

   In the Top-Down Design Process we have stressed writing
each function in SOP form (sum of products form), the form
most used in industry. A function can also be written in
SOS form (sum of sums form), POS form (product of sums
form), or POP form (product of products form).

6-38.

$$F = \overline{A \cdot B \cdot C \cdot D \cdot E \cdot G}$$

$$\bar{F} = (A \cdot B \cdot C \cdot D) \cdot (E \cdot G)$$
Signal list: F,A,B,C,D,E,G

LLP: F = (A·B·C·D)·(E·G)
NIP: 0   1 1 1 1 1 1

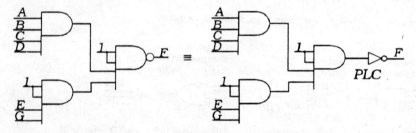

   In the Top-Down Design Process we have stressed writing
each function in SOP form (sum of products form), the form
most used in industry. A function can also be written in
SOS form (sum of sums form), POS form (product of sums
form), or POP form (product of products form).

6-39.

$$F = A \cdot B \cdot C \cdot D \cdot E \cdot G$$

$$F = (A \cdot B \cdot C) \cdot (D \cdot E \cdot G)$$
Signal list: F,A,B,C,D,E,G

LLP: $F = (A \cdot B \cdot C) \cdot (D \cdot E \cdot G)$
NIP: 1   1 1 1   1 1 1

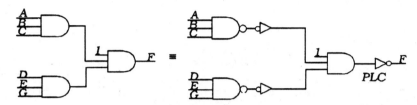

      In the Top-Down Design Process we have stressed writing each function in SOP form (sum of products form), the form most used in industry. A function can also be written in SOS form (sum of sums form), POS form (product of sums form), or POP form (product of products form).

6-40.

$$F = A + B + C$$

$$F = (A + B) + C$$
Signal list: F,A,B,C

LLP: $F = (A + B) + C$
NIP: 1   1   1   1

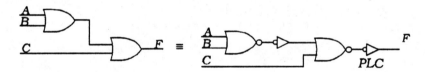

      In the Top-Down Design Process we have stressed writing each function in SOP form (sum of products form), the form most used in industry. A function can also be written in SOS form (sum of sums form), POS form (product of sums form), or POP form (product of products form).

6-41.

$$F = \overline{A + B + C + D}$$

$$\overline{F} = (A + B) + (C + D)$$
Signal list: F,A,B,C,D

LLP: $F = (A + B) + (C + D)$
NIP: 0   1 1   1 1

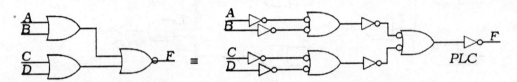

    In the Top-Down Design Process we have stressed writing each function in SOP form (sum of products form), the form most used in industry. A function can also be written in SOS form (sum of sums form), POS form (product of sums form), or POP form (product of products form).

6-42.

$$F = A \bullet B \bullet C \bullet D \bullet E$$

$$F = (A \bullet B) \bullet (C \bullet D) \bullet E$$
Signal list: F,A,B,C,D,E

LLP: $F = (A \bullet B) \bullet (C \bullet D) \bullet E$
NIP: 1   1 1 1 1   1

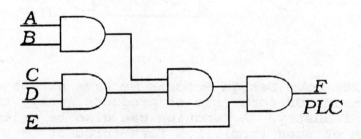

    In the Top-Down Design Process we have stressed writing each function in SOP form (sum of products form), the form most used in industry. A function can also be written in SOS form (sum of sums form), POS form (product of sums form), or POP form (product of products form).

6-43.

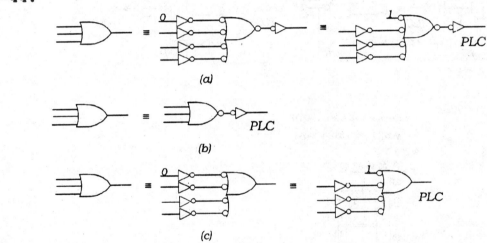

6-44.

159

6-45.

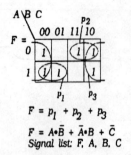

$$F = p_1 + p_2 + p_3$$

$$F = A \cdot \bar{B} + \bar{A} \cdot B + \bar{C}$$
Signal list: F, A, B, C

LLP: $F = A \cdot B + A \cdot B + C \cdot 1$ *Thinking ahead, each*
NTP: 1   1 0   0 1   0 1   *distributed connection*
                                             needs an AND element.

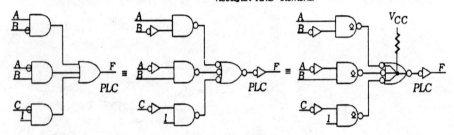

6-46.

$$F = W \cdot X \cdot \bar{Y} + X \cdot \neg \bar{Z} + X \cdot \bar{Y} + W \cdot Y \cdot \bar{Z}$$

$$F = W \cdot X \cdot \bar{Y} + X \cdot Z + X \cdot \bar{Y} + W \cdot Y \cdot \bar{Z}$$

$$F = p_1 + p_2 + p_3$$

$$F = X \cdot \bar{Y} + X \cdot Z + W \cdot Y \cdot \bar{Z}$$

$$F = X \cdot \bar{Y} + X \cdot \neg \bar{Z} + W \cdot X \cdot \neg Z$$
Signal list: F, W, X, Y, ~Z

LLP: $F = X \cdot Y + X \cdot \neg Z + W \cdot Y \cdot \neg Z$
PIP: H H L   H L   H H H

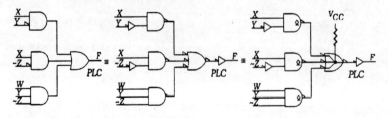

160

6-47.

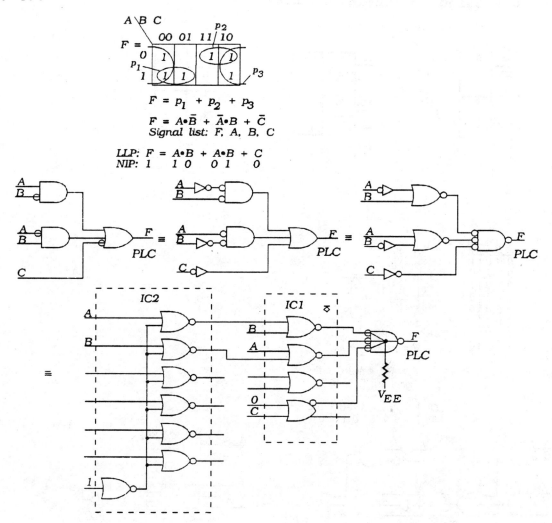

$$F = p_1 + p_2 + p_3$$

$$F = A \cdot \bar{B} + \bar{A} \cdot B + \bar{C}$$

Signal list: F, A, B, C

LLP: F = A•B + A•B + C
NIP: 1   1 0   0 1   0

Section 6-5 Using AND-OR-Invert Elements

6-48.

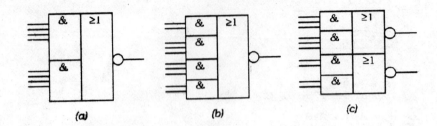

      (a)   (b)   (c)

6-49.

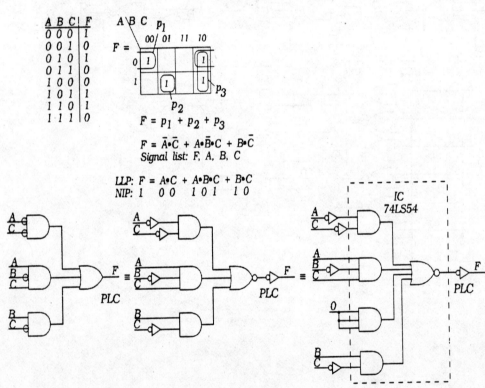

$F = p_1 + p_2 + p_3$

$F = \bar{A} \cdot \bar{C} + A \cdot \bar{B} \cdot C + B \cdot \bar{C}$

Signal list: F, A, B, C

LLP: $F = A \cdot C + A \cdot B \cdot C + B \cdot C$
NIP: 1   0 0   1 0 1   1 0

6-50.

The function is represented as

A	B	C	D	F
0	0	0	0	0
1	1	1	1	1
X	X	X	X	0

$$F = \bar{A} \cdot \bar{B} \cdot \bar{C} \cdot \bar{D} + A \cdot B \cdot C \cdot D$$
Signal list: F, A, B, C, D

LLP: $F = A \cdot B \cdot C \cdot D + A \cdot B \cdot C \cdot D$
PIP: H   L L L L   H H H H

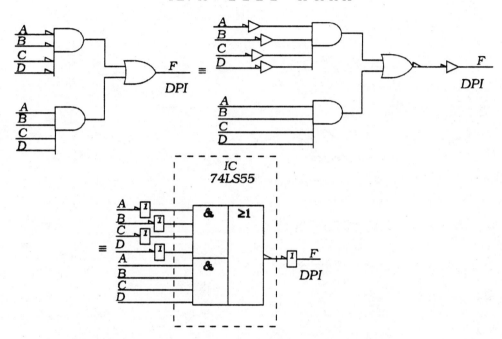

6-51.

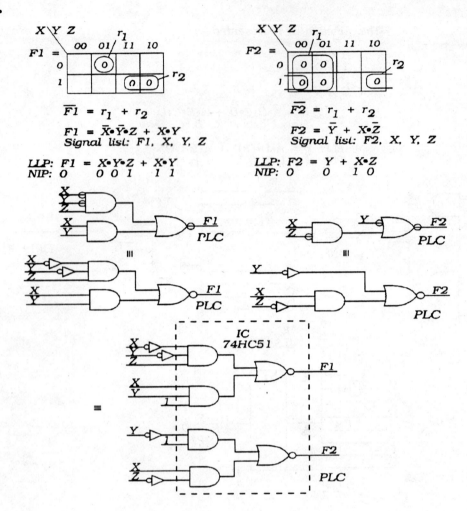

$\overline{F1} = r_1 + r_2$

$F1 = \overline{X} \cdot \overline{Y} \cdot Z + X \cdot Y$
Signal list: F1, X, Y, Z

$\overline{F2} = r_1 + r_2$

$F2 = \overline{Y} + X \cdot \overline{Z}$
Signal list: F2, X, Y, Z

LLP: $F1 = X \cdot Y \cdot Z + X \cdot Y$
NIP: 0 0 0 1 1 1

LLP: $F2 = Y + X \cdot Z$
NIP: 0 0 1 0

164

6-52. (a) $F = \overline{B} + A$
Signal list: F, A, B

    (b) $F = \overline{\overline{A} \cdot B}$

        $\overline{F} = \overline{A} \cdot B$
Signal list: F, A, B

    (c) $F = \overline{Z} + Y \cdot \overline{Z} + X \cdot Y$
Signal list: F, X, Y, Z

    (d) $F = \overline{\overline{X} \cdot Z + \overline{Y} \cdot Z}$

        $\overline{F} = \overline{X} \cdot Z + \overline{Y} \cdot Z$
Signal list: F, X, Y, Z

6-53. (a) $F = \overline{\sim A \cdot \overline{C} + \overline{B} \cdot C}$
Signal list: F, ~A, B, C

    (b) $F = \overline{B \cdot C + A \cdot \overline{C}}$

        $\overline{F} = B \cdot C + A \cdot \overline{C}$
Signal list: F, A, B, C

    (c) $\sim F = \overline{\overline{W} \cdot \overline{Z} + \overline{W} \cdot X \cdot Y + W \cdot \overline{Y} \cdot Z}$

        $\sim F = \overline{W} \cdot \overline{Z} + \overline{W} \cdot X \cdot Y + W \cdot \overline{Y} \cdot Z$
Signal list: ~F[NL], W, X, Y, Z

Alternate method: F is available just to the left
of the dotted line in the PLC.

so, $F = \overline{\overline{W} \cdot \overline{Z} + \overline{W} \cdot X \cdot Y + W \cdot \overline{Y} \cdot Z}$

      $\overline{F} = \overline{W} \cdot \overline{Z} + \overline{W} \cdot X \cdot Y + W \cdot \overline{Y} \cdot Z$

but, $\sim F = F$ inverted or $\overline{F}$

thus, $\sim F = \overline{W} \cdot \overline{Z} + \overline{W} \cdot X \cdot Y + W \cdot \overline{Y} \cdot Z$
Signal list: ~F[NL], W, X, Y, Z

    (d) $\overline{F}$ or $\sim F = \overline{\overline{W} \cdot \overline{Y} \cdot Z + \overline{W} \cdot X \cdot Y + W \cdot \overline{Z} + W \cdot Y}$

        $\overline{\sim F} = \overline{W} \cdot \overline{Y} \cdot Z + \overline{W} \cdot X \cdot Y + W \cdot Z + W \cdot Y$
Signal list: ~F[NL], W, X, Y, Z

Alternate method: F is available just to the left
of the dotted line in the PLC.

so, $F = \overline{W} \cdot \overline{Y} \cdot Z + \overline{W} \cdot X \cdot Y + W \cdot \overline{Z} + W \cdot Y$

but, $\overline{F}$ or $\tilde{F}$ = F inverted or $\overline{F}$

thus, $\tilde{F} = \overline{\overline{W} \cdot \overline{Y} \cdot Z + \overline{W} \cdot X \cdot Y + W \cdot \overline{Z} + W \cdot Y}$

$\overline{\tilde{F}} = \overline{W} \cdot \overline{Y} \cdot Z + \overline{W} \cdot X \cdot Y + W \cdot \overline{Z} + W \cdot Y$
Signal list: $\tilde{F}$[NL], W, X, Y, Z

6-54. (a) $\tilde{F} = \overline{W + Z} + \overline{W + X + \overline{Y}} + \overline{\overline{W} + Y + \overline{Z}}$

$\tilde{F} = \overline{W} \cdot \overline{Z} + \overline{W} \cdot \overline{X} \cdot Y + W \cdot \overline{Y} \cdot Z$
Signal list: $\tilde{F}$, W[NL], X, Y, Z

(b) $\overline{F}$ or $\tilde{F} = \overline{\overline{W} \cdot \overline{Y} \cdot Z} \cdot \overline{\overline{W} \cdot X \cdot Z} \cdot \overline{W \cdot \overline{Z}} \cdot \overline{W \cdot Y}$

$\tilde{F} = \overline{\overline{W} \cdot \overline{Y} \cdot Z} \cdot \overline{\overline{W} \cdot X \cdot Z} \cdot \overline{W \cdot \overline{Z}} \cdot \overline{W \cdot Y}$

Put the equation in SOP form by inverting
both sides of the equation and then apply
DeMorgan's Theorem.

$\overline{\tilde{F}} = \overline{\overline{\overline{W} \cdot \overline{Y} \cdot Z} \cdot \overline{\overline{W} \cdot X \cdot Z} \cdot \overline{W \cdot \overline{Z}} \cdot \overline{W \cdot Y}}$

$\overline{\tilde{F}} = \overline{W} \cdot \overline{Y} \cdot Z + \overline{W} \cdot X \cdot Z + W \cdot \overline{Z} + W \cdot Y$
Signal list: $\tilde{F}$, W, X, Y[NL], Z

Section 6-7 Obtaining Boolean Functions from Diagrams Using Direct Polarity Indication

6-55. For signal name $\tilde{B}$(H): If the signal state $\tilde{B}$ = 1
then the signal is high, else the signal is not high.

For signal name $\overline{\tilde{B}}$(L): If the signal state $\overline{\tilde{B}}$ = 1
($\tilde{B}$ = 0), then the signal is low, else the signal is
not low.

	Signal level high	Signal level low
$\tilde{B}$(H)	$\tilde{B}$ = 1	$\tilde{B}$ = 0
$\overline{\tilde{B}}$(L)	$\overline{\tilde{B}}$ = 0 ($\tilde{B}$ = 1)	$\overline{\tilde{B}}$ = 1 ($\tilde{B}$ = 0)

6-56. For signal name C(L): If the signal state C = 1
then the signal is low, else the signal is not low.

For signal name $\overline{C}$(H): If the signal state $\overline{C}$ = 1
(C = 0), then the signal is high, else the signal is
not high.

166

	Signal level high	Signal level low
C(L)	C = 0	C = 1
$\overline{C}$(H)	$\overline{C}$ = 1 (C = 0)	$\overline{C}$ = 0 (C = 1)

6-57. Show that $\tilde{}A$(H) $\neq$ $\tilde{}A$(L)

For $\tilde{}A$(H):
  $\tilde{}A$ = 1 when high
  $\tilde{}A$ = 0 when not high

For $\tilde{}A$(L):
  $\tilde{}A$ = 1 when low
  $\tilde{}A$ = 0 when not low

Comparing these results:
  When $\tilde{}A$ = 1 the first signal is high and the second signal is low.

  When $\tilde{}A$ = 0 the first signal is low and the second signal is high.

This analysis shows that the signals are obviously not equivalent.

6-58. Show that $\overline{\sim D}$(H) = ~D(L)

For $\overline{\tilde{}D}$(H):

  $\overline{\tilde{}D}$ = 1 ($\tilde{}D$ = 0) when high

  $\overline{\tilde{}D}$ = 0 ($\tilde{}D$ = 1) when not high

For $\tilde{}D$(L):

  $\tilde{}D$ = 1 when low

  $\tilde{}D$ = 0 when not low

Comparing these results:
  When $\tilde{}D$ = 1 both signals are low.

  When $\tilde{}D$ = 0 both signals are high.

This analysis shows that the signals are therefore equivalent.

6-59. (a) $\overline{F}$(L) = B·$\overline{A}$(L)

  $\overline{F}$ = B·$\overline{A}$
  Signal list: F, A, B

(b) F(H) = ($\overline{B}$ + A)(H)

  F   = $\overline{B}$ + A
  Signal list: F, A, B

(c) $\overline{F}$(L) = ($\overline{X}$·Z)(L)

  $\overline{F}$ = $\overline{X}$·Z
  Signal list: F, X, Z

Note that the circuit specified is more

complicated than required.

    (d) $F(H) = (\tilde{Z} + Y \cdot \tilde{Z} + X \cdot Y)(H)$

              $F = \tilde{Z} + Y \cdot \tilde{Z} + X \cdot Y$
              Signal list: F, X, Y, Z

6-60. (a) $\overline{F}(L) = (B \cdot C + A \cdot \overline{C})(L)$

           $\overline{F} = B \cdot C + A \cdot \overline{C}$
           Signal list: F, A, B, C

    (b) $F(H) = (\overline{\tilde{A} \cdot C} + \overline{B} \cdot C)(H)$

           $F = \overline{\tilde{A} \cdot C} + \overline{B} \cdot C$
           Signal list: F, ˜A, B, C

    (c) $\overline{\tilde{F}}(H) = (\overline{W} \cdot \overline{Y} \cdot \overline{Z} + \overline{W} \cdot X \cdot Z + W \cdot \overline{Z})(H)$

           $\overline{\tilde{F}} = \overline{W} \cdot \overline{Y} \cdot \overline{Z} + \overline{W} \cdot X \cdot Z + W \cdot \overline{Z}$
           Signal list: ˜F[NL], W, X, Y, Z

    (d) $\overline{F}(L)$ or $\tilde{F}(L) = (\overline{W} \cdot \overline{X} \cdot Y + W \cdot \overline{Y} \cdot Z)(L)$

                $\tilde{F} = \overline{W} \cdot \overline{X} \cdot Y + W \cdot \overline{Y} \cdot Z$
                    Signal list: ˜F[NL], W, X, Y, Z

6-61. (a) $\overline{\tilde{F}}(L) = (\overline{(W + Y + \overline{Z})} + \overline{(W + \overline{X} + \overline{Z})}$

               $+ \overline{(\overline{W} + Z)} + \overline{(\overline{W} + \overline{Y})})(L)$

           $\tilde{F}(L) = (\overline{W} \cdot \overline{Y} \cdot X + \overline{W} \cdot X \cdot Z + W \cdot \overline{Z} + W \cdot Y)(L)$

           $\tilde{F} = \overline{W} \cdot \overline{Y} \cdot X + \overline{W} \cdot X \cdot Z + W \cdot \overline{Z} + W \cdot Y$
           Signal list: ~F, W, X, Y[NL], Z

    (b) $F(L) = ((\overline{\overline{W} \cdot \overline{Z}}) \cdot (\overline{\overline{W} \cdot \overline{X} \cdot Y}) \cdot (W \cdot \overline{Y} \cdot Z))(L)$

           $\overline{F}(H) = (\overline{(\overline{\overline{W} \cdot \overline{Z}}) \cdot (\overline{\overline{W} \cdot \overline{X} \cdot Y}) \cdot (W \cdot \overline{Y} \cdot Z)})(H)$

           $\overline{F} = \overline{W} \cdot \overline{Z} + \overline{W} \cdot \overline{X} \cdot Y + W \cdot \overline{Y} \cdot Z$
           Signal list: F[NL], W[NL], X, Y, Z

Section 7-2 Implementing Logic Functions
Using MSI Multiplexers

7-1. (a) $F(A,B) = A \cdot B$ with respect to variable B
Signal list: F, A, B

$F(A,B) = F(A,0) \cdot \overline{B} + F(A,1) \cdot B$

for $F(A,B) = A \cdot B$

$I(0) = F(A,0) = 0$

$I(1) = F(A,1) = A$

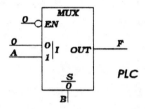

(b) $F(A,B) = A + B$ with respect to variable A
Signal list: F, A, B

$F(A,B) = F(0,B) \cdot \overline{A} + F(1,B) \cdot A$

for $F(A,B) = A + B$

$I(0) = F(0,B) = B$

$I(1) = F(1,B) = 1$

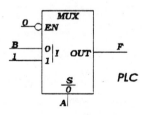

(c) $F(A,B) = A + B$ with respect to variable B
Signal list: F, A, B

$F(A,B) = F(A,0) \cdot \overline{B} + F(A,1) \cdot B$

for $F(A,B) = A + B$

$I(0) = F(A,0) = A$

$$I(1) = F(A,1) = 1$$

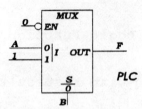

(d) $F(A,B) = A \oplus B$ with respect to variable A
Signal list: F, A, B

$$F(A,B) = F(0,B) \cdot \overline{A} + F(1,B) \cdot A$$

for $F(A,B) = \overline{A} \cdot B + A \cdot \overline{B}$

$$I(0) = F(0,B) = B$$

$$I(1) = F(1,B) = \overline{B}$$

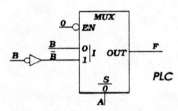

7-2. (a) $F(A,B) = A \cdot B$ with respect to variables A and B
Signal list: F, A, B

$$F(A,B) = F(0,0) \cdot \overline{A} \cdot \overline{B} + F(0,1) \cdot \overline{A} \cdot B$$

$$+ F(1,0) \cdot A \cdot \overline{B} + F(1,1) \cdot A \cdot B$$

for $F(A,B) = A \cdot B$

$$I(0) = F(0,0) = 0$$

$$I(1) = F(0,1) = 0$$

$$I(2) = F(1,0) = 0$$

$$I(3) = F(1,1) = 1$$

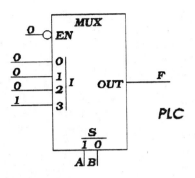

(b) $F(A,B) = A + B$ with respect to variables A and B
Signal list: F, A, B

$$F(A,B) = F(0,0) \cdot \overline{A} \cdot \overline{B} + F(0,1) \cdot \overline{A} \cdot B$$

$$+ F(1,0) \cdot A \cdot \overline{B} + F(1,1) \cdot A \cdot B$$

for $F(A,B) = A + B$

$I(0) = F(0,0) = 0$

$I(1) = F(0,1) = 1$

$I(2) = F(1,0) = 1$

$I(3) = F(1,1) = 1$

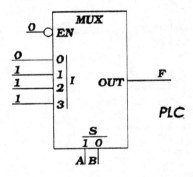

7-3. (a) $F(X,Y,Z) = X + \overline{Y} \cdot Z$ with respect to variable X
Signal list: F, X, Y, Z

$$F(X,Y,Z) = F(0,Y,Z) \cdot \overline{X} + F(1,Y,Z) \cdot X$$

for $F(X,Y,Z) = X + \overline{Y} \cdot Z$

$I(0) = F(0,Y,Z) = \overline{Y} \cdot Z$

$I(1) = F(1,Y,Z) = 1$

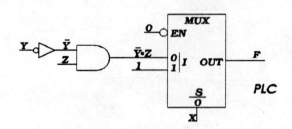

(b) $F(X,Y,Z) = X \cdot Y + \overline{X} \cdot Z$ with respect to variable X
Signal list: F, X, Y, Z

$F(X,Y,Z) = F(0,Y,Z) \cdot \overline{X} + F(1,Y,Z) \cdot X$

for $F(X,Y,Z) = X \cdot Y + \overline{X} \cdot Z$

$I(0) = F(0,Y,Z) = Z$

$I(1) = F(1,Y,Z) = Y$

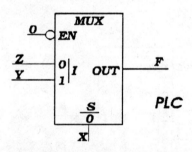

(c) $F(X,Y,Z) = \Sigma\ m(0,2,4,5)$ with respect to variable X
Signal list: F, X, Y, Z

$F =$

X\Y Z	00	01	11	10
0	1	0	0	1
1	1	1	0	0

p_1

p_1

$F(X,Y,Z) = p_1 + p_2$

$F(X,Y,Z) = \overline{X} \cdot \overline{Z} + X \cdot \overline{Y}$

$F(X,Y,Z) = F(0,Y,Z) \cdot \overline{X} + F(1,Y,Z) \cdot X$

for $F(X,Y,Z) = \overline{X} \cdot \overline{Z} + X \cdot \overline{Y}$

172

$I(0) = F(0,Y,Z) = \overline{Z}$

$I(1) = F(1,Y,Z) = \overline{Y}$

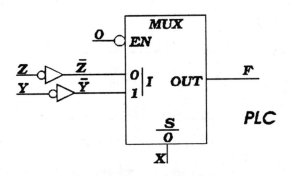

(d) $F(X,Y,Z) = \overline{X} + \overline{Y} \cdot \overline{Z}$ with respect to variable X
Signal list: F, X, Y, Z

$F(X,Y,Z) = F(0,Y,Z) \cdot \overline{X} + F(1,Y,Z) \cdot X$

for $F(X,Y,Z) = \overline{X} + \overline{Y} \cdot \overline{Z}$

$I(0) = F(0,Y,Z) = 1$

$I(1) = F(1,Y,Z) = \overline{Y} \cdot \overline{Z}$

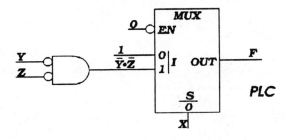

7-4. (a) $F(X,Y,Z) = X + \overline{Y} \cdot Z$ with respect to variables X and Y
Signal list: F, X, Y, Z

$F(X,Y,Z) = F(0,0,Z) \cdot \overline{X} \cdot \overline{Y} + F(0,1,Z) \cdot \overline{X} \cdot Y$

$+ F(1,0,Z) \cdot X \cdot \overline{Y} + F(1,1,Z) \cdot X \cdot Y$

for $F(X,Y,Z) = X + \overline{Y} \cdot Z$

$I(0) = F(0,0,Z) = Z$

$I(1) = F(0,1,Z) = 0$

$I(2) = F(1,0,Z) = 1$

$$I(3) = F(1,1,Z) = 1$$

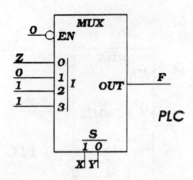

(b) $F(X,Y,Z) = X \cdot Y + \overline{X} \cdot Z$ with respect to variables X and Y

Signal list: F, X, Y, Z

$$F(X,Y,Z) = F(0,0,Z) \cdot \overline{X} \cdot \overline{Y} + F(0,1,Z) \cdot \overline{X} \cdot Y$$

$$+ F(1,0,Z) \cdot X \cdot \overline{Y} + F(1,1,Z) \cdot X \cdot Y$$

for $F(X,Y,Z) = X \cdot Y + \overline{X} \cdot Z$

$$I(0) = F(0,0,Z) = Z$$

$$I(1) = F(0,1,Z) = Z$$

$$I(2) = F(1,0,Z) = 0$$

$$I(3) = F(1,1,Z) = 1$$

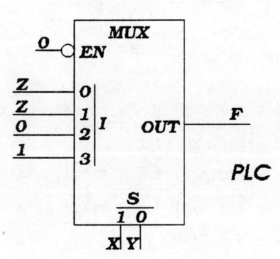

(c) From 7-3c, $F(X,Y,Z) = \overline{X} \cdot \overline{Z} + X \cdot \overline{Y}$ with respect to variables X and Y

Signal list: F, X, Y, Z

174

$$F(X,Y,Z) = F(0,0,Z) \cdot \overline{X} \cdot \overline{Y} + F(0,1,Z) \cdot \overline{X} \cdot Y$$
$$+ F(1,0,Z) \cdot X \cdot \overline{Y} + F(1,1,Z) \cdot X \cdot Y$$

for $F(X,Y,Z) = \overline{X} \cdot \overline{Z} + X \cdot \overline{Y}$

$I(0) = F(0,0,Z) = \overline{Z}$

$I(1) = F(0,1,Z) = \overline{Z}$

$I(2) = F(1,0,Z) = 1$

$I(3) = F(1,1,Z) = 0$

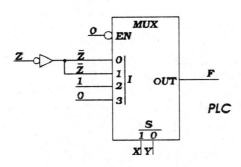

(d) $F(X,Y,Z) = \overline{X} + \overline{Y} \cdot \overline{Z}$ with respect to variables X and Y

Signal list: F, X, Y, Z

$$F(X,Y,Z) = F(0,0,Z) \cdot \overline{X} \cdot \overline{Y} + F(0,1,Z) \cdot \overline{X} \cdot Y$$
$$+ F(1,0,Z) \cdot X \cdot \overline{Y} + F(1,1,Z) \cdot X \cdot Y$$

for $F(X,Y,Z) = \overline{X} + \overline{Y} \cdot \overline{Z}$

$I(0) = F(0,0,Z) = 1$

$I(1) = F(0,1,Z) = 1$

$I(2) = F(1,0,Z) = \overline{Z}$

$I(3) = F(1,1,Z) = 0$

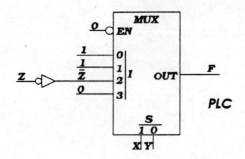

PLC

7-5. $F(A,B,C,D) = \Sigma\ m(0,2,8,9,10)$
        Signal list: F, A, B, C, D

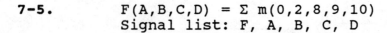

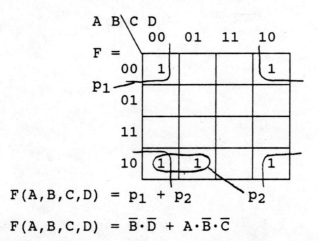

$F(A,B,C,D) = p_1 + p_2$

$F(A,B,C,D) = \overline{B}\cdot\overline{D} + A\cdot\overline{B}\cdot\overline{C}$

A 8-to-1 line multiplexer has three select inputs and we chose to write Shannon's Expansion Theorem with respect to variables A, B, and C.

$$F(A,B,C,D) = F(0,0,0,D)\cdot\overline{A}\cdot\overline{B}\cdot\overline{C} + F(0,0,1,D)\cdot\overline{A}\cdot\overline{B}\cdot C$$

$$+ F(0,1,0,D)\cdot\overline{A}\cdot B\cdot\overline{C} + F(0,1,1,D)\cdot\overline{A}\cdot B\cdot C$$

$$+ F(1,0,0,D)\cdot A\cdot\overline{B}\cdot\overline{C} + F(1,0,1,D)\cdot A\cdot\overline{B}\cdot C$$

$$+ F(1,1,0,D)\cdot A\cdot B\cdot\overline{C} + F(1,1,1,D)\cdot A\cdot B\cdot C$$

for $F(A,B,C,D) = \overline{B}\cdot\overline{D} + A\cdot\overline{B}\cdot\overline{C}$

$I(0) = F(0,0,0,D) = \overline{D}$

$I(1) = F(0,0,1,D) = \overline{D}$

176

$$I(2) = F(0,1,0,D) = 0$$

$$I(3) = F(0,1,1,D) = 0$$

$$I(4) = F(1,0,0,D) = 1$$

$$I(5) = F(1,0,1,D) = \overline{D}$$

$$I(6) = F(1,1,0,D) = 0$$

$$I(7) = F(1,1,1,D) = 0$$

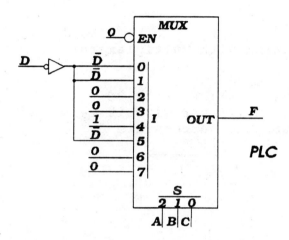

7-6.   from 7-5, $F(A,B,C,D) = \overline{B} \cdot \overline{D} + A \cdot \overline{B} \cdot \overline{C}$

A 4-to-1 line multiplexer has two select inputs and we choose to write Shannon's Expansion Theorem with respect to variables A and B.

$$F(A,B,C,D) = F(0,0,C,D) \cdot \overline{A} \cdot \overline{B} + F(0,1,C,D) \cdot \overline{A} \cdot B$$

$$+ F(1,0,C,D) \cdot A \cdot \overline{B} + F(1,1,C,D) \cdot A \cdot B$$

for $F(A,B,C,D) = \overline{B} \cdot \overline{D} + A \cdot \overline{B} \cdot \overline{C}$

$$I(0) = F(0,0,C,D) = \overline{D}$$

$$I(1) = F(0,1,C,D) = 0$$

$$I(2) = F(1,0,C,D) = \overline{D} + \overline{C}$$

$$I(3) = F(1,1,C,D) = 0$$

177

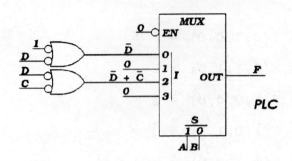

Section 7-3 Designing with Multiplexers

7-7. Given: $F(A,B,C) = \Sigma m(2,5,7)$
Signal list: F, A, B, C
For a type 0 Muliplexer circuit $f_2 = f_5 = f_7 = 1$ and all other characteristic numbers are 0.

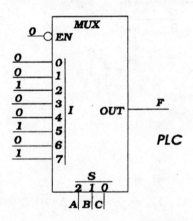

7-8. Given: $F(A,B,C) = \Sigma m(2,5,7)$
Signal list: F, A, B, C
For a type 1 Muliplexer circuit implementation one input variable is partitioned off as illustrated. Any one of the three input variables can be partitioned off, but the most convenient one to partition off is C (the least significant bit).

A	B	C	F	F
0	0	0	0	
0	0	1	0	0
0	1	0	1	
0	1	1	0	$\overline{C}$
1	0	0	0	
1	0	1	1	C
1	1	0	0	
1	1	1	1	C

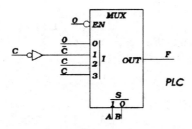

7-9. Given: $F(A,B,C) = \Sigma m(2,5,7)$

Signal list: F, A, B, C

For a type 2 Muliplexer circuit implementation two input variables are partitioned off as illustrated. Any two of the three input variables can be partitioned off, but the most convenient two to partition off are B C (the least significant bits).

A	B	C	F	F
0	0	0	0	
0	0	1	0	
0	1	0	1	$B \cdot \overline{C}$
0	1	1	0	
1	0	0	0	
1	0	1	1	
1	1	0	0	C
1	1	1	1	

179

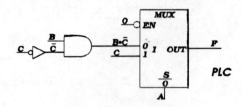

7-10. Given: F(A,B,C,D) = Σm(0,2,8,9,10)
              Signal list: F, A, B, C, D
        (a) for a gate-level design

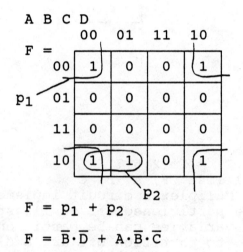

$$F = p_1 + p_2$$

$$F = \bar{B}\cdot\bar{D} + A\cdot\bar{B}\cdot\bar{C}$$

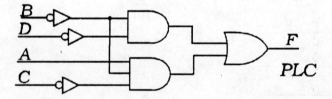

      (b) For a type 3 MUX design three input variables are partitioned off as illustrated. Any three of the four input variables can be partitioned off, but the most convenient three to partition off are B C D (the least significant bits).

180

A	B	C	D	F	F
0	0	0	0	1	
0	0	0	1	0	
0	0	1	0	1	
0	0	1	1	0	
0	1	0	0	0	$\overline{B}\cdot\overline{C}$
0	1	0	1	0	
0	1	1	0	0	
0	1	1	1	0	
1	0	0	0	1	
1	0	0	1	1	
1	0	1	0	1	
1	0	1	1	0	
1	1	0	0	0	
1	1	0	1	0	
1	1	1	0	0	
1	1	1	1	0	

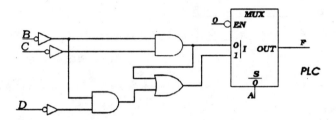

$F_{subfunction} =$

$$F_{subfunction} = p_1 + p_2$$

$$F_{subfunction} = \overline{B}\cdot\overline{C} + \overline{B}\cdot\overline{D}$$

The gate-level design requires fewer symbols to draw using the same type of gates as the MUX design.

7-11. Given: $F(A,B,C,D) = \Sigma m(0,2,8,9,10)$
          Signal list: F, A, B, C, D
  (a) See Problem 7-10a for the gate-level design.

  (b) For a type 2 MUX design two input variables are partitioned off as illustrated. Any two of the four input variables can be partitioned off, but the most convenient two to partition off are C D (the least significant bits).

181

A	B	C	D	F	F
0	0	0	0	1	
0	0	0	1	0	
0	0	1	0	1	
0	0	1	1	0	$\overline{D}$
0	1	0	0	0	
0	1	0	1	0	
0	1	1	0	0	
0	1	1	1	0	0
1	0	0	0	1	
1	0	0	1	1	
1	0	1	0	1	
1	0	1	1	0	$\overline{C \cdot D} = \overline{C} + \overline{D}$
1	1	0	0	0	
1	1	0	1	0	
1	1	1	0	0	
1	1	1	1	0	0

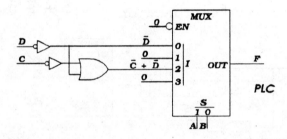

The type 2 MUX design requires fewer symbols to draw.

7-12. Given: $F(A,B,C,D) = \Sigma m(0,2,8,9,10)$
          Signal list: F, A, B, C, D
  (a) See Problem 7-10a for the gate-level design.

    (b) For a type 1 MUX design one input variable is
partitioned off as illustrated. Any one of the four input
variables can be partitioned off, but the most convenient
one to partition off is D (the least significant bit).

182

A	B	C	D	F	F
0	0	0	0	1	
0	0	0	1	0	$\overline{D}$
0	0	1	0	1	
0	0	1	1	0	$\overline{D}$
0	1	0	0	0	
0	1	0	1	0	0
0	1	1	0	0	
0	1	1	1	0	0
1	0	0	0	1	
1	0	0	1	1	1
1	0	1	0	1	
1	0	1	1	0	$\overline{D}$
1	1	0	0	0	
1	1	0	1	0	0
1	1	1	0	0	
1	1	1	1	0	0

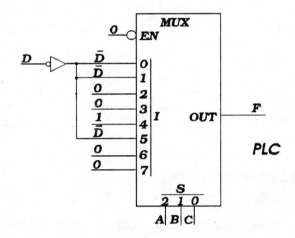

The type 1 MUX design requires fewer symbols to draw. Type 1 MUX designs are often preferred to gate-level circuits since they are considered both simpler to design and easier to understand by many engineers.

7-13. (a) Given: F1(A,B,C) = Σm(1,3,4,5,6,)
                     Signal list: F1, A, B, C

For a type 1 MUX design one input variable is partitioned off as illustrated. Any one of the three input variables can be partitioned off, but the most convenient one to partition off is C (the least significant bit).

A B	C	F1	F1
0 0	0	0	
0 0	1	1	C
0 1	0	0	
0 1	1	1	C
1 0	0	1	
1 0	1	1	1
1 1	0	1	
1 1	1	0	$\overline{C}$

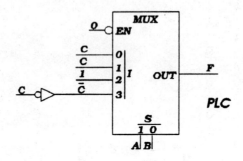

(b) Given: F2(X,Y,Z) = Σm(0,2,4,5,7)
              Signal list: F2, X, Y, Z

For a type 1 MUX design one input variable is
partitioned off as illustrated. Any one of the three input
variables can be partitioned off, but the most convenient
one to partition off is Z (the least significant bit).

X Y	Z	F2	F2
0 0	0	1	
0 0	1	0	$\overline{Z}$
0 1	0	1	
0 1	1	0	$\overline{Z}$
1 0	0	1	
1 0	1	1	1
1 1	0	0	
1 1	1	1	Z

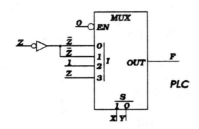

(c) Given: F3(A,B,C,D) = $\Sigma m(2,6,7,8,9,12,13,14,15)$
     Signal list: F3, A, B, C, D

     For a type 1 MUX design one input variable is
partitioned off as illustrated. Any one of the three input
variables can be partitioned off, but the most convenient
one to partition off is D (the least significant bit).

A	B	C	D	F3	F3
0	0	0	0	0	
0	0	0	1	0	0
0	0	1	0	1	
0	0	1	1	0	$\overline{D}$
0	1	0	0	0	
0	1	0	1	0	0
0	1	1	0	1	
0	1	1	1	1	1
1	0	0	0	1	
1	0	0	1	1	1
1	0	1	0	0	
1	0	1	1	0	0
1	1	0	0	1	
1	1	0	1	1	1
1	1	1	0	1	
1	1	1	1	1	1

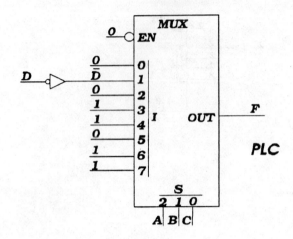

7-14. (a) Given: F1(A,B,C) = $\Sigma m(1,3,4,5,6,)$
            Signal list: F1, A, B, C

        For a type 0 Muliplexer circuit $f_1 = f_3 = f_4 = f_5 = f_6$ = 1 and all other characteristic numbers are 0.

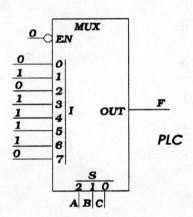

        (b) Given: F2(X,Y,Z) = $\Sigma m(0,2,4,5,7)$
            Signal list: F2, X, Y, Z

        For a type 0 Muliplexer circuit $f_0 = f_2 = f_4 = f_5 = f_7$ = 1 and all other characteristic numbers are 0.

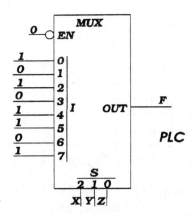

(c) Given: $F3(A,B,C,D) = \Sigma m(2,6,7,8,9,12,13,14,15)$
          Signal list: F3, A, B, C, D

For a type 0 Multiplexer circuit $f_2 = f_6 = f_7 = f_8 = f_9 = f_{12} = f_{13} = f_{14} = f_{15} = 1$ and all other characteristic numbers are 0.

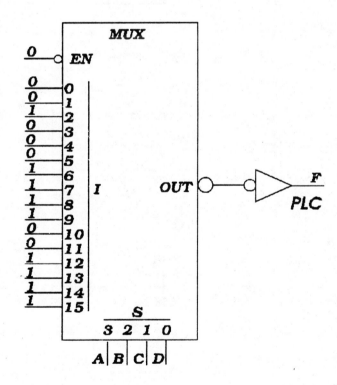

7-15. (a)   Given the following Karnaugh map:

```
      A\B C
         00   01   11   10
F =
      0 | 0  | 0  | 1  | 0 |
      1 | 1  | 0  | 1  | 1 |
```

Signal list: F, A, B, C

Using a Karnaugh map it is easy to partition off any one of the three input variables A, B, or C to obtain a type 1 MUX design. The number of different solutions, that is, the number of combinations is $C(n,r) = n!/((n-r)!r!)$, so for $n = 3$, the total number of input variables, and $r = 1$, the number of map-entered variables, $C(3,1) = 3$.

```
  A\B C                            B C
     00   01   11   10                00        01        11        10
F =                            F =
  0 | 0 | 0 | 1 | 0 |               | A     |  0      |  1      |  A     |
  1 | 1 | 0 | 1 | 1 |    =
```

```
  A\B C                            A\C
     00   01   11   10                 0         1
F =                            F =
  0 | 0 | 0 | 1 | 0 |               0 | 0    |  B    |
  1 | 1 | 0 | 1 | 1 |    =         1 | 1    |  B    |
```

```
  A\B C                            A\B
     00   01   11   10                 0         1
F =                            F =
  0 | 0 | 0 | 1 | 0 |               0 | 0    |  C    |
  1 | 1 | 0 | 1 | 1 |    =         1 | C̄    |  1    |
```

188

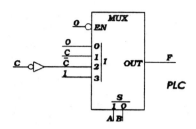

PLC

(b)   Given the following Karnaugh map:

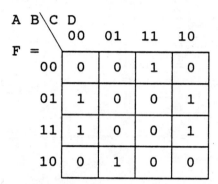

Signal list: F, A, B, C, D

     Using a Karnaugh map it is easy to partition off any
one of the four input variables A, B, C, or D to obtain a
type 1 MUX design as illustrated below for variables A and
D. The number of different solutions, that is, the number of
combinations is $C(n,r) = n!/((n-r)!r!)$, so for n = 4, the
total number of input variables, and r = 1, the number of
map-entered variables, $C(4,1) = 4$.

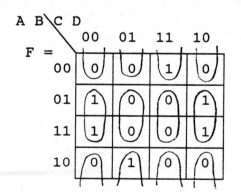 =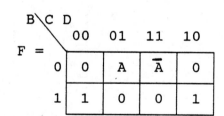

189

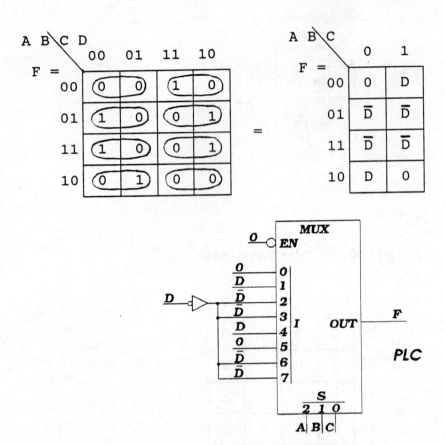

7-16. (a)   Given the following Karnaugh map:

$$
\begin{array}{c}
A\backslash B\ C \\
\end{array}
$$

	00	01	11	10
F = 0	0	0	1	0
1	1	0	1	1

Signal list: F, A, B, C

One possible type 2 MUX design is shown as follows.

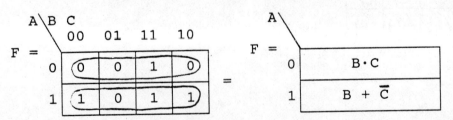

190

Two other solutions are possible, one where B is the input variable connected to the single select input of the MUX and the other where C is the input variable connected to the single select input of the MUX. Recall, the number of combinations $C(n,r) = n!/((n-r)!r!)$, and for $n = 3$, the total number of input variables, and $r = 2$, the number of map-entered variables $C(3,2) = 3$.

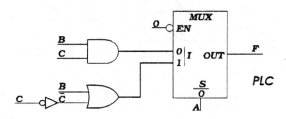

(b)   Given the following Karnaugh map:

A B\C D

F =	00	01	11	10
00	0	0	1	0
01	1	0	0	1
11	1	0	0	1
10	0	1	0	0

Signal list: F, A, B, C, D

One possible type 2 MUX design is shown as follows.

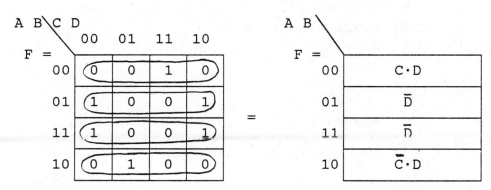

Five other solutions are possible, where input variable pairs A C, A D, B C, B D, and C D are connected to select inputs of the MUX, that is, the number of combinations $C(n,r) = n!/((n-r)!r!)$, and for $n = 4$, the total number of input variables, and $r = 2$, the number of map-entered variables $C(4,2) = 6$.

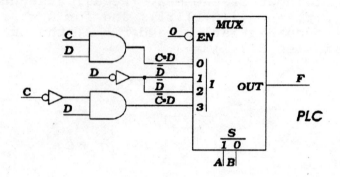

7-17. (a) $F(W,X,Y,Z) = OUT = \sum_{i=0}^{2^n-1} I_i \cdot m_i \cdot EN$

where $m_i(W,X,Y,Z)$, that is, $n = 4$

$F(W,X,Y,Z) = \sum_{i=0}^{15} I_i \cdot m_i,$ for $EN = 1$

$= 0 \cdot m_0 + 0 \cdot m_1 + 1 \cdot m_2 + 0 \cdot m_3 + 0 \cdot m_4$
$+ 0 \cdot m_5 + 1 \cdot m_6 + 1 \cdot m_7 + 1 \cdot m_8 + 0 \cdot m_9$
$+ 1 \cdot m_{10} + 0 \cdot m_{11} + 0 \cdot m_{12} + 1 \cdot m_{13}$
$+ 1 \cdot m_{14} + 1 \cdot m_{15}$

$= m_2 + m_6 + m_7 + m_8 + m_{10} + m_{13}$
$+ m_{14} + m_{15}$

$= \Sigma m(2,6,7,8,10,13,14,15)$
where $m = m(W,X,Y,Z)$

(b) $F(A,B,C,D) = OUT = \sum_{i=0}^{2^n-1} I_i \cdot m_i \cdot EN$

where $m_i(A,B,C)$, that is, $n = 3$

$F(A,B,C,D) = \sum_{i=0}^{7} I_i \cdot m_i,$ for $EN = 1$

$= 0 \cdot m_0 + 1 \cdot m_1 + \bar{D} \cdot m_2 + D \cdot m_3 + 0 \cdot m_4$

$+ 1 \cdot m_5 + \bar{D} \cdot m_6 + D \cdot m_7$

$= m_1 + \bar{D} \cdot m_2 + D \cdot m_3 + m_5$

$+ \bar{D} \cdot m_6 + D \cdot m_7$

192

$$= \overline{A} \cdot \overline{B} \cdot C + \overline{D} \cdot (\overline{A} \cdot B \cdot \overline{C}) + D \cdot (\overline{A} \cdot B \cdot C)$$

$$+ A \cdot \overline{B} \cdot C + \overline{D} \cdot (A \cdot B \cdot \overline{C}) + D \cdot (A \cdot B \cdot C)$$

$$A(MSB), \quad B, \quad C, \quad D(LSB)$$

$$= \overline{A} \cdot \overline{B} \cdot C \cdot (D + \overline{D}) + \overline{A} \cdot B \cdot \overline{C} \cdot \overline{D}$$

$$+ \overline{A} \cdot B \cdot C \cdot D + A \cdot \overline{B} \cdot C \cdot (D + \overline{D})$$

$$+ A \cdot B \cdot \overline{C} \cdot \overline{D} + A \cdot B \cdot C \cdot D$$

$$= m_3 + m_2 + m_4 + m_7 + m_{11} + m_{10}$$
$$+ m_{12} + m_{15}$$

$$= \Sigma m(2,3,4,7,10,11,12,15)$$
$$where \; m = m(A,B,C,D)$$

7-18. (a) $\quad F(A,B,C) = OUT = \sum\limits_{i=0}^{2^n-1} I_i \cdot m_i \cdot EN$

$\quad$ where $m_i(A,B)$, that is, $n = 2$

$\quad F(A,B,C) = \sum\limits_{i=0}^{4} I_i \cdot m_i, \qquad$ for $EN = 1$

$$= C \cdot m_0 + C \cdot m_1 + 0 \cdot m_2 + 1 \cdot m_3$$

$$= C \cdot m_0 + C \cdot m_1 + m_3$$

$$= C \cdot \overline{A} \cdot \overline{B} + C \cdot \overline{A} \cdot B + A \cdot B$$
$$A(MSB), \quad B, \quad C(LSB)$$

$$= \overline{A} \cdot \overline{B} \cdot C + \overline{A} \cdot B \cdot C + A \cdot B \cdot (C + \overline{C})$$

$$= m_1 + m_3 + m_7 + m_6$$

$$= \Sigma m(1,3,6,7)$$
$$where \; m = m(A,B,C)$$

(b) $\quad F(X,Y,Z) = OUT = \sum\limits_{i=0}^{2^n-1} I_i \cdot m_i \cdot EN$

$\quad$ where $m_i(X)$, that is, $n = 1$

$\quad F(X,Y,Z) = \sum\limits_{i=0}^{1} I_i \cdot m_i, \qquad$ for $EN = 1$

$$= Z \cdot m_0 + Y \cdot m_1$$

$$= Z \cdot \overline{X} + Y \cdot X$$
$$X(MSB), \quad Y(LSB)$$

$$= \overline{X} \cdot (Y + \overline{Y}) \cdot Z + X \cdot Y \cdot (Z + \overline{Z})$$

$$= m_3 + m_1 + m_7 + m_6$$

$$= \Sigma m(1,3,6,7)$$
$$\text{where } m = m(X,Y,Z)$$

Section 7-4 Additional techniques for Designing with Multiplexers

7-19. Given $F(A,B,C,D,E) = \Sigma\, m(0,5,7,11,15,16,18,25,29)$
  Signal list: F, A, B, C, D, E
  (a) Design a circuit using a 16-to 1 line Multiplexer to implement the function specified.

m	E F F	m	E F F	m	E F F	m	E F F
0	0 1	8	0 0	16	0 1	24	0 0
1	1 0 $\overline{E}$	9	1 0 0	17	1 0 $\overline{E}$	25	1 1 E
2	0 0	10	0 0	18	0 1	26	0 0
3	1 0 0	11	1 1 E	19	1 0 $\overline{E}$	27	1 0 0
4	0 0	12	0 0	20	0 0	28	0 0
5	1 1 E	13	1 0 0	21	1 0 0	29	1 1 E
6	0 0	14	0 0	22	0 0	30	0 0
7	1 1 E	15	1 1 E	23	1 0 0	31	1 0 0

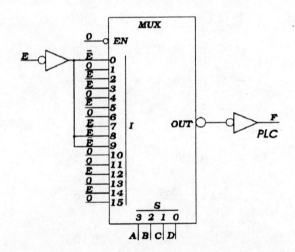

(b) Design a circuit using two 8-to-1 line MUXs and one 2-to-1 line MUX to construct a 16-to-1 MUX tree to implement the function specified.

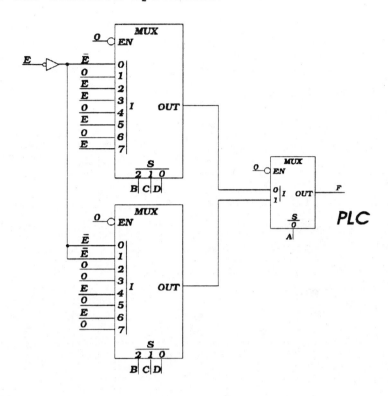

7-20. Use four 2-to-1 line MUXs and one 4-to-1 line MUX to design a type 1 MUX design for the following function.

$$F(W,X,Y,Z) = \Sigma\ m(5,7,13,14,15)$$
Signal list: F, W, X, Y, Z

W	X	Y	Z	F	F
0	0	0	0	0	
0	0	0	1	0	0
0	0	1	0	0	
0	0	1	1	0	0
0	1	0	0	0	
0	1	0	1	1	Z
0	1	1	0	0	
0	1	1	1	1	Z
1	0	0	0	0	
1	0	0	1	0	0
1	0	1	0	0	
1	0	1	1	0	0
1	1	0	0	0	
1	1	0	1	1	Z
1	1	1	0	1	
1	1	1	1	1	1

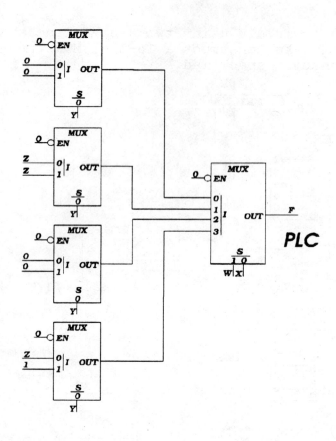

PLC

7-21. $F1(X,Y,Z) = \Sigma\ m(1,3,4,5,7)$
$F2(X,Y,Z) = \Sigma\ m(1,2,3,4,6)$
Signal list: F1, F2, X[NL], Y, Z

(a) Method 1 (assume all positive logic signals then
add Inverters to negative logic signal
lines)

X	Y	Z	F1	F2	F1	F2
0	0	0	0	0		
0	0	1	1	1	Z	Z
0	1	0	0	1		
0	1	1	1	1	Z	1
1	0	0	1	1		
1	0	1	1	0	1	$\overline{Z}$
1	1	0	0	1		
1	1	1	1	0	Z	$\overline{Z}$

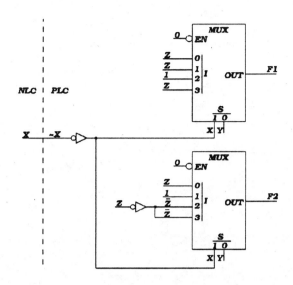

(b)   Method 2   (fix truth table by complementing
                  negative logic signals)

m	~X	Y	Z	F1	F2	F1	F2
	1	0	0	0	0		
2	1	0	1	1	1	Z	Z
	1	1	0	0	1		
3	1	1	1	1	1	Z	1
	0	0	0	1	1		
0	0	0	1	1	0	1	$\overline{Z}$
	0	1	0	0	1		
1	0	1	1	1	0	Z	$\overline{Z}$

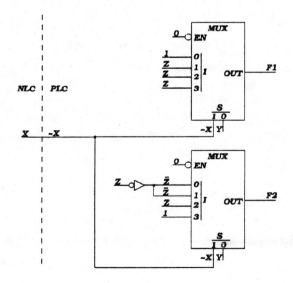

7-22. (a) $F(A,B,C) = \overline{A}\cdot\overline{B}\cdot(C + \overline{C}) + (A + \overline{A})\cdot\overline{B}\cdot\overline{C}$

$\qquad\qquad + \overline{A}\cdot(B + \overline{B})\cdot\overline{C}$

$\qquad\quad = m_1 + m_0 + m_4 + m_0 + m_2 + m_0$

$\qquad\quad = m_0 + m_1 + m_2 + m_4$

$\qquad\quad = \Sigma\, m(0,1,2,4)$

(b) $F(A,B,C) = \overline{A}\cdot\overline{B} + \overline{B}\cdot\overline{C} + \overline{A}\cdot\overline{C}$

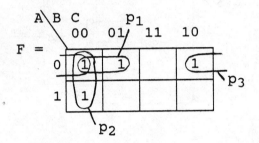

$\qquad F(A,B,C) = m_0 + m_1 + m_2 + m_4$

$\qquad\qquad\qquad = \Sigma\, m(0,1,2,4)$

(c) $F(A,B,C) = \overline{A}\cdot\overline{B} + \overline{B}\cdot\overline{C} + \overline{A}\cdot\overline{C}$

A	B	C	F
0	0	0	1
0	0	1	1
0	1	0	1
0	1	1	
1	0	0	1
1	0	1	
1	1	0	
1	1	1	

$\qquad F(A,B,C) = m_0 + m_1 + m_2 + m_4$

$\qquad\qquad\qquad = \Sigma\, m(0,1,2,4)$

7-23. (a) $F(A,B,C) = B\cdot C + A\cdot C + A\cdot B$

$\qquad$ Signal list: F, A, B, C

A	B	C	F	F
0	0	0	0	
0	0	1	0	0
0	1	0	0	
0	1	1	1	C
1	0	0	0	
1	0	1	1	C
1	1	0	1	
1	1	1	1	1

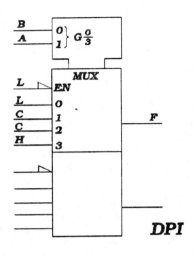

DPI

(b) $F(A,B) = A \cdot B = A \cdot B + \overline{A} \cdot \overline{B}$
     Signal list: F, A, B[NL]

A	B	F
0	0	1
0	1	0
1	0	0
1	1	1

A	~B	F	F
0	1	1	
0	0	0	~B
1	1	0	
1	0	1	$\overline{\sim B}$

by method 2

Recall: $B[NL] = B(L) = \overline{B}(H) = \overline{B} = \,\sim B$

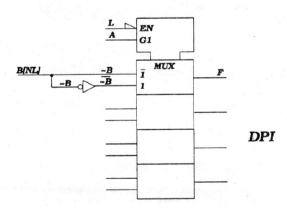

DPI

(c) $F(W,X,Y,Z) = W \cdot \overline{X} \cdot Y + Z$
     Signal list: F[NL], W, X, Y, Z

199

W	X	Y	Z	F	F
0	0	0	0	0	
0	0	0	1	1	Z
0	0	1	0	0	
0	0	1	1	1	Z
0	1	0	0	0	
0	1	0	1	1	Z
0	1	1	0	0	
0	1	1	1	1	Z
1	0	0	0	0	
1	0	0	1	1	Z
1	0	1	0	1	
1	0	1	1	1	1
1	1	0	0	0	
1	1	0	1	1	Z
1	1	1	0	0	
1	1	1	1	1	Z

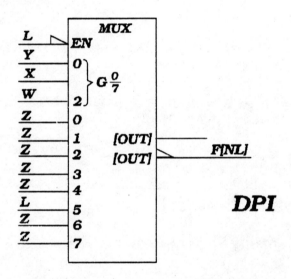

Note: Since method 1 was used, the output
needs to be inverted hence the reason
for using the active low output.

Section 7-5 Implementing Logic Functions Using MSI Decoders

7-24. (a) F1(A,B,C) = Σ m(2,5,7), Signal list: F1, A, B, C

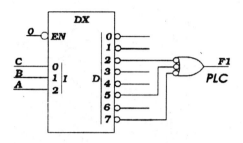

(b) F2(A,B,C,D) = Σ m(3,6,9,12,14,15)
Signal list: F2, A, B, C, D

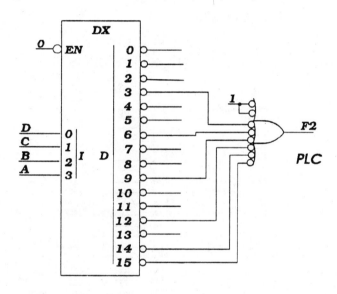

(c) F3(W,X,Y,Z) = Σ m(0,5,8,9,10,11)
Signal list: F3, W, X, Y, Z

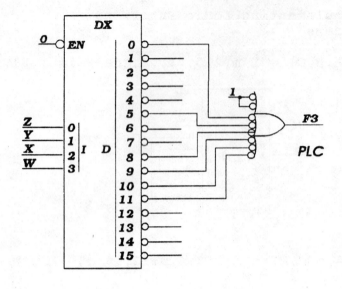

(d) F4(V,W,X,Y,Z) = Σ m(1,5,19,23,31)
Signal list: F4, V, W, X, Y, Z

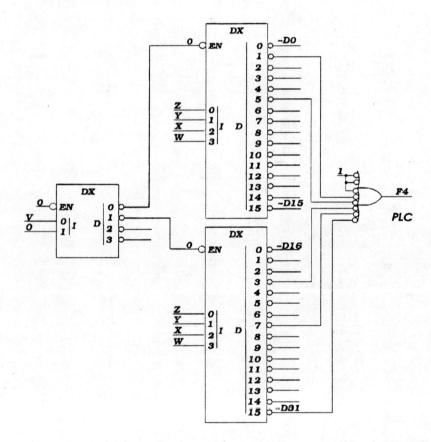

7-25. $F1(A,B,C) = \Sigma\, m(0,1,2)$ The function contains three 1s and five 0s (there are fewer 1s therefore implement the function with the 1s).

$F1 = m_0 + m_1 + m_2$

$Di = m_i$ for $i = 0$ to 7, and $EN = 1$

$F1 = D0 + D1 + D2$
Signal list: F1, ~D0, ~D1, ~D2
(For the gate level design around the decoder -- assume decoder has negated outputs which is true for all TTL general purpose n-to-2^n decoders.)

$F1 = \overline{~D0} + \overline{~D1} + \overline{~D2}$ (function written in design form or available form, i. e., the form expressing the signals that are available at the output of the decoder).

LLP: $F1 =$ ~D0 + ~D1 + ~D2
NIP: 1   0   0   0

$F2(A,B,C) = \Sigma\, m(1,2,3,4,5,7)$ The function contains six 1s and two 0s (there are fewer 0s therefore implement the function with the 0s).

therefore, $\overline{F2}(A,B,C) = \Sigma\, m(0,6)$

$\overline{F2} = m_0 + m_6$

$Di = m_i$ for $i = 0$ to 7, and $EN = 1$

$\overline{F2} = D0 + D6$
Signal list: F2, ~D0, ~D6
(For the gate level design around the decoder -- assume decoder has negated outputs which is true for all TTL general purpose n-to-2^n decoders.)

$\overline{F2} = \overline{~D0} + \overline{~D6}$ (available form)

LLP: $F2 =$ ~D0 + ~D6
NIP: 0   0   0

$F3(A,B,C) = \Sigma\, m(2,3,5,7)$ The function contains four 1s and four 0s (same number of 1s as 0s therefore choose either 1s or 0s to implement the function).

$F3(A,B,C) = m_2 + m_3 + m_5 + m_7$

$Di = m_i$ for $i = 0$ to 7, and $EN = 1$

F3 = D2 + D3 + D5 + D7
Signal list: F3, ~D2, ~D3, ~D5, ~D7
(For the gate level design around the decoder --
 assume decoder has negated outputs which is true for
 all TTL general purpose n-to-2^n decoders.)

F3 = $\overline{\text{~D2}}$ + $\overline{\text{~D3}}$ + $\overline{\text{~D5}}$ + $\overline{\text{~D7}}$ (available form)

LLP: F3 = ~D2 + ~D3 + ~D5 + ~D7
NIP: 1   0   0   0   0

Using the design documentation the logic diagram can
be drawn as follows.

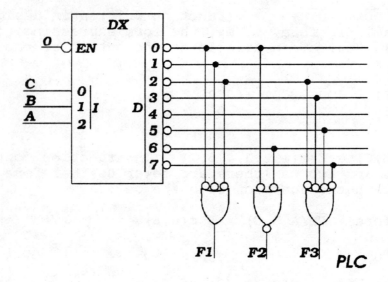

7-26. F3(I3,I2,I1,I0) = Σ m(11,12)
      F2(I3,I2,I1,I0) = Σ m(7,8,9,10)
      F1(I3,I2,I1,I0) = Σ m(5,6,9,10)

      F0(I3,I2,I1,I0) = $\overline{\text{I0}}$
      Signal list: F3, F2, F1, F0, I3, I2, I1, I0

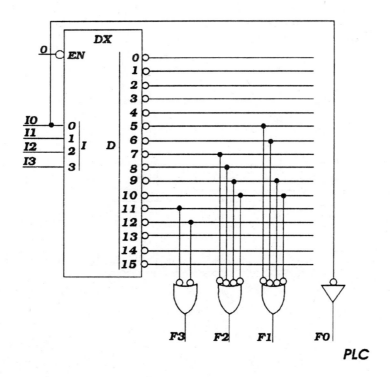

PLC

7-27. F3(I3,I2,I1,I0) = Σ m(5,6,7,8,9)
F2(I3,I2,I1,I0) = Σ m(1,2,3,4,9)
F1(I3,I2,I1,I0) = Σ m(0,3,4,7,8)

F0(I3,I2,I1,I0) = $\overline{I0}$
Signal list: F3, F2, F1, F0, I3, I2, I1, I0

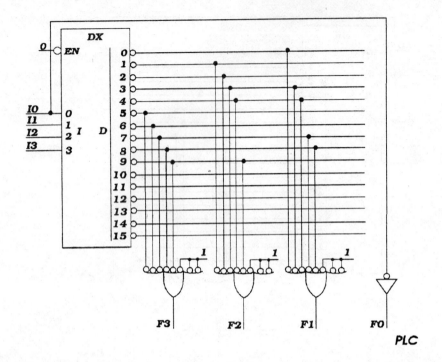

PLC

7-28. $F2(C,B,A) = \Sigma\ m(1,4,7)$

$\overline{F1}(C,B,A) = \Sigma\ m(3,6)$

$\overline{F0}(C,B,A) = \Sigma\ m(0,4,7)$

7-29. $F1(X,Y,Z) = X\cdot\overline{Y} + X\cdot Z$

$F2(X,Y,Z) = Y\cdot Z + X\cdot\overline{Y}\cdot Z$
Signal list: F1[NL],F2,X,Y[NL],Z

$F1(X,Y,Z) = X\cdot\overline{Y}\cdot(\ \overline{Z} + Z) + X\cdot(\overline{Y} + Y)\cdot Z$
$F1(X,Y,Z) = \Sigma\ m(4,5,7)$

$F2(X,Y,Z) = (\overline{X} + X)\cdot Y\cdot Z + X\cdot\overline{Y}\cdot Z$
$F2(X,Y,Z) = \Sigma\ m(3,7,5) = \Sigma\ m(3,5,7)$
Signal list: F1[NL],F2,X,Y[NL],Z

   (a) Method 1 (assume all positive logic signals then
               add Inverters to negative logic signal
               lines)

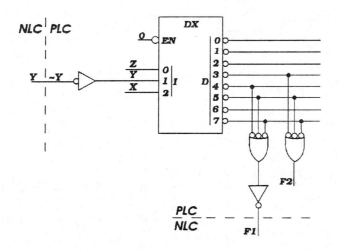

(b)   Method 2 (fix truth table by complementing
                negative logic signals)

X Y Z	F1 F2		m	X ~Y Z	~F1 F2
0 0 0	0 0		2	0 1 0	1 0
0 0 1	0 0		3	0 1 1	1 0
0 1 0	0 0		0	0 0 0	1 0
0 1 1	0 1		1	0 0 1	1 1
1 0 0	1 0		6	1 1 0	0 0
1 0 1	1 1		7	1 1 1	0 1
1 1 0	0 0		4	1 0 0	1 0
1 1 1	1 1		5	1 0 1	0 1

$$\overline{\text{~F1}}(X,\text{~Y},Z) = \Sigma\ m(5,6,7)$$
$$F2(X,\text{~Y},Z) = \Sigma\ m(1,5,7)$$
Signal list: F1[NL],F2,X,Y[NL],Z

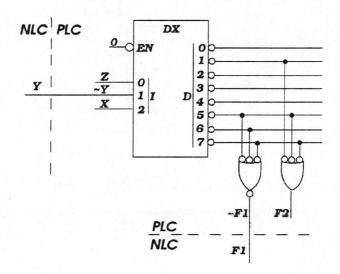

7-30. Using method 2 (fix truth table by complementing negative logic signals)

m	2421				BCD							
	I3	I2	I1	I0	F3	F2	F1	F0	~F3	~F2	~F1	~F0
0	0	0	0	0	0	0	0	0	1	1	1	1
1	0	0	0	1	0	0	0	1	1	1	1	0
2	0	0	1	0	0	0	1	0	1	1	0	1
3	0	0	1	1	0	0	1	1	1	1	0	0
4	0	1	0	0	0	1	0	0	1	0	1	1
11	1	0	1	1	0	1	0	1	1	0	1	0
12	1	1	0	0	0	1	1	0	1	0	0	1
13	1	1	0	1	0	1	1	1	1	0	0	0
14	1	1	1	0	1	0	0	0	0	1	1	1
15	1	1	1	1	1	0	0	1	0	1	1	0

$\overline{\tilde{F3}}(I3,I2,I1,I0) = \Sigma\ m(14,15)$

$\overline{\tilde{F2}}(I3,I2,I1,I0)\cdot = \Sigma\ m(4,11,12,13)$

$\overline{\tilde{F1}}(I3,I2,I1,I0) = \Sigma\ m(2,3,12,13)$

$\tilde{F0}(I3,I2,I1,I0) = \overline{I0}$

Signal list: F3[NL],F2[NL],F1[NL],F0[NL],I3,I2,I1,I0

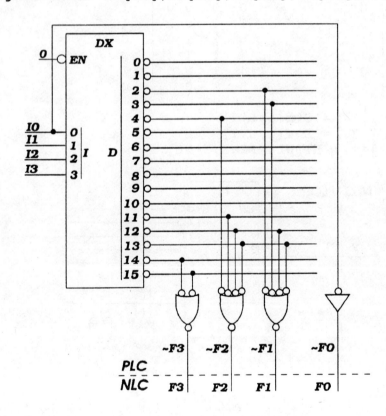

7-31. (a)   A15 A14 A13 = 101

Decoder design

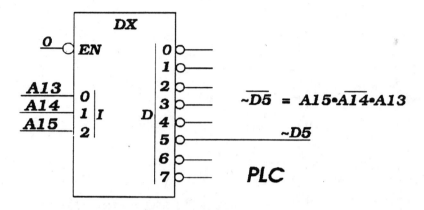

$$\overline{\sim D5} = A15 \cdot \overline{A14} \cdot A13$$

~D5

PLC

Gate-level design

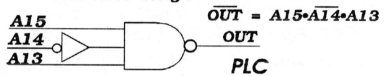

$$\overline{OUT} = A15 \cdot \overline{A14} \cdot A13$$

OUT

PLC

The decoder design is preferred since other addresses can be decoded without using additional hardware.

(b)   A15 A14 A13 A12 A11 = 11011

Decoder design

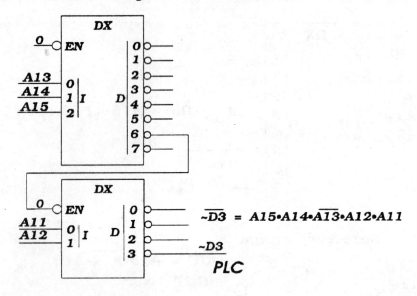

$$\overline{\sim D3} = A15 \cdot A14 \cdot \overline{A13} \cdot A12 \cdot A11$$

~D3

PLC

Gate-level design

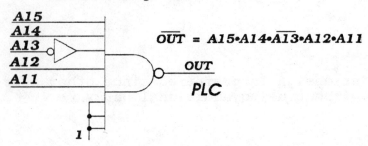

$$\overline{OUT} = A15 \cdot A14 \cdot \overline{A13} \cdot A12 \cdot A11$$

OUT

PLC

    The decoder design is preferred since other addresses
can be decoded without using additional hardware.

(c) A15 A14 A13 A12 A11 A10 = 001001

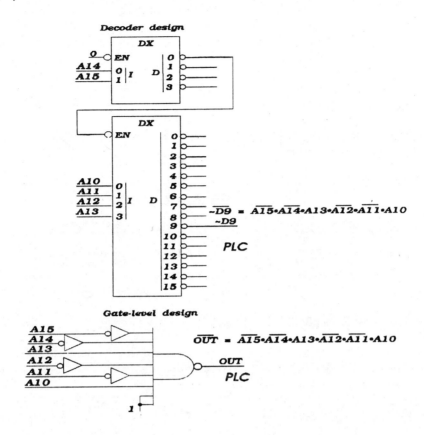

Decoder design

$\overline{D9} = \overline{A15} \cdot \overline{A14} \cdot A13 \cdot \overline{A12} \cdot \overline{A11} \cdot A10$

~D9

PLC

Gate-level design

$\overline{OUT} = \overline{A15} \cdot \overline{A14} \cdot A13 \cdot \overline{A12} \cdot \overline{A11} \cdot A10$

OUT

PLC

   The decoder design is preferred since other addresses
can be decoded without using additional hardware.

Section 7-6 Implementing Logic Functions Using Exclusive OR and Exclusive NOR Elements

7-32. $FO(A,B,C,D) = \Sigma\ m(1,2,4,7,8,11,13,14)$

A B\C D	00	01	11	10
FO = 00	0	1	0	1
01	1	0	1	0
11	0	1	0	1
10	1	0	1	0

<u>Observe that</u>

$\overline{C} \cdot D + C \cdot \overline{D} = C \oplus D$

$\overline{C} \cdot \overline{D} + C \cdot D = C \odot D = \overline{C \oplus D}$

211

FO =	
00	$C \oplus D$
01	$\overline{C \oplus D}$
11	$C \oplus D$
10	$\overline{C \oplus D}$

Observe that

$$\bar{B}\cdot(C \oplus D) + B\cdot(\overline{C \oplus D}) = B \oplus C \oplus D$$
$$B\cdot(C \oplus D) + \bar{B}\cdot(\overline{C \oplus D}) = B \odot C \oplus D$$
$$= \overline{B \oplus C \oplus D}$$

A

FO =	
0	$B \oplus C \oplus D$
1	$\overline{B \oplus C \oplus D}$

Observe that

$$FO = \bar{A}\cdot(B \oplus C \oplus C)$$
$$+ A\cdot(\overline{B \oplus C \oplus D})$$
$$= A \oplus B \oplus C \oplus D$$

7-33. $FO(X,Y,Z) = \Sigma\ m(0,3,5,6)$

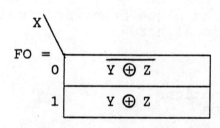

FE =	00	01	11	10
0	1	0	1	0
1	0	1	0	1

Observe that

$$\bar{Y}\cdot\bar{Z} + Y\cdot Z = Y \odot Z = \overline{Y \oplus Z}$$
$$\bar{Y}\cdot Z + Y\cdot\bar{Z} = Y \oplus Z$$

X

FO =	
0	$\overline{Y \oplus Z}$
1	$Y \oplus Z$

Observe that

$$FE = \bar{X}\cdot(\overline{Y \oplus Z}) + X\cdot(Y \oplus Z)$$
$$= X \odot (Y \oplus Z)$$
$$= \overline{X \oplus Y \oplus Z}$$

7-34. (a) $F1 = X \oplus Y \oplus Z$, Signal list: F1, X, Y, Z

$F1 = X \odot Y \odot Z$ by P8 from 1st expression

$F1 = \bar{X} \odot Y \oplus Z$ by P5 and P11 from 1st expression

212

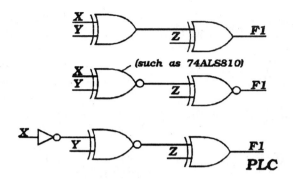

(such as 74ALS810)

PLC

Using the Exclusive OR and Exclusive NOR operator properties P1 through P13 additional valid circuits are possible.

(b) $F2 = W \oplus X \oplus Y \oplus Z$,   Signal list: F2, W, X, Y, Z

$F2 = \overline{W \oplus X \oplus Y \odot Z}$ by P5 and P1 from 1st expression

$F2 = W \odot X \oplus Y \odot Z$ by P7 and P2 from 2nd expression

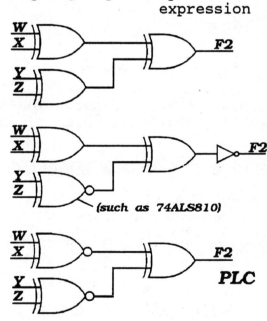

(such as 74ALS810)

PLC

Using the Exclusive OR and Exclusive NOR operator properties P1 through P13 additional valid circuits are possible.

(c) F3 = $\overline{A \oplus B \oplus C \oplus D}$,   Signal list: F3,A,B,C,D

F3 = $A \odot B \oplus C \oplus D$ by P5 and P2 from 1st expression

F3 = $A \odot B \oplus C \odot \overline{D}$ by P7 and P11 from 2nd expression

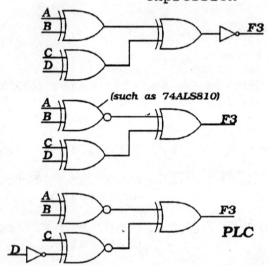

(such as 74ALS810)

PLC

Using the Exclusive OR and Exclusive NOR operator properties P1 through P13 additional valid circuits are possible.

7-35. (a) F1 = $(A \odot B) \oplus (C \odot D)$ = $\overline{A \oplus B} \oplus \overline{C \odot D}$

= $A \oplus B \oplus C \oplus D$
This is an odd function.

(b) F2 = $(W \oplus X) \odot (Y \oplus Z)$ = $\overline{W \oplus X \oplus Y \oplus Z}$
This is an even function.

(c) F3 = $((A \odot B) \odot (C \oplus D)) \oplus (E \odot F)$

= $\overline{A \odot B \odot C \odot D \oplus E \odot F}$

= $A \odot B \odot C \odot D \odot E \odot F$

= $\overline{A \oplus B \oplus C \oplus D \oplus E \oplus F}$ (since n is even)
This is an even function.

214

(d) $F4 = (((T \oplus U) \odot \bar{V}) \odot ((W \odot X) \odot \bar{Y})) \odot \bar{Z}$

$\qquad = (T \oplus U \oplus V) \odot (W \odot X \oplus Y) \oplus Z$

$\qquad = \overline{T \oplus U \oplus V \oplus W \odot X \oplus Y \oplus Z}$

$\qquad = T \oplus U \oplus V \oplus W \oplus X \oplus Y \oplus Z$
    This is an odd function.

7-36. Three information bits sent from the source, and a 1 is used to signal an error condition at the destination for a single parity bit error.

    (a) For a transmission bit pattern that is even, the output PG from the Parity Generator must be an odd function as illustrated below.

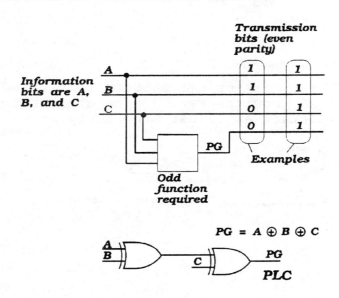

For a transmission bit pattern that is even, the output PC from the Parity Checker must be an odd function as illustrated below since 1 is used as the error signal.

215

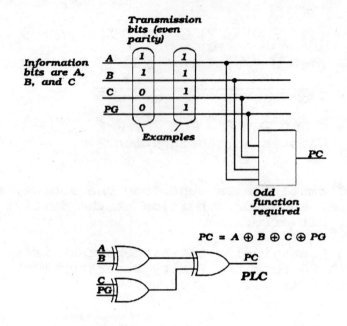

Information bits are A, B, and C

Transmission bits (even parity)

	A	1	1
	B	1	1
	C	0	1
	PG	0	1

Examples

PC

Odd function required

$$PC = A \oplus B \oplus C \oplus PG$$

PC

PLC

(b) For a transmission bit pattern that is odd, the output PG from the Parity Generator must be an even function as illustrated below.

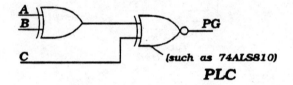

PG

(such as 74ALS810)

PLC

For a transmission bit pattern that is odd, the output PC from the Parity Checker must be an even function as illustrated below since 1 is used as the an error signal.

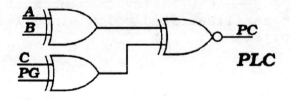

PC

PLC

216

Notice we could provide an Inverter at the output
of either Parity Generator circuit above and
provide either an odd or even parity generator
using the same circuit depending on the connection
made. The same is true with either Parity
Checker circuit as illustrated in the circuits
shown below.

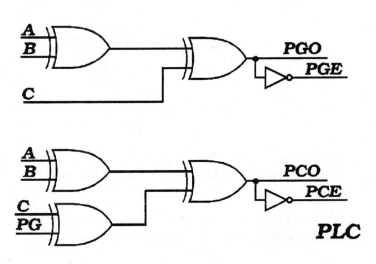

7-37. Design a Parity Generator/Parity Checker circuit that
can be use to provide either an odd or even parity bit in a
single-bit error detection system that utilizes seven
information bits.

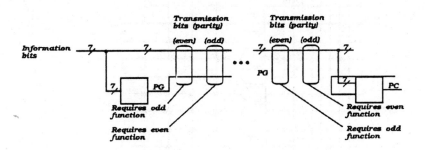

For input signals A,B,C,D,E,F,G

$PGO = A \oplus B \oplus C \oplus D \oplus E \oplus F \oplus G$ (odd function)

$PGE = \overline{A \oplus B \oplus C \oplus D \oplus E \oplus F \oplus G}$ (even function)

$$PCO = A \oplus B \oplus C \oplus D \oplus E \oplus F \oplus G \oplus PG \quad \text{(odd function)}$$

$$PCE = \overline{A \oplus B \oplus C \oplus D \oplus E \oplus F \oplus G \oplus PG} \quad \text{(even function)}$$

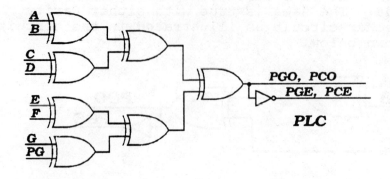

To use the circuit as a Parity Generator set PG = 0 so the Exclusive OR acts like a buffer for signal G, i.e., $G \oplus 0 = G$.

To use the circuit as a Parity Checker connect PG to the parity bit represented by the signal PG.

Section 7-7   Implementing Logic Functions Using Programmable Devices

7-38.

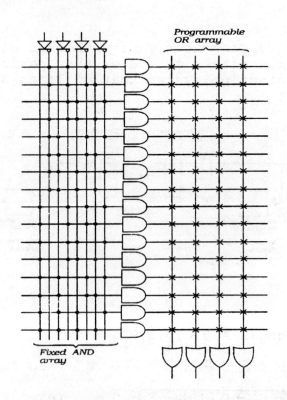

There are 32 fuses required for a simple 3-input, 4-output PROM; however, 64 fuses (twice as many) are required for a simple 4-input, 4-output PROM.

7-39.

XS3				BCD			
I3	I2	I1	I0	F3	F2	F1	F0
0	0	1	1	0	0	0	0
0	1	0	0	0	0	0	1
0	1	0	1	0	0	1	0
0	1	1	0	0	0	1	1
0	1	1	1	0	1	0	0
1	0	0	0	0	1	0	1
1	0	0	1	0	1	1	0
1	0	1	0	0	1	1	1
1	0	1	1	1	0	0	0
1	1	0	0	1	0	0	1

ADDR (HEX) INPUT	DATA (HEX) OUTPUT
00	0F
01	0F
02	0F
03	0F
04	1F
05	2F
06	3F
07	4F
08	5F
09	6F
0A	7F
0B	8F
0C	9F
0D	0F
0E	0F
0F	0F

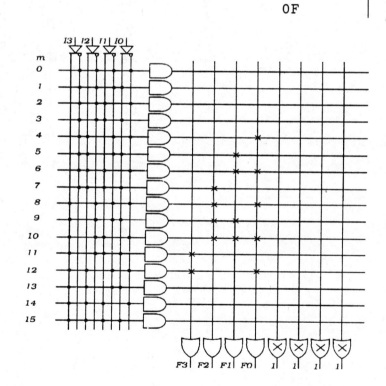

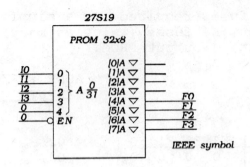

27S19

PROM 32x8

IEEE symbol

7-40.

BCD				XS3			
I3	I2	I1	I0	F3	F2	F1	F0
0	0	0	0	0	0	1	1
0	0	0	1	0	1	0	0
0	0	1	0	0	1	0	1
0	0	1	1	0	1	1	0
0	1	0	0	0	1	1	1
0	1	0	1	1	0	0	0
0	1	1	0	1	0	0	1
0	1	1	1	1	0	1	0
1	0	0	0	1	0	1	1
1	0	0	1	1	1	0	0

ADDR (HEX) INPUT	DATA (HEX) OUTPUT
00	3F
01	4F
02	5F
03	6F
04	7F
05	8F
06	9F
07	AF
08	BF
09	CF
0A	0F
0B	0F
0C	0F
0D	0F
0E	0F
0F	0F

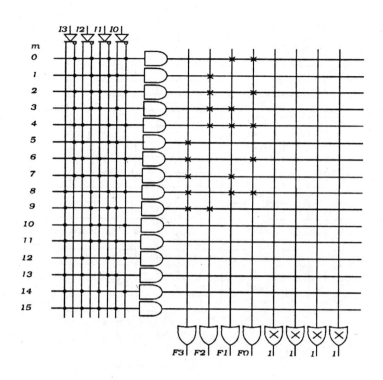

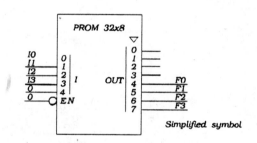

PROM 32x8

Simplified symbol

7-41.

2421				BCD							
I3	I2	I1	I0	F3	F2	F1	F0	~F3	~F2	~F1	~F0
0	0	0	0	0	0	0	0	1	1	1	1
0	0	0	1	0	0	0	1	1	1	1	0
0	0	1	0	0	0	1	0	1	1	0	1
0	0	1	1	0	0	1	1	1	1	0	0
0	1	0	0	0	1	0	0	1	0	1	1
1	0	1	1	0	1	0	1	1	0	1	0
1	1	0	0	0	1	1	0	1	0	0	1
1	1	0	1	0	1	1	1	1	0	0	0
1	1	1	0	1	0	0	0	0	1	1	1
1	1	1	1	1	0	0	1	0	1	1	0

ADDR (HEX) INPUT	DATA (HEX) OUTPUT
00	FF
01	EF
02	DF
03	CF
04	BF
05	0F
06	0F
07	0F
08	0F
09	0F
0A	0F
0B	AF
0C	9F
0D	8F
0E	7F
0F	6F

221

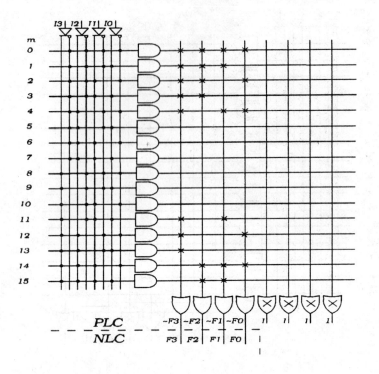

PLC
NLC

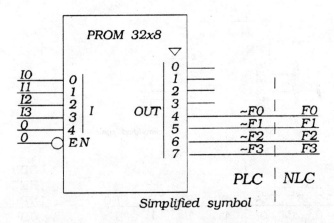

Simplified symbol

7-42. Since the simple 4-input, 4-output PAL in Fig. P7-42 has nonnegated or active high outputs, we need to obtain the SOP expressions for the 1s of the functions to conform to the PAL's AND/OR architecture.

F1(A,B,C,D) = Σ m(6,7,9,11,12,13)
F2(A,B,C,D) = Σ m(0,2,3,4,5,10,11,13,15)
F3(A,B,C,D) = Σ m(2,3,6,7,10,11,14,15)
Signal list: F1, F2, F3, A, B, C, D

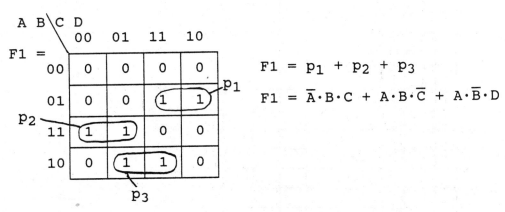

$F1 = p_1 + p_2 + p_3$

$F1 = \overline{A} \cdot B \cdot C + A \cdot B \cdot \overline{C} + A \cdot \overline{B} \cdot D$

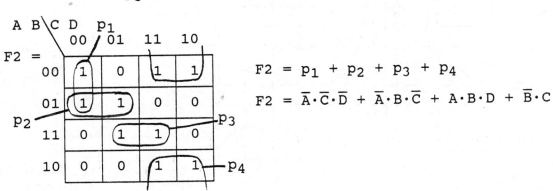

$F2 = p_1 + p_2 + p_3 + p_4$

$F2 = \overline{A} \cdot \overline{C} \cdot \overline{D} + \overline{A} \cdot B \cdot \overline{C} + A \cdot B \cdot D + \overline{B} \cdot C$

A B\C D
        00   01   11   10
F3 =
00 | 0 | 0 | 1 | 1 | p1
01 | 0 | 0 | 1 | 1 |
11 | 0 | 0 | 1 | 1 |
10 | 0 | 0 | 1 | 1 |

$F3 = p_1$

$F3 = C$

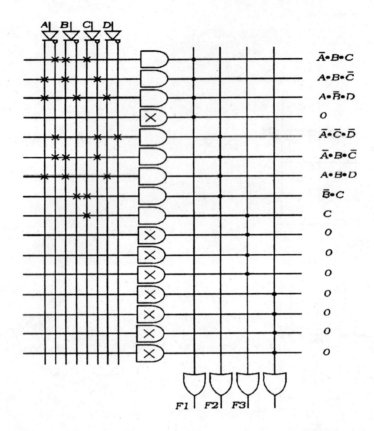

If this simple PAL were commercially available is would be described as a 4H4.

7-43. The simple PAL in Fig. P7-43 has negated or active low outputs, thus we need to obtain the SOP expressions for the 0s of the functions in order to conform to the PAL's AOI (AND/OR Invert) architecture. Using the 0s in the Karnaugh maps we can write the functions as follows.

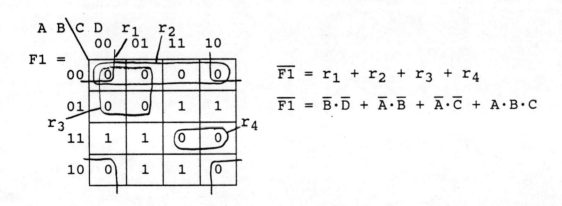

$$\overline{F1} = r_1 + r_2 + r_3 + r_4$$

$$\overline{F1} = \overline{B} \cdot \overline{D} + \overline{A} \cdot \overline{B} + \overline{A} \cdot \overline{C} + A \cdot B \cdot C$$

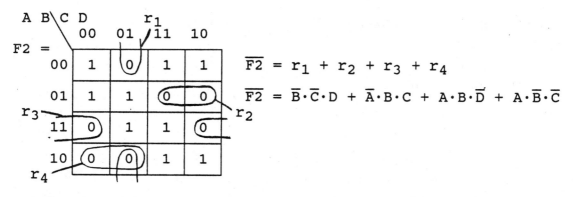

$\overline{F2} = r_1 + r_2 + r_3 + r_4$

$\overline{F2} = \overline{B} \cdot \overline{C} \cdot D + \overline{A} \cdot B \cdot C + A \cdot B \cdot \overline{D} + A \cdot \overline{B} \cdot \overline{C}$

$\overline{F3} = r_1$

$\overline{F3} = \overline{C}$

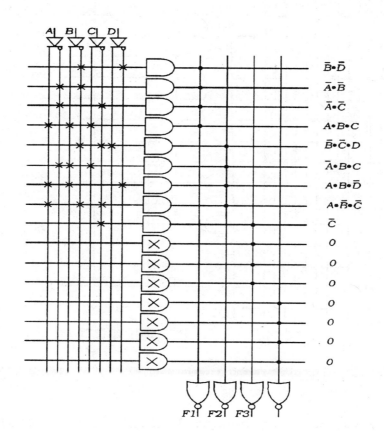

225

If this simple PAL were commercially available is would described as a 4L4.

7-44. Given the following functions to implement in the simple PAL shown in Fig. P7-44:

$$\overline{F1}(X,Y,Z) = X + \overline{Y} \cdot Z$$

$$F2(W,X,Y,Z) = W \cdot \overline{X} \cdot Y + Z$$
$$F3(A,B,C) = A \cdot C + B \cdot C + A \cdot B$$
Signal list = F1,F2,F3,W,X,Y,Z,A,B,C

The simple PAL in Fig. P7-44 has nonnegated or active high outputs, thus we need to obtain the SOP expressions for the 1s of the functions in order to conform to the PAL's AND/OR architecture. Inverting the function F1 we obtain

$$F1 = \overline{X + \overline{Y} \cdot Z} = \overline{X} \cdot (\overline{\overline{Y} \cdot Z}) = \overline{X} \cdot (Y + \overline{Z}) = \overline{X} \cdot Y + \overline{X} \cdot \overline{Z}.$$

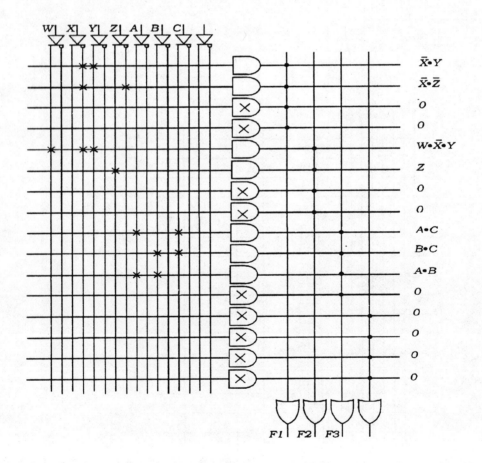

If this simple PAL were commercially available is would be described as a 8H4.

7-45. The simple PAL in Fig. P7-43 has negated or active low outputs, thus we need to obtain the SOP expressions for the 0s of the functions in order to conform to the PAL's AOI (AND/OR Invert) architecture.

BCD				2421			
I3	I2	I1	I0	F3	F2	F1	F0
0	0	0	0	0	0	0	0
0	0	0	1	0	0	0	1
0	0	1	0	0	0	1	0
0	0	1	1	0	0	1	1
0	1	0	0	0	1	0	0
0	1	0	1	1	0	1	1
0	1	1	0	1	1	0	0
0	1	1	1	1	1	0	1
1	0	0	0	1	1	1	0
1	0	0	1	1	1	1	1

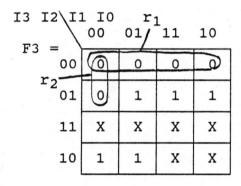

$$\overline{F3} = r_1 + r_2$$

$$\overline{F3} = \overline{I3}\cdot\overline{I2} + \overline{I3}\cdot\overline{I1}\cdot\overline{I0}$$

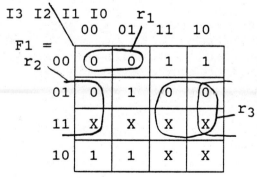

$$\overline{F2} = r_1 + r_2$$

$$\overline{F2} = \overline{I3}\cdot\overline{I2} + \overline{I3}\cdot\overline{I1}\cdot I0$$

F1 = (K-map)

$$\overline{F1} = r_1 + r_2 + r_3$$

$$\overline{F1} = \overline{I3}\cdot\overline{I2}\cdot\overline{I1} + I2\cdot\overline{I0} + I2\cdot I1$$

$I3\ I2 \backslash I1\ I0$

FO =

	00	01	11	10	r_1
00	0	1	1	0	
01	0	1	1	0	
11	X	X	X	X	
10	0	1	X	X	

$\overline{FO} = r_1$

$\overline{FO} = \overline{IO}$

Signal list: F3, F2, F1, F0, I3, I2, I1, I0

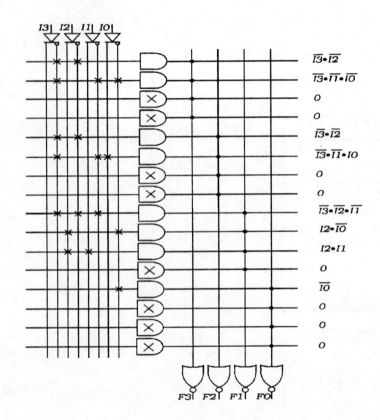

$\overline{I3} \cdot \overline{I2}$

$\overline{I3} \cdot \overline{I1} \cdot \overline{I0}$

o

o

$\overline{I3} \cdot \overline{I2}$

$\overline{I3} \cdot \overline{I1} \cdot I0$

o

o

$\overline{I3} \cdot \overline{I2} \cdot \overline{I1}$

$I2 \cdot \overline{I0}$

$I2 \cdot I1$

o

$\overline{I0}$

o

o

o

7-46. (a) Since PAL 10H8 and PAL 12H6 (see pages 714 and 715 in the text for the PAL circuit diagrams) both have nonnegated or active high outputs, we need to obtain the SOP expressions for the 1s of the functions to conform to the PALs' AND/OR architectures. The Boolean functions to be implemented are

$F1(A,B,C,D) = \Sigma\ m(6,7,9,11,12,13)$
$F2(A,B,C,D) = \Sigma\ m(0,2,3,4,5,10,11,13,15)$
$F3(A,B,C,D) = \Sigma\ m(2,3,6,7,10,11,14,15)$
Signal list: F1, F2, F3, A, B, C, D.

228

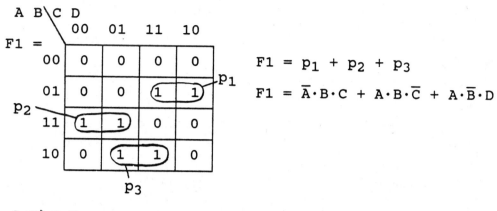

$F1 = p_1 + p_2 + p_3$

$F1 = \overline{A} \cdot B \cdot C + A \cdot B \cdot \overline{C} + A \cdot \overline{B} \cdot D$

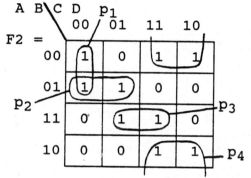

$F2 = p_1 + p_2 + p_3 + p_4$

$F2 = \overline{A} \cdot \overline{C} \cdot \overline{D} + \overline{A} \cdot B \cdot \overline{C} + A \cdot B \cdot D + \overline{B} \cdot C$

A B\C D

	00	01	11	10
F3 =				
00	0	0	1	1
01	0	0	1	1
11	0	0	1	1
10	0	0	1	1

$F3 = p_1$

$F3 = C$

Only Boolean functions with two product terms or less can be implemented by the circuit for the PAL 10H8. Since the reduced function for F1 contains three product terms and the reduced function for F2 contain four product terms the circuit for the PAL 12H6 must be used since it can implement Boolean functions with four product terms or less. The implementation using the PAL 12H6 is as follows.

229

Logic Diagram　　　　　　　　**12H6**

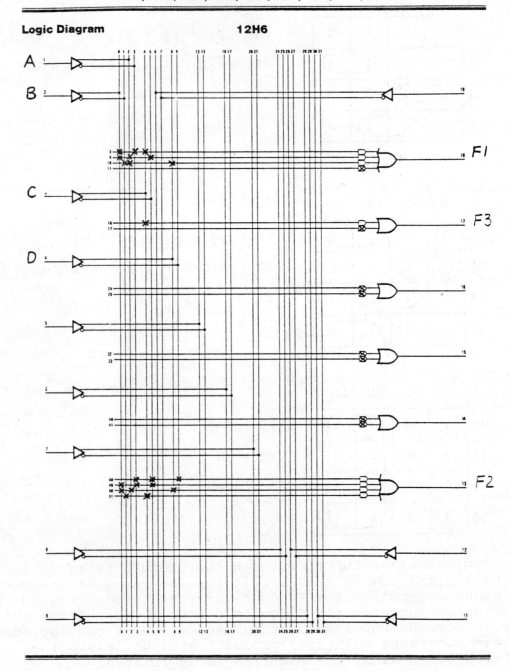

230

(b)   Since PAL 10L8 and PAL 14L4 (see pages 717 and 721 in the text for the PAL circuit diagrams) both have negated or active low outputs, we need to obtain the SOP expressions for the 0s of the functions to conform to the PALs' AND/OR Invert architectures. Using the 0s in the Karnaugh maps we can write the functions as follows.

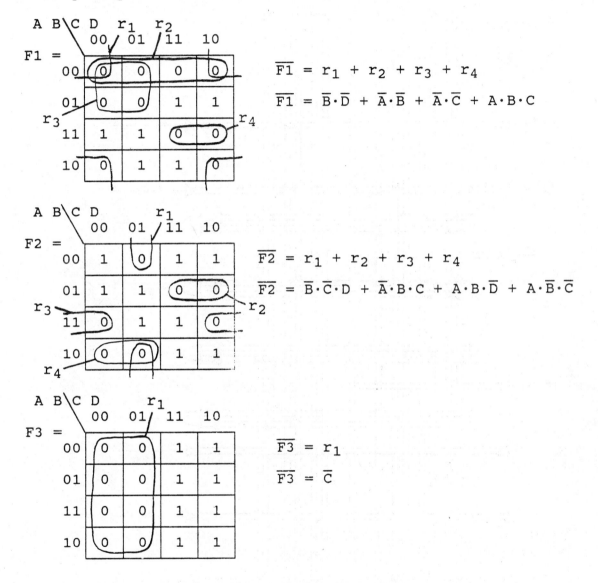

$$\overline{F1} = r_1 + r_2 + r_3 + r_4$$

$$\overline{F1} = \overline{B}\cdot\overline{D} + \overline{A}\cdot\overline{B} + \overline{A}\cdot\overline{C} + A\cdot B\cdot C$$

$$\overline{F2} = r_1 + r_2 + r_3 + r_4$$

$$\overline{F2} = \overline{B}\cdot\overline{C}\cdot D + \overline{A}\cdot B\cdot C + A\cdot B\cdot\overline{D} + A\cdot\overline{B}\cdot\overline{C}$$

$$\overline{F3} = r_1$$

$$\overline{F3} = \overline{C}$$

Only Boolean functions with two product terms or less can be implemented by the circuit for the PAL 10L8. Since the reduced functions for F1 and F2 each contain four product terms the circuit for the PAL 14L4 must be used since it can implement Boolean functions with four product terms or less. The implementation using the PAL 14L4 is shown as follows.

Logic Diagram   14L4

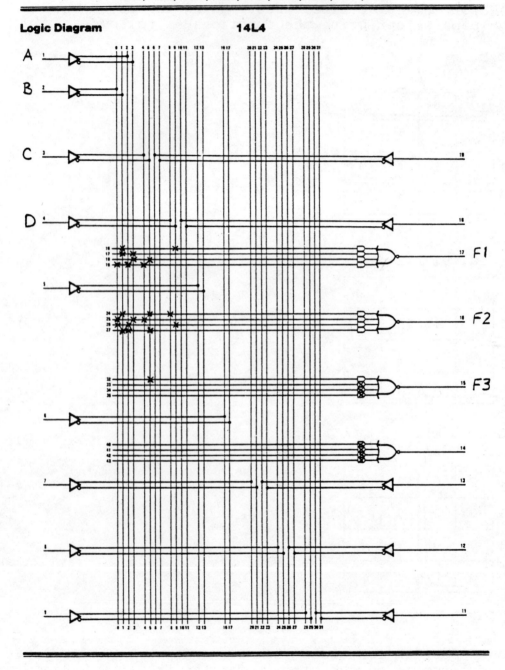

7-47. The 2421 to BCD code converter signal list is F3[NL], F2[NL], F1[NL], F0[NL], I3, I2, I1, I0. Using method 2 (fix truth table by complementing negative logic signals)

	2421				BCD							
m	I3	I2	I1	I0	F3	F2	F1	F0	~F3	~F2	~F1	~F0
0	0	0	0	0	0	0	0	0	1	1	1	1
1	0	0	0	1	0	0	0	1	1	1	1	0
2	0	0	1	0	0	0	1	0	1	1	0	1
3	0	0	1	1	0	0	1	1	1	1	0	0
4	0	1	0	0	0	1	0	0	1	0	1	1
11	1	0	1	1	0	1	0	1	1	0	1	0
12	1	1	0	0	0	1	1	0	1	0	0	1
13	1	1	0	1	0	1	1	1	1	0	0	0
14	1	1	1	0	1	0	0	0	0	1	1	1
15	1	1	1	1	1	0	0	1	0	1	1	0

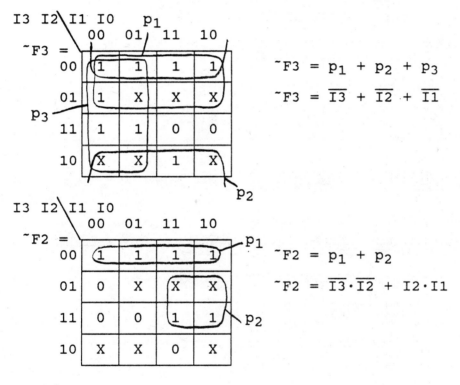

$$\text{~F3} = p_1 + p_2 + p_3$$
$$\text{~F3} = \overline{I3} + \overline{I2} + \overline{I1}$$

$$\text{~F2} = p_1 + p_2$$
$$\text{~F2} = \overline{I3}\cdot\overline{I2} + I2\cdot I1$$

$$\text{~F1} = p_1 + p_2$$
$$\text{~F1} = \overline{I3}\cdot\overline{I1} + I3\cdot I1$$

$$\text{~F0} = \overline{I0}$$

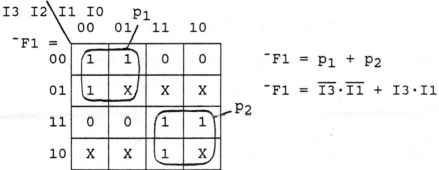

Signal list: F3[NL],F2[NL],F1[NL],F0[NL],I3,I2,I1,I0

233

Logic Diagram   14H4

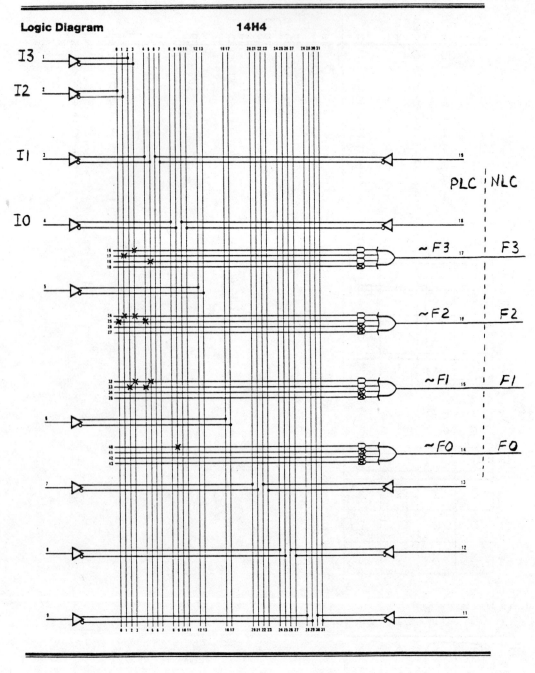

PLC | NLC

~F3   F3

~F2   F2

~F1   F1

~F0   F0

7-48. Implement the following equations using the simple PLA shown in Fig. P7-48. Program each polarity fuse to fit the SOP form of the output function with the fewest number of product terms.

$$F1(A,B,C) = \Sigma\ m(2,5,6)$$
$$F2(A,B,C,D) = \Sigma\ m(1,4,5,6,7,8,9,12,13,14,15)$$
$$F3(A,B,C,D) = \Sigma\ m(0,1,2,3,4,6,8,9,10,11,12,14) + d(13,15)$$
$$F4(A,B,C,D) = \Sigma\ m(0,2,4,5,12,13)$$
Signal list: F1, F1, F3, F4, A, B, C, D

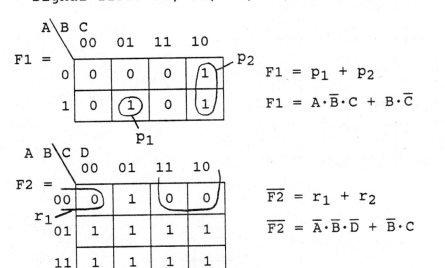

$$F1 = p_1 + p_2$$
$$F1 = A \cdot \overline{B} \cdot C + B \cdot \overline{C}$$

$$\overline{F2} = r_1 + r_2$$
$$\overline{F2} = \overline{A} \cdot \overline{B} \cdot \overline{D} + \overline{B} \cdot C$$

$$\overline{F3} = r_1$$
$$\overline{F3} = B \cdot D$$

$$F4 = p_1 + p_2$$
$$F4 = \overline{A} \cdot \overline{B} \cdot \overline{D} + B \cdot \overline{C}$$

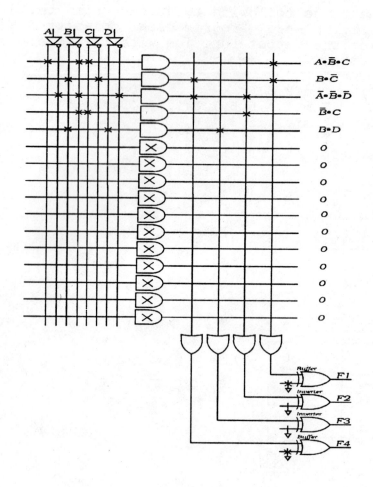

The architectural size of this simple PLA is 4x16x4.

7-49.

	BCD				84-2-1		
I3	I2	I1	I0	F3	F2	F1	F0
0	0	0	0	0	0	0	0
0	0	0	1	0	1	1	1
0	0	1	0	0	1	1	0
0	0	1	1	0	1	0	1
0	1	0	0	0	1	0	0
0	1	0	1	1	0	1	1
0	1	1	0	1	0	1	0
0	1	1	1	1	0	0	1
1	0	0	0	1	0	0	0
1	0	0	1	1	1	1	1

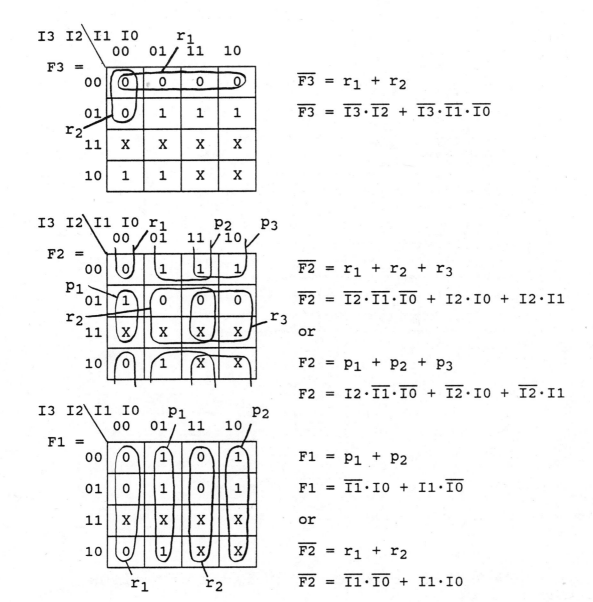

$$\overline{F3} = r_1 + r_2$$

$$\overline{F3} = \overline{I3} \cdot \overline{I2} + \overline{I3} \cdot \overline{I1} \cdot \overline{I0}$$

$$\overline{F2} = r_1 + r_2 + r_3$$

$$\overline{F2} = \overline{I2} \cdot \overline{I1} \cdot \overline{I0} + I2 \cdot I0 + I2 \cdot I1$$

or

$$F2 = p_1 + p_2 + p_3$$

$$F2 = I2 \cdot \overline{I1} \cdot \overline{I0} + \overline{I2} \cdot I0 + \overline{I2} \cdot I1$$

$$F1 = p_1 + p_2$$

$$F1 = \overline{I1} \cdot I0 + I1 \cdot \overline{I0}$$

or

$$\overline{F2} = r_1 + r_2$$

$$\overline{F2} = \overline{I1} \cdot \overline{I0} + I1 \cdot I0$$

$$F0 = I0$$

Signal list: F3, F2, F1, F0, I3, I2, I1, I0

237

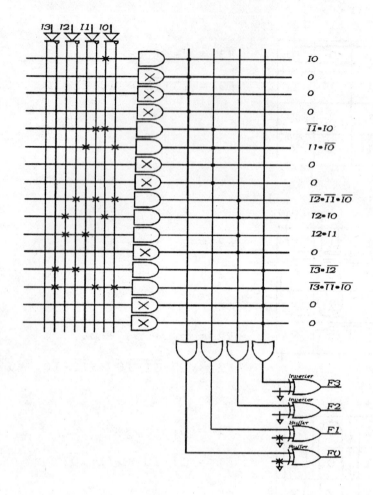

If this simple PAL were commercially available is would be described as a 4P4.

Section 7-8 Programming PALs
Using PALASM

Examples 7-13 through 7-16 are worked in the text on page 398 through page 412.

Section 7-9 Programming PALs
Using PALASM

7-50. 1. Obtain a minimum covering of the 1s (0s) of the function.
    2. Add product terms to cover each occurrence of adjacent 1s (0s) that are not already contained in the same

p-subcube (r-subcube) in the minimum covering of the function.

7-51. (a) $F(A,B,C) = \Sigma\, m(1,3,4,5)$, Signal list: F, A, B, C

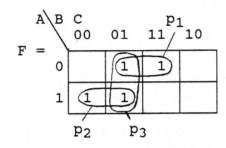

Minimum covering
$F = p_1 + p_2$

$F = \overline{A}\cdot C + A\cdot\overline{B}$

Logic hazard-free covering
$F = p_1 + p_2 + p_3$

$F = \overline{A}\cdot C + A\cdot\overline{B} + \overline{B}\cdot C$

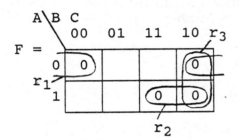

Minimum covering

$\overline{F} = r_1 + r_2$

$\overline{F} = \overline{A}\cdot\overline{C} + A\cdot B$

Logic hazard-free covering

$\overline{F} = r_1 + r_2 + r_3$

$\overline{F} = \overline{A}\cdot\overline{C} + A\cdot B + B\cdot\overline{C}$

(b) $F(X,Y,Z) = \Sigma\, m(1,2,3,6)$, Signal list: F, X, Y, Z

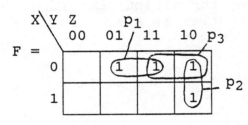

Minimum covering
$F = p_1 + p_2$

$F = \overline{X}\cdot Z + Y\cdot\overline{Z}$

Logic hazard-free covering
$F = p_1 + p_2 + p_3$

$F = \overline{X}\cdot Z + Y\cdot\overline{Z} + \overline{X}\cdot Y$

Minimum covering

$\overline{F} = r_1 + r_2$

$\overline{F} - \overline{Y}\cdot\overline{Z} + X\cdot Z$

Logic hazard-free covering

$\overline{F} = r_1 + r_2 + r_3$

$\overline{F} = \overline{Y}\cdot\overline{Z} + X\cdot Z + X\cdot\overline{Y}$

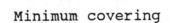

(c) $F(X,Y,Z) = \Sigma\, m(0,2,3,4,6)$, Signal list: F,X,Y,Z

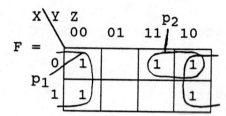

Minimum covering
$F = p_1 + p_2$

$F = \overline{Z} + \overline{X}\cdot Y$

This function does not have any adjacent 1s that are not already contained in the same p-subcube required for a minimum covering of the function; therefore, the minimum covering of the function is also a logic hazard-free covering.

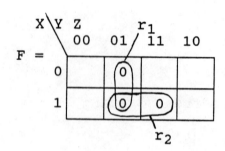

Minimum covering

$\overline{F} = r_1 + r_2$

$\overline{F} = \overline{Y}\cdot Z + X\cdot Z$

This function does not have any adjacent 0s that are not already contained in the same r-subcube required for a minimum covering of the function; therefore, the minimum covering of the function is also a logic hazard-free covering.

7-52. $F = \overline{B}\cdot\overline{D} + A\cdot B\cdot\overline{C} + A\cdot C\cdot D$

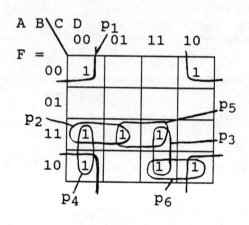

The circuit potentially contains three logic hazards. Product terms p_4, p_5, and p_6 will eliminate these hazards.

Logic hazard-free function
$F = p_1 + p_2 + p_3 + p_4 + p_5 + p_6$

$F = \overline{B}\cdot\overline{D} + A\cdot B\cdot\overline{C} + A\cdot C\cdot D$

$\quad + A\cdot\overline{C}\cdot\overline{D} + A\cdot B\cdot D + A\cdot\overline{B}\cdot C$

7-53. (a) $F(A,B,C,D) = \Sigma\, m(0,1,2,5,6,7,15)$
Signal list: F, A, B, C, D

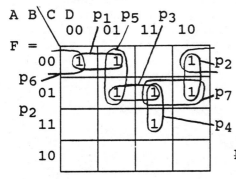

A circuit for the 1s potentially contains <u>three</u> logic hazards. Product terms p_5, p_6, and p_7 will eliminate these hazards.

Logic hazard-free function

$$F = p_1 + p_2 + p_3 + p_4 + p_5 + p_6 + p_7$$

$$F = \overline{B}\cdot\overline{D} + A\cdot B\cdot\overline{C} + A\cdot C\cdot D$$

$$+ A\cdot\overline{C}\cdot\overline{D} + A\cdot B\cdot D + A\cdot\overline{B}\cdot C$$

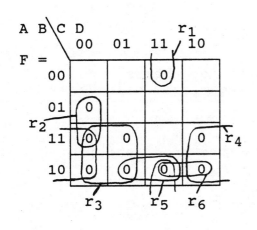

A circuit for the 0s potentially contains <u>two</u> logic hazards. Product terms r_5, and r_6 will eliminate these hazards.

Logic hazard-free function

$$\overline{F} = r_1 + r_2 + r_3 + r_4 + r_5 + r_6$$

$$\overline{F} = \overline{B}\cdot C\cdot D + B\cdot\overline{C}\cdot\overline{D} + A\cdot\overline{C} + A\cdot C\cdot\overline{D}$$

$$+ A\cdot\overline{B}\cdot D + A\cdot\overline{B}\cdot C$$

(b) $F(W,X,Y,Z) = \Sigma\, m(0,1,5,7,8,9,14,15)$
Signal list: F, W, X, Y, Z

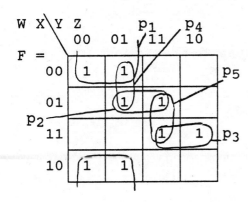

A circuit for the 1s potentially contains <u>two</u> logic hazards. Product terms p_4, and p_5 will eliminate these hazards.

Logic hazard-free function
$$F = p_1 + p_2 + p_3 + p_4 + p_5$$

$$F = \overline{X}\cdot\overline{Y} + \overline{W}\cdot X\cdot Z + W\cdot X\cdot Y$$

$$+ \overline{W}\cdot\overline{Y}\cdot Z + X\cdot Y\cdot Z$$

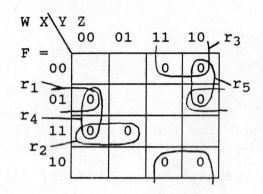

A circuit for the 0s potentially contains <u>two</u> logic hazards. Product terms r_5, and r_6 will eliminate these hazards.

Logic hazard-free function

$$\overline{F} = r_1 + r_2 + r_3 + r_4 + r_5$$

$$\overline{F} = \overline{W} \cdot X \cdot \overline{Z} + W \cdot X \cdot \overline{Y} + \overline{X} \cdot Y$$

$$+ X \cdot \overline{Y} \cdot \overline{Z} + \overline{W} \cdot Y \cdot \overline{Z}$$

(c) $F(A,B,C,D) = \Sigma\ m(2,3,8,9,10,11)$
       Signal list: F, A, B, C, D

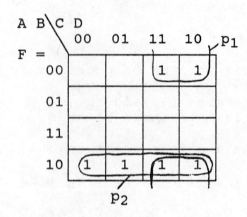

A circuit for the 1s contains <u>no</u> logic hazards.

Logic hazard-free function
$$F = p_1 + p_2$$

$$F = \overline{B} \cdot C + A \cdot \overline{B}$$

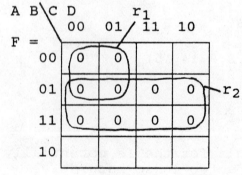

A circuit for the 0s contains <u>no</u> logic hazards.

Logic hazard-free function

$$\overline{F} = r_1 + r_2$$

$$\overline{F} = \overline{A} \cdot \overline{C} + B$$

Section 8-2 Combinational Logic Circuits
Versus Sequential Logic Circuits

8-1. By adding delays from the input to each output and recording the worst case delay for each output.

8-2.

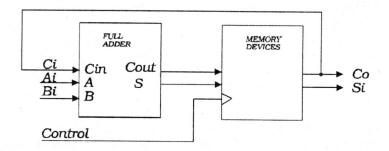

8-3.

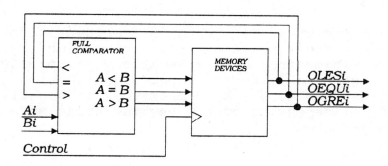

8-4. An asynchronous circuit called a system clock.

Synchronous and asynchronous sequential circuits have a memory property while combinational circuits do not.

8-5.

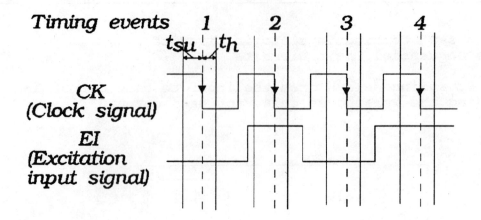

Timing events

CK
(Clock signal)

EI
(Excitation
input signal)

8-6. (a) Timing events are marked in a sequential
synchronous circuit by a system clock signal
making a transition from 0 to 1 (or from 1 to 0)
each time the clock ticks after one period of the
system clock.

(b) Timing events are marked in a sequential
fundamental mode circuit by an external input
signal making a transition from 0 to 1 or from 1
to 0.

(c) Timing events are marked in a sequential pulse
mode circuit by an external input signal making a
transition from 0 to 1 back to 0 (a positive
pulse) or an external input signal making a
transition from 1 to 0 back to 1 (a negative
pulse) depending on the way the pulse mode circuit
is designed.

**Section 8-3   The Basic Bistable Memory
Devices**

8-7.

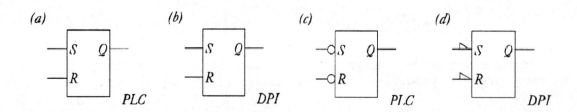

8-8.   (a)

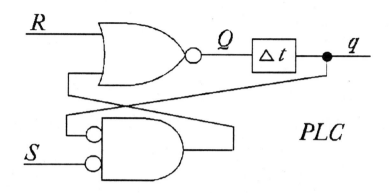

(b) $Q = \overline{R + \overline{q \cdot \overline{S}}} = \overline{R} \cdot (q + S) = \overline{R} \cdot q + \overline{R} \cdot S$

(c)

Present state (PS) output signal	External input signals	Next state (NS) output signal	
q	S R	Q	Comment
0	0 0	0	reset state (Q = 0)
0	0 1	0	reset state (Q = 0)
0	1 0	1	set state (Q = 1)
0	1 1	0	S R = 11 not normally allowed
1	0 0	1	set state (Q = 1)
1	0 1	0	reset state (Q = 0)
1	1 0	1	set state (Q = 1)
1	1 1	0	S R = 11 not normally allowed

(d)

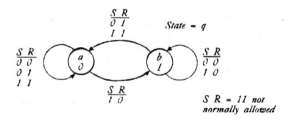

(e)

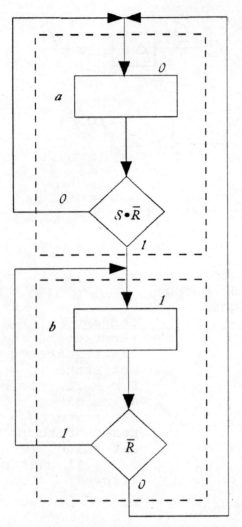

State - q

$S\ R\ =\ 11\ not$
$normally\ allowed$

(f)

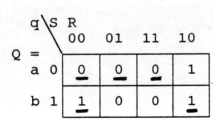

S R = 11 not
normally allowed

(g)

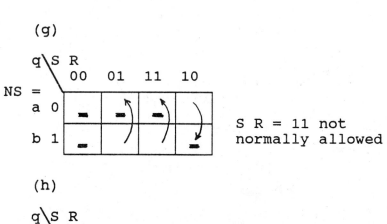

S R = 11 not
normally allowed

(h)

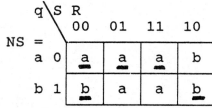

S R = 11 not
normally allowed

(i)

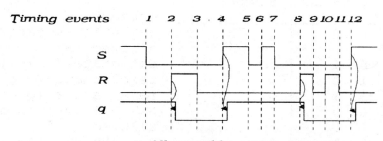

Timing events   1 2   3   4   5 6 7   8 9 10 11 12

All possible states are not
represented in this timing diagram.

247

8-9. (a)

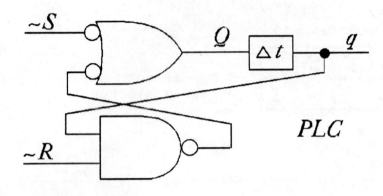

PLC

(b) $Q = \overline{\tilde{S}} + \overline{\overline{q \cdot \tilde{R}}} = \overline{\tilde{S}} + q \cdot \tilde{R}$

(c)

Present state (PS) output signal	External input signals		Next state (NS) output signal	
q	~S	~R	Q	Comment
0	0	0	1	S R = 11 not normally allowed
0	0	1	1	set state (Q = 1)
0	1	0	0	reset state (Q = 0)
0	1	1	0	reset state (Q = 0)
1	0	0	1	S R = 11 not normally allowed
1	0	1	1	set state (Q = 1)
1	1	0	0	reset state (Q = 0)
1	1	1	1	set state (Q = 1)

(d)

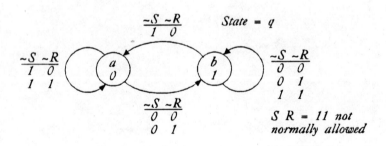

(e)

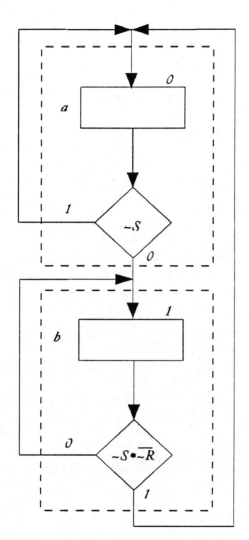

State = q

S R = 11 not
normally allowed

(f)

q \ ~S ~R	00	01	11	10
Q =				
a 0	1	1	0	0
b 1	1	1	1	0

S R = 11 not
normally allowed

249

(g)

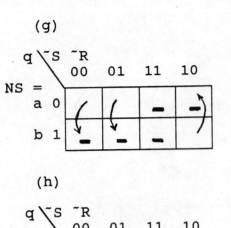

S R = 11 not
normally allowed

(h)

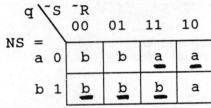

S R = 11 not
normally allowed

(i)

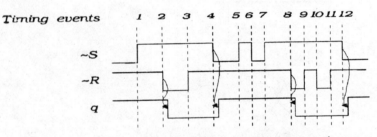

All possible states are not
represented in this timing diagram.

8-10. (a)

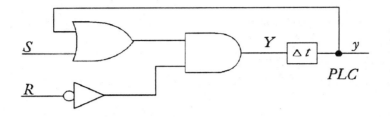

(b) $Y = (y + S) \cdot \overline{R} = y \cdot \overline{R} + S \cdot \overline{R}$

(c)

Present state (PS) output signal y	External input signals S R	Next state (NS) output signal Y	Comment
0	0 0	0	reset state (Y = 0)
0	0 1	0	reset state (Y = 0)
0	1 0	1	set state (Y = 1)
0	1 1	0	S R = 11 not normally allowed
1	0 0	1	set state (Y = 1)
1	0 1	0	reset state (Y = 0)
1	1 0	1	set state (Y = 1)
1	1 1	0	S R = 11 not normally allowed

(d)

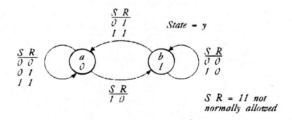

(e)

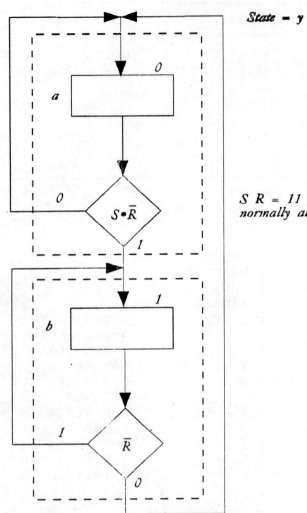

State − y

S R = 11 not
normally allowed

(f)

y \ S R	00	01	11	10
a 0	0	0	0	1
b 1	1	0	0	1

Y =

S R = 11 not
normally allowed

252

(g)

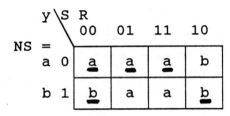

NS =

S R = 11 not
normally allowed

(h)

y\S R	00	01	11	10
a 0	a	a	a	b
b 1	b	a	a	b

NS =

S R = 11 not
normally allowed

(i)

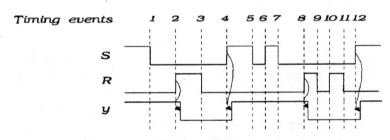

All possible states are not
represented in this timing diagram.

8-11.

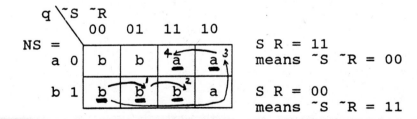

NS =

S R = 11
means ~S ~R = 00

S R = 00
means ~S ~R = 11

Since transitions 1, 2 and transitions 3, 4
end up in different states the race is critical.

8-12.

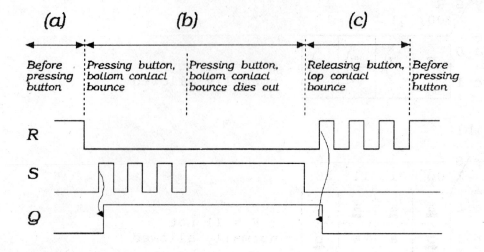

Section 8-4 Additional Bistable Memory Devices

8-13. 1. Latches
2. Master-slave (pulse triggered) flip-flops with or without data-lockout
3. Edge-triggered flip-flops

8-14.

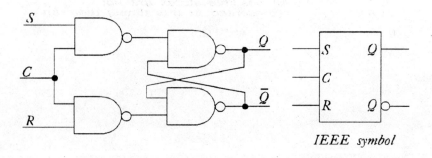

IEEE symbol

8-15. (a)

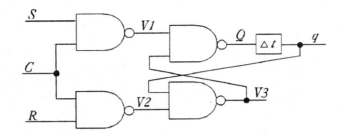

(b) $V1 = \overline{C \cdot S}$

$V2 = \overline{C \cdot R}$

$V3 = \overline{q \cdot V2}$

$Q = \overline{V1 \cdot V3} = \overline{\overline{C \cdot S} \cdot \overline{q \cdot V2}} = C \cdot S + q \cdot V2$

$= C \cdot S + q \cdot \overline{C \cdot R} = C \cdot S + q \cdot (\overline{C} + \overline{R})$

$= C \cdot S + q \cdot \overline{C} + q \cdot \overline{R}$

(c)

q \ C S R	000	001	011	010	100	101	111	110
Q= a 0	0	0	0	0	0	0	1	1
b 1	1	1	1	1	1	0	1	1

S R = 11 not normally allowed

q \ C S R	000	001	011	010	100	101	111	110
NS= a 0	—	—	—	—	—	↗	↓	↓
b 1	—	—	—	—	—	↶	—	—

(d)

q \ C S R	000	001	011	010	100	101	111	110
NS= a 0	a	a	a	a	a	a	b	b
b 1	b	b	b	b	b	a	b	b

(e)

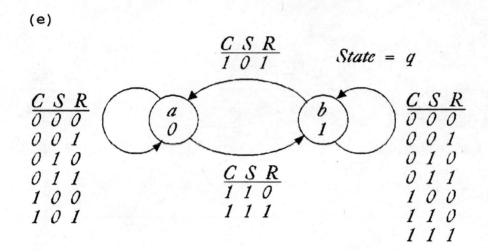

$$\frac{C \; S \; R}{1 \; 0 \; 1}$$

State = *q*

$$\begin{array}{ccc} C & S & R \\ \hline 0 & 0 & 0 \\ 0 & 0 & 1 \\ 0 & 1 & 0 \\ 0 & 1 & 1 \\ 1 & 0 & 0 \\ 1 & 0 & 1 \end{array}$$

$$\frac{C \; S \; R}{1 \; 1 \; 0}$$
$$1 \; 1 \; 1$$

$$\begin{array}{ccc} C & S & R \\ \hline 0 & 0 & 0 \\ 0 & 0 & 1 \\ 0 & 1 & 0 \\ 0 & 1 & 1 \\ 1 & 0 & 0 \\ 1 & 1 & 0 \\ 1 & 1 & 1 \end{array}$$

(f)

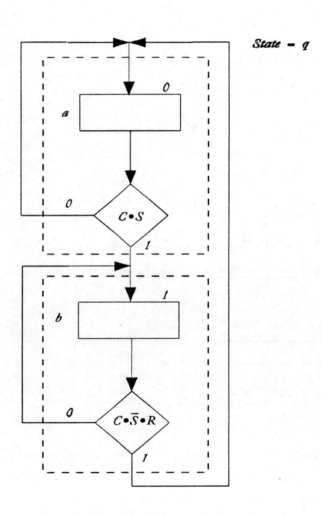

State = *q*

(g)

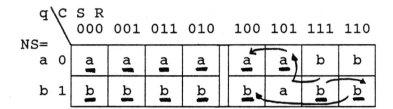

From the flow map (observe the arrows for the sequence of events) it can be observed that C S R = 111 are the input conditions normally not allowed, since these conditions results in a critical race condition when S and R both change to 0 simultaneously while C = 1, i.e., the circuit may end up in either state a or state b depending on the delay paths in the circuit..

8-16.

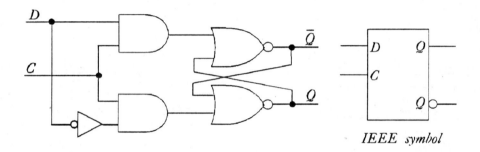

IEEE symbol

8-17. (a)

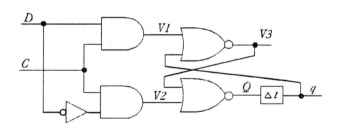

(b)   V1 = C·D

V2 = C·$\overline{D}$

V3 = $\overline{q + V1}$

Q = $\overline{V3 + V2}$ = $\overline{\overline{q + V1} + C·\overline{D}}$ = (q + V1)·$\overline{C·\overline{D}}$

$$Q = (q + C \cdot D) \cdot (\overline{C} + D) = q \cdot \overline{C} + q \cdot D + C \cdot D$$

(c)

q\C D	00	01	11	10
Q = a 0	0	0	1	0
b 1	1	1	1	0

q\C D	00	01	11	10
NS= a 0	—	—	—	—
b 1	—	—	—	

(d)

q\C D	00	01	11	10
NS= a 0	a	a	b	a
b 1	b	b	b	. a

(e)

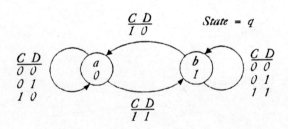

(f)

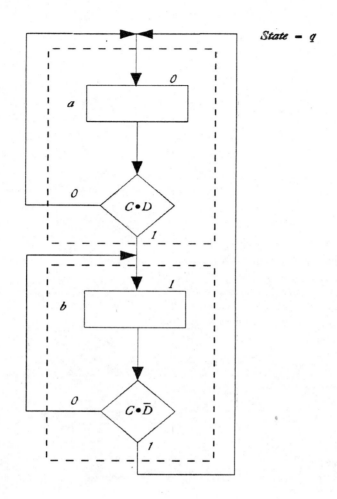

State — q

(g)

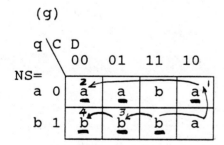

C D changing from 11 to 10 to 00, arrows 1 and 2, result in state a (q = 0) while C D changing from 11 to 01 to 00, arrows 3 and 4, result in state b (q = 1). The race is obviously critical and should not be allowed to happen.

8-18.

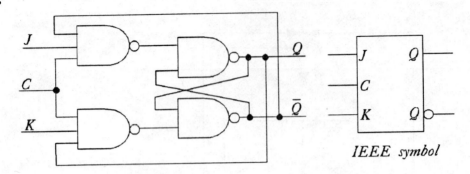

IEEE symbol

8-19. (a)

There is no timing break in this feedback path as a result of the delay model; therefore, use the characteristic equation for the basic latch to determine the characteristic equation of the circuit.

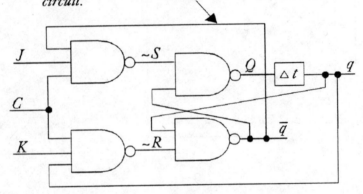

(b)

$$Q = \overline{{}^{\sim}S} + q \cdot {}^{\sim}R$$

but

$$^{\sim}S = \overline{\overline{q} \cdot J \cdot C}$$

$$^{\sim}R = \overline{C \cdot K \cdot q}$$

$$Q = \overline{q} \cdot J \cdot C + q \cdot \overline{C \cdot K \cdot q}$$

$$= \overline{q} \cdot J \cdot C + q \cdot (\overline{C} + \overline{K} + \overline{q})$$

$$= \overline{q} \cdot C \cdot J + q \cdot \overline{C} + q \cdot \overline{K}$$

(c)

q\C J K	000	001	011	010	100	101	111	110
Q=								
a 0	0	0	0	0	0	0	1	1
b 1	1	1	1	1	1	0	0	1

q\C J K	000	001	011	010	100	101	111	110
NS=								
a 0	a	a	a	a	a	a	b	b
b 1	b	b	b	b	b	a	a	b

arrows show
where circuit
is unstable

(d) From the Karnaugh map or the flow map it can be observed that the circuit is unstable for the input conditions C J K = 111.

(e) The circuit can be made to operate properly by supplying a positive control pulse that is narrower that the propagation delay through the circuit.

8-20.

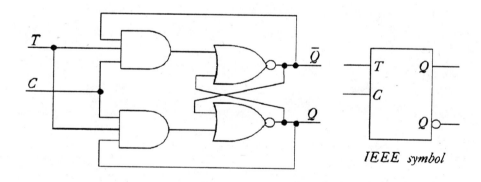

IEEE symbol

8-21. (a)

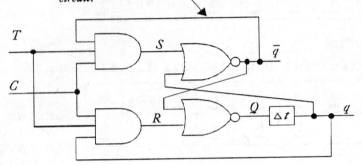

There is no timing break in this feedback path as a result of the delay model; therefore, use the characteristic equation for the basic latch to determine the characteristic equation of the circuit.

(b)

$$Q = S \cdot \overline{R} + q \cdot \overline{R}$$

but $S = \overline{q} \cdot T \cdot C$

$$R = q \cdot T \cdot C$$

$$Q = \overline{q} \cdot T \cdot C \cdot \overline{q \cdot T \cdot C} + q \cdot \overline{q \cdot T \cdot C}$$

$$Q = (\overline{q} \cdot T \cdot C + q) \cdot \overline{q \cdot T \cdot C}$$

$$Q = (\overline{q} \cdot T \cdot C + q) \cdot (\overline{q} + \overline{T} + \overline{C})$$

$$Q = \overline{q} \cdot C \cdot T + q \cdot \overline{T} + q \cdot \overline{C}$$

(c)

q\C T

	00	01	11	10
Q=				
a 0	0	0	1	0
b 1	1	1	0	1

q\C T

	00	01	11	10
NS=				
a 0	a	a	b	a
b 1	b	b	a	b

arrows show
where circuit
is unstable

262

(d) The circuit is unstable when CT = 11. To make the circuit operate properly a positive control pulse must be used that has a pulse width that is narrower than the propagation delay through the circuit.

8-22.

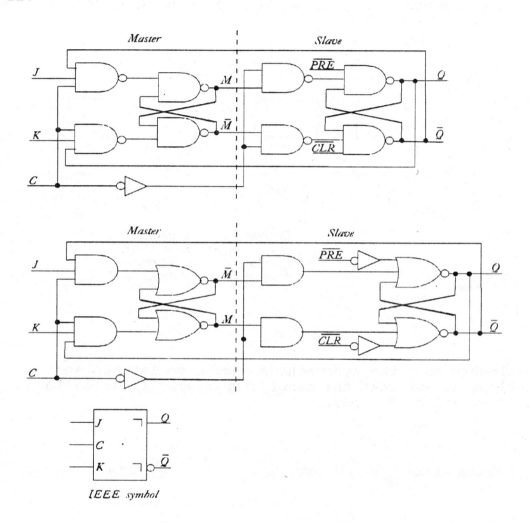

IEEE symbol

8-23. In a master slave J-K flip-flop the feedback path is broken, i. e., interrupted, when the control input is 1 and also when the control input is 0; therefore when J K = 11 and C = 1 the circuit does not oscillate but toggles as required without the need for a pulse narrowing circuit at the control input. In a gated J-K latch the feedback path is never broken and thus the circuit can oscillate when J K = 11 and C = 1 unless the width of pulse at the control input is restricted to a narrow pulse via a pulse narrowing circuit. The width of the pulse must be narrower than the propagation delay path through the latch circuit in order to prevent the circuit from oscillating.

8-24. Master-slave (pulse triggered) flip-flops and edge triggered flip-flops.

8-25.

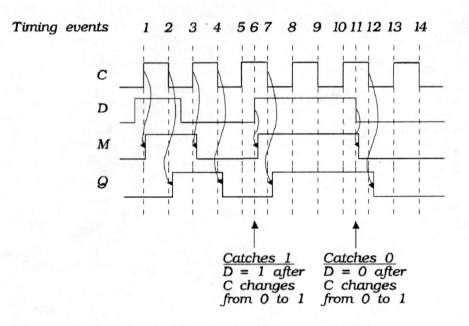

Timing events   1 2   3 4 5 6 7   8   9 10 11 12 13 14

C
D
M
Q

Catches 1
D = 1 after
C changes
from 0 to 1

Catches 0
D = 0 after
C changes
from 0 to 1

8-26. Insure that the synchronous inputs to latches and flip-flops always meet the manufactures specification for data setup and hold times.

8-27.

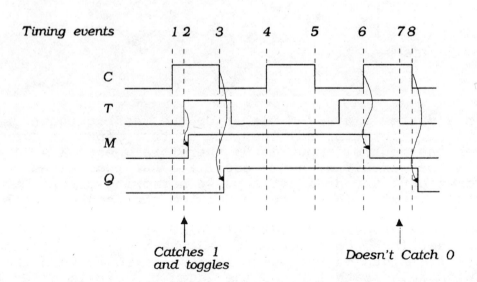

Timing events   1 2   3   4   5   6   7 8

C
T
M
Q

Catches 1
and toggles

Doesn't Catch 0

Yes the T input signal violates both the data setup and the data hold time for a master-slave (pulse-triggered) T flip-flop. The T signal should be stable before the 0 to 1 transition of C and remain stable until after the 1 to 0 transition of C for a master-slave (pulse-triggered) flip-flop without data lockout. Notice the flip-flop catches 1s but not 0s.

8-28.

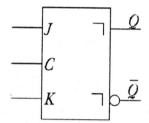
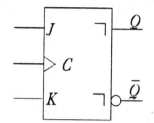

IEEE symbol (master-slave J-K flip-flop without data-lockout)

IEEE symbol (master-slave J-K flip-flop with data-lockout)

8-29.

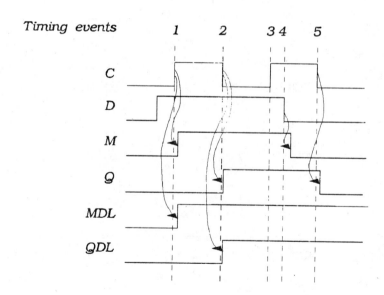

Q ends up as a 0 after the last 1 to 0 transition of C (timing event 5) because of the 0 caught by M at timing event 4. QDL ends up as a 1 since the change in D at timing event 4 is locked out.

8-30.

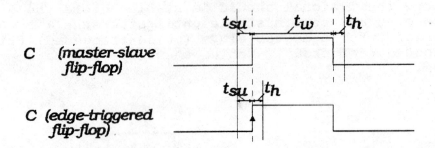

C *(master-slave flip-flop)*

C *(edge-triggered flip-flop)*

8-31.

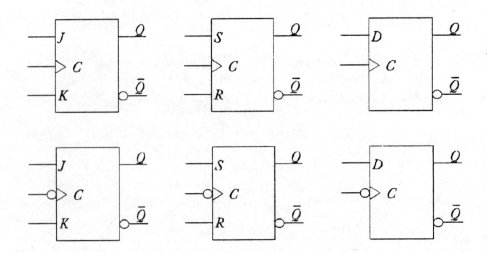

8-32.

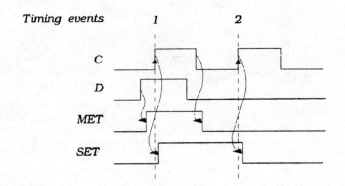

8-33.

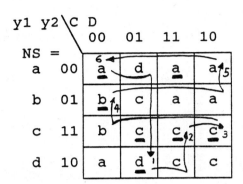

8-34.

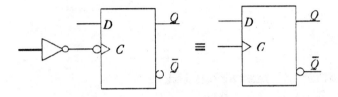

8-35.

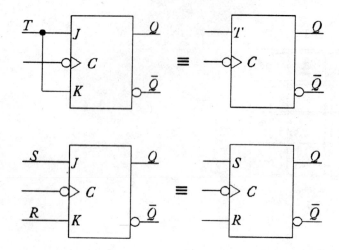

Section 8-5   Reduced Characteristic and Excitation Tables for Bistable Devices

8-37. (a) $Q(q,J,K) = m_0 \cdot q + m_1 \cdot 0 + m_2 \cdot 1 + m_3 \cdot \bar{q}$

where $m = m(J,K)$

$$= \bar{J} \cdot \bar{K} \cdot q + J \cdot \bar{K} + J \cdot K \cdot \bar{q}$$

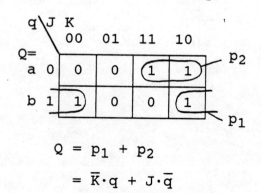

$Q = p_1 + p_2$

$\quad = \bar{K} \cdot q + J \cdot \bar{q}$

(b)

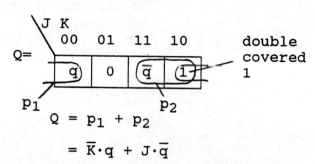

$Q = p_1 + p_2$

$\quad = \bar{K} \cdot q + J \cdot \bar{q}$

8-38. (a) $Q(q,T) = m_0 \cdot q + m_1 \cdot \bar{q}$

where $m = m(T)$

$$= \bar{T} \cdot q + T \cdot \bar{q}$$

(b)

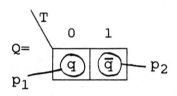

$Q = p_1 + p_2$

$$= \bar{T} \cdot q + T \cdot \bar{q}$$

8-39.

S R	Q
0 0	q
0 1	0
1 0	1
1 1	X

D	Q
0	0
1	1

J K	Q
0 0	q
0 1	0
1 0	1
1 1	$\bar{q}$

T	Q
0	q
1	$\bar{q}$

8-40.

S R	Q
0 0	q
0 1	0
1 0	1
1 1	X

=>

S R q	Q
0 0 0	0
0 0 1	1
0 1 0	0
0 1 1	0
1 0 0	1
1 0 1	1
1 1 0	X
1 1 1	X

=>

q Q	S R
0 0	0 X
0 1	1 0
1 0	0 1
1 1	X 0

D	Q
0	0
1	1

=>

D q	Q
0 0	0
0 1	0
1 0	1
1 1	1

=>

q Q	D
0 0	0
0 1	1
1 0	0
1 1	1

J K	Q
0 0	q
0 1	0
1 0	1
1 1	$\bar{q}$

=>

J K q	Q
0 0 0	0
0 0 1	1
0 1 0	0
0 1 1	0
1 0 0	1
1 0 1	1
1 1 0	1
1 1 1	0

=>

q Q	J K
0 0	0 X
0 1	1 X
1 0	X 1
1 1	X 0

269

T	Q
0	q
1	$\bar{q}$

=>

T	q	Q
0	0	0
0	1	1
1	0	1
1	1	0

=>

q	Q	T
0	0	0
0	1	1
1	0	1
1	1	0

Section 8-6 Metastability and Synchronization Using Bistables

8-41. (a) The circuit in Fig. P8-41 is a positive pulse synchronizer.

(b) The multiple-stage synchronizer circuit has the advantage that it can catch narrow asynchronous input pulses that occur between clock timing events that the single-stage synchronizer circuit cannot.

(c) To reset the S-R latch when Y1·SYNOUT = 1 since the Y1·SYNOUT signal is applied to the R input of the latch (recall S R = 11 causes Y1 = 0)..

(d)

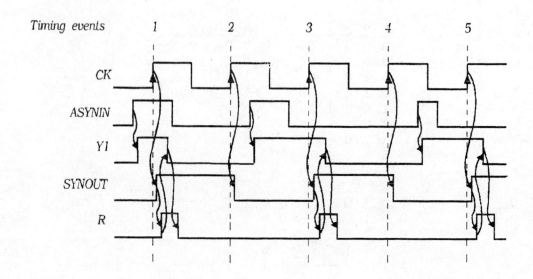

8-42. (a) The circuit in Fig. P8-42 is a negative pulse synchronizer.

(b) The multiple-stage synchronizer circuit can catch narrow asynchronous input pulses that occur between clock timing events that the single-stage synchronizer circuit cannot.

(c)

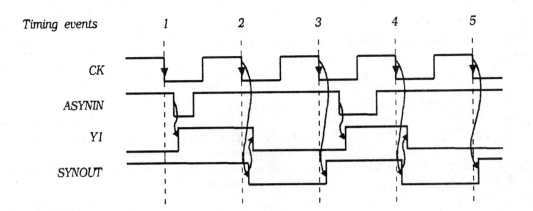

SOLUTIONS TO PROBLEMS IN CHAPTER 9

Section 9-2 Moore, Mealy, and Mixed Type Synchronous State Machines

9-1. Moore type outputs: Z2, Z3, Z5, Z7
Mealy type outputs: Z1, Z4, Z6, Z8

9-2. Moore type outputs are dependent only on the present state of the circuit and are independent of the external inputs to the circuit. Mealy type outputs are dependent on both the external inputs and the present state of the circuit.

9-3. Design a Moore type synchronous state machine with two external inputs X1 and X2 and one output Z. When $\overline{X1} \cdot X2 = 1$ at the next clock timing event, output Z goes to 1. Output Z then goes to 0 unless X2 = 1 causing the output to stay at 1.

(a)

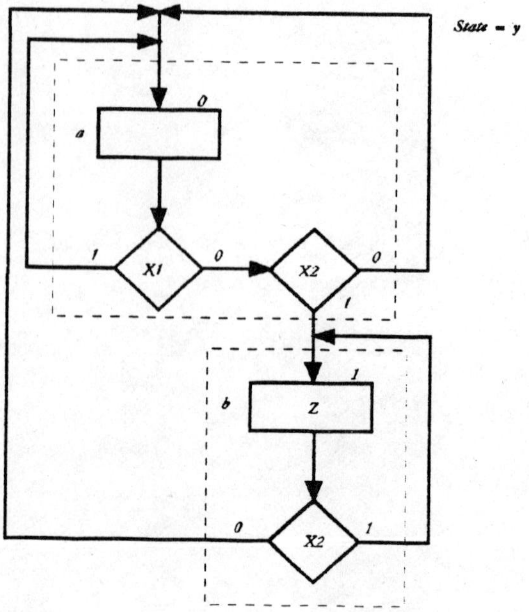

(b)
$$Y = \overline{y} \cdot \overline{X1} \cdot X2 \quad \text{set operation from state a to b}$$
$$+\; y \cdot X2 \quad \text{hold operation from state b to b}$$
$$= (\overline{y} \cdot \overline{X1} + y) \cdot X2$$
$$= (\overline{X1} + y) \cdot X2$$
$$= \overline{X1} \cdot X2 + y \cdot X2$$
$$Z = y$$

(c) D = Y

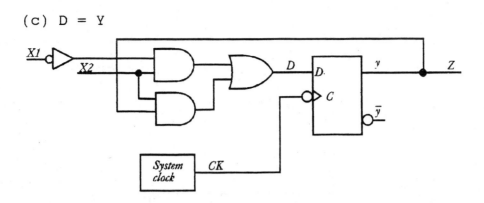

9-4. Modify the design specification in Problem 9-3

such that $\overline{X1} \cdot X2 = 0$ causes the output Z to stay at 1. $\overline{X1} = 1$
causes the output Z to **stay at** 0; otherwise, output Z **goes to** 1.

(a)

State = y

State/Output = y/PS Z
Inputs = X1 X2

b   Z   1

0   $\overline{X1} \cdot X2$ 1

a   0

1   $\overline{X1}$ 0

0 0
1 0
1 1
b
1/1

0 1   1 0
             1 1

a
0/0

0 0
0 1

273

(b) $Y = y \cdot (X1 + \overline{X2})$ hold operation from state b to b

       $+ \ \overline{y} \cdot X1$ set operation from state a to b

       $= y \cdot X1 + y \cdot \overline{X2} + \overline{y} \cdot X1$

       $= X1 + y \cdot \overline{X2}$

   $Z = y$

(c) $D = Y$

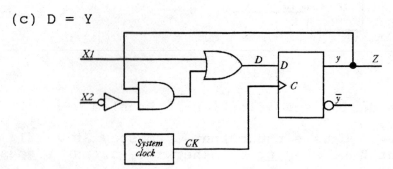

9-5.   (a)

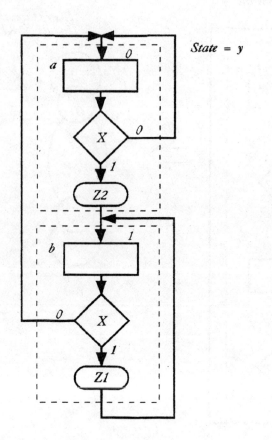

274

(b)

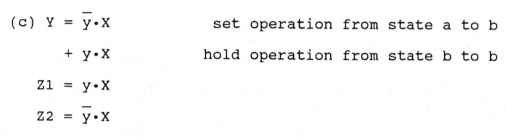

(c) $Y = \bar{y} \cdot X$ set operation from state a to b

     $+ \, y \cdot X$ hold operation from state b to b

   $Z1 = y \cdot X$

   $Z2 = \bar{y} \cdot X$

(d) $D = Y = X, \quad Z1 = y \cdot X, \quad Z2 = \bar{y} \cdot X$

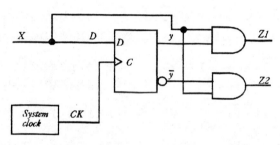

9-6.   (a) $Y1 = \overline{y1} \cdot y2 \cdot X$ set operation from state b to d

        $+ \, y1 \cdot \overline{y2} \cdot X$ hold operation from state d to d

        $+ \, \overline{y1} \cdot \overline{y2} \cdot X$ set operation from state a to c

        $+ \, y1 \cdot y2 \cdot \bar{X}$ hold operation from state c to d

     $Y2 = \overline{y1} \cdot \overline{y2} \cdot \bar{X}$ set operation from state a to b

        $+ \, \overline{y1} \cdot y2 \cdot \bar{X}$ hold operation from state b to b

        $+ \, \overline{y1} \cdot \overline{y2} \cdot X$ set operation from state a to c

   $\overline{Z1} = \overline{y1} \cdot \overline{y2}$ SOP expression for the 0s of the function

   $\overline{Z2} = y1 \cdot y2$ SOP expression for the 0s of the function

(b) $Y1 = \overline{y1} \cdot y2 \cdot X$ set operation from state b to c

        $+ \; y1 \cdot y2 \cdot \overline{X}$ hold operation from state d to c

        $+ \; y1 \cdot \overline{y2} \cdot X$ hold operation from state c to d

        $+ \; \overline{y1} \cdot y2 \cdot \overline{X}$ set operation from state b to d

        $+ \; \overline{y1} \cdot \overline{y2} \cdot \overline{X}$ set operation from state a to d

   $Y2 = \overline{y1} \cdot \overline{y2} \cdot X$ set operation from state a to b

        $+ \; \overline{y1} \cdot y2 \cdot \overline{X}$ hold operation from state b to d

        $+ \; y1 \cdot \overline{y2} \cdot X$ set operation from state c to d

        $+ \; \overline{y1} \cdot \overline{y2} \cdot \overline{X}$ set operation from state a to d

   $Z1 = \overline{y1} \cdot y2 \cdot X + y1 \cdot y2 \cdot X$ SOP expression for the 1s of the function

   $Z2 = \overline{y1} \cdot \overline{y2} \cdot X + y1 \cdot y2 \cdot X$ SOP expression for the 1s of the function

(c) $Y1 = \overline{y1} \cdot \overline{y2} \cdot X1 \cdot X2$ set operation from state a to b

        $+ \; y1 \cdot y2 \cdot \overline{X1}$ hold operation from state b to b

        $+ \; y1 \cdot y2 \cdot X1$ hold operation from state b to c

        $+ \; y1 \cdot \overline{y2} \cdot \overline{X2}$ hold operation from state c to c

   $Y2 = \overline{y1} \cdot \overline{y2} \cdot \overline{X1} \cdot X2$ set operation from state a to b

        $+ \; y1 \cdot y2 \cdot \overline{X1}$ hold operation from state b to b

        $+ \; y1 \cdot \overline{y2} \cdot X2$ set operation from state c to d

   $Z1 = \overline{y1} \cdot \overline{y2} + y1 \cdot \overline{y2} + \overline{y1} \cdot y2$ SOP expression for the 1s of the function

or $\overline{Z1} = y1 \cdot y2$ SOP expression for the 0s of the function

   $Z2 = y1 \cdot \overline{y2}$ SOP expression for the 1s of the function

9-7. (a) $D = \overline{y} \cdot \overline{X1} \cdot X2 + y \cdot X2$

       $Z = y$

(b)   The bistable equation for a D flip-flop is

$$Y = D, \text{ so } Y = D = \overline{y} \cdot \overline{X1} \cdot X2 + y \cdot X2$$

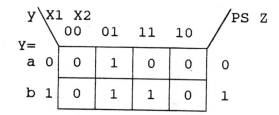

	00	01	11	10	PS Z
Y= a 0	0	1	0	0	0
b 1	0	1	1	0	1

y\X1 X2 ... /PS Z

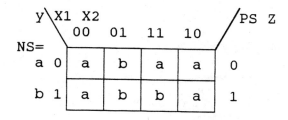

	00	01	11	10	PS Z
NS= a 0	a	b	a	a	0
b 1	a	b	b	a	1

y\X1 X2 ... /PS Z

(c)

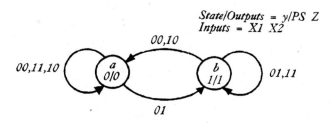

State/Outputs = y/PS Z
Inputs = X1 X2

9-8.   (a) D1 = y1 $\oplus$ y2

D2 = $\overline{y2}$

Z = y1

(b) The bistable equation for a D flip-flop is

Yi = Di

so   Y1 = D1 = y1 $\oplus$ y2

Y2 = D2 = $\overline{y2}$

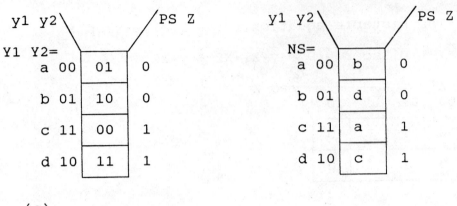

Y1 Y2=	y1 y2	Y1 Y2	PS Z
a	00	01	0
b	01	10	0
c	11	00	1
d	10	11	1

NS=	y1 y2	NS	PS Z
a	00	b	0
b	01	d	0
c	11	a	1
d	10	c	1

(c)

$State/Output = y1\ y2/PS\ Z$

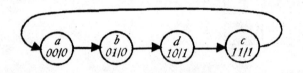

9-9.   (a) T1 = y2

   T2 = 1

   $Z = \overline{y1}$

   (b) The bistable equation for a T flip-flop is

   $Yi = yi \oplus Ti$

so $Y1 = y1 \oplus y2$

   $Y2 = y2 \oplus 1 = \overline{y2}\cdot 1 + y2\cdot\overline{1} = \overline{y2}$

Y1 Y2=	y1 y2	Y1 Y2	PS Z
a	00	01	1
b	01	10	1
c	11	00	0
d	10	11	0

NS=	y1 y2	NS	PS Z
a	00	b	1
b	01	d	1
c	11	a	0
d	10	c	0

(c)

State/Output = y1 y2/PS Z

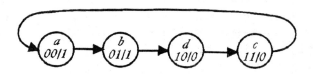

9-10. (a) $J1 = X$ $\quad K1 = \overline{y2} \cdot \overline{X} + y2 \cdot X$ $\qquad Z1 = y1 \cdot y2$

$\quad J2 = \overline{y1}$ $\quad K2 = y1 + X$ $\qquad Z2 = \overline{X} + \overline{y1} \cdot y2$

(b) The bistable equation for a J-K flip-flop is

$Yi = yi \cdot \overline{Ki} + \overline{yi} \cdot Ji$

$Y1 = y1 \cdot \overline{K1} + \overline{y1} \cdot J1$

$\quad = y1 \cdot (\overline{\overline{y2} \cdot \overline{X} + y2 \cdot X}) + \overline{y1} \cdot X$

$\quad = y1 \cdot (y2 + X) \cdot (\overline{y2} + \overline{X}) + \overline{y1} \cdot X$

$\quad = y1 \cdot (y2 \cdot \overline{X} + \overline{y2} \cdot X) + \overline{y1} \cdot X$

$\quad = y1 \cdot y2 \cdot \overline{X} + y1 \cdot \overline{y2} \cdot X + \overline{y1} \cdot X$

$Y2 = y2 \cdot \overline{K2} + \overline{y2} \cdot J2$

$\quad = y2 \cdot (\overline{y1 + X}) + \overline{y2} \cdot \overline{y1}$

$\quad = y2 \cdot (\overline{y1} \cdot \overline{X}) + \overline{y2} \cdot \overline{y1}$

$\quad = \overline{y1} \cdot y2 \cdot \overline{X} + \overline{y1} \cdot \overline{y2}$

y1 y2\X	0	1	/PS Z1
Y1 Y2,PS Z2=			
a 00	01, 1	11, 0	0
b 01	01, 1	10, 1	0
c 11	10, 1	00, 0	1
d 10	00, 1	10, 0	0

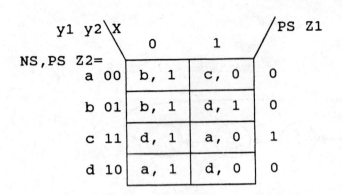

y1 y2 \ X	0	1	PS Z1
NS,PS Z2=			
a 00	b, 1	c, 0	0
b 01	b, 1	d, 1	0
c 11	d, 1	a, 0	1
d 10	a, 1	d, 0	0

State/Output = y1 y2/PS Z1
Input/Output = X/PS Z2

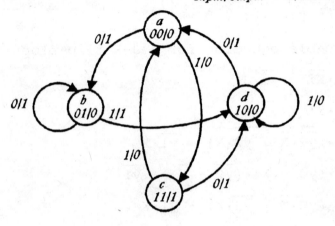

(c)

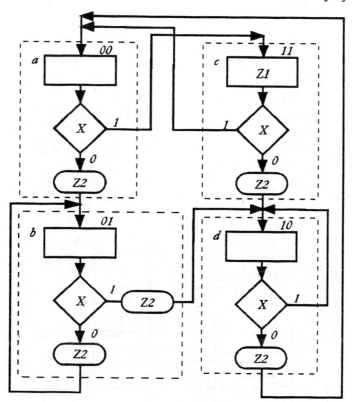

State = y1 y2

Section 9-3   Synchronous Sequential Design
of Moore and Mealy Machines

9-11. (a)

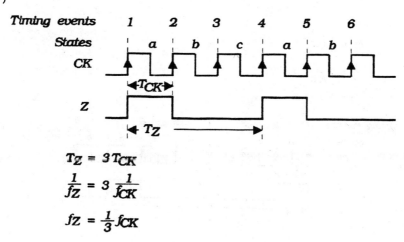

$$T_Z = 3 T_{CK}$$

$$\frac{1}{f_Z} = 3 \frac{1}{f_{CK}}$$

$$f_Z = \frac{1}{3} f_{CK}$$

(b)

PS	NS	PS Z
a	b	1
b	c	0
c	a	0

(c)

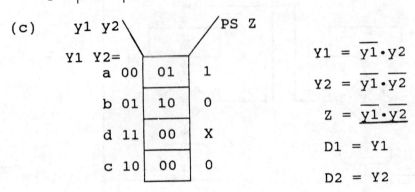

y1 y2   PS Z

Y1 Y2=

a 00	01	1
b 01	10	0
d 11	00	X
c 10	00	0

$Y1 = \overline{y1} \cdot y2$

$Y2 = \overline{y1 \cdot \overline{y2}}$

$Z = \overline{y1 \cdot \overline{y2}}$

$D1 = Y1$

$D2 = Y2$

(d)

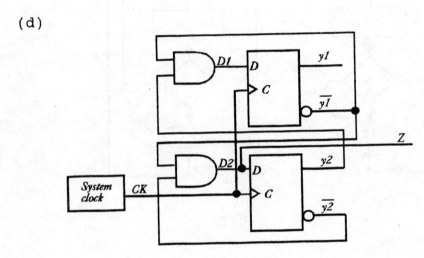

9-12. (a)

$$fz = \tfrac{1}{5} f_{CK}$$

$$\frac{1}{fz} = 5 \frac{1}{f_{CK}}$$

$$T_Z = 5 T_{CK}$$

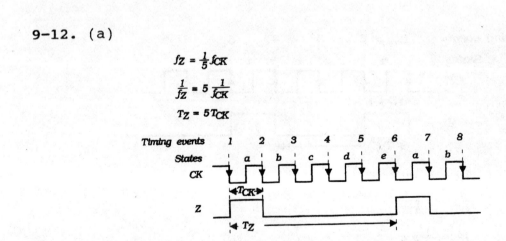

PS	NS	PS Z
a	b	1
b	c	0
c	d	0
d	e	0
e	a	0

(c)

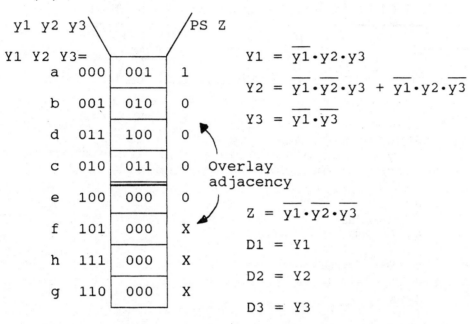

y1 y2 y3 ⟍   ⟋ PS Z

Y1 Y2 Y3=

a	000	001	1
b	001	010	0
d	011	100	0
c	010	011	0
e	100	000	0
f	101	000	X
h	111	000	X
g	110	000	X

$Y1 = \overline{\overline{y1} \cdot y2 \cdot y3}$

$Y2 = \overline{y1} \cdot \overline{y2} \cdot y3 + \overline{y1} \cdot y2 \cdot \overline{y3}$

$Y3 = \overline{y1} \cdot \overline{y3}$

Overlay adjacency

$Z = \overline{y1} \cdot \overline{y2} \cdot \overline{y3}$

$D1 = Y1$

$D2 = Y2$

$D3 = Y3$

(d)

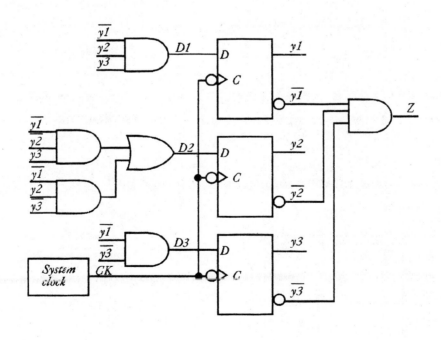

283

9-13.

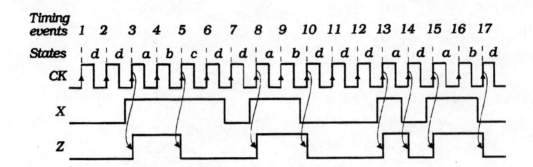

Timing events	1	2	3	4	5	6	7	8	9	10	11	12	13	14	15	16	17
States	d	d	a	b	c	d	d	a	b	d	d	d	a	d	a	b	d

A Moore output can change only after a clock timing event while a Mealy output can change either after a clock timing event or after an external input change.

9-14. (a)

	NS, PS Z	
PS	X = 0	X = 1
a	a, 0	b, 0
b	c, 0	b, 0
c	a, 0	b, 1
d	a, 0	b, 0

(b)

b	a=c X		
c	X	X	
d	√	a=c X	X
	a	b	c

a = d

therefore   d is a redundant state and is not necessary
and   since $2^n \geq s$
                s = 3
                n = 2
the number of flip-flops cannot be reduced

9-15. (a)

	NS			
PS	X = 0	X = 1	PS Z	
a	d	b	0	
b	e	c	0	
c	d	f	1	
d	a	b	0	
e	b	d	1	
f	d	c	1	

284

(b)

	a	b	c	d	e
b	d=e b=c X				
c	X	X			
d	✓	a=e b=c X	X		
e	X	X	b=d d=f X	X	
f	X	X	✓	X	b=d c=d X

$$a = d \qquad c = f$$

therefore   d and f are redundant states and are not necessary

<u>before reduction</u>

$s_{before} = s_b = 6$

since $2^{n_{before}} = 2^{nb} \geq s_b$

$nb = 3$

<u>after reduction</u>

$s_{after} = s_a = 4$

since $2^{n_{after}} = 2^{na} \geq s_a$

$na = 2$

Yes, the number of flip-flops
can be reduces for this design.

9-16. 1. m state variables can be permuted m! different way.
     2. m state variables can be complemented 2^m different ways.

9-17.

r (rows)	v (state var.)	C_{DFF}	C_{JKFF}
2	1	2	1
3	2	12	3
4	2	12	3
5	3	1120	140
6	3	3360	420
7	3	6720	840
8	3	6720	840
9	4	172,972,800	10,810,800

9-18. (a)

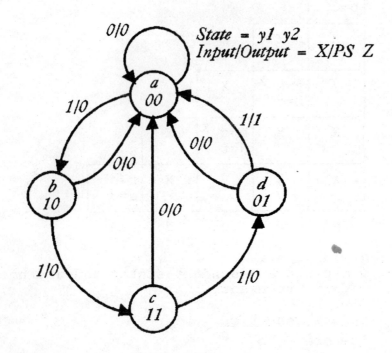

(b)

PS	NS, PS Z	
	X = 0	X = 1
a	a, 0	b, 0
b	a, 0	c, 0
c	a, 0	d, 0
d	a, 0	a, 1

(c)

y1 y2\X

Y1 Y2, PS Z=

	0	1
a 00	00, 0	10, 0
d 01	00, 0	00, 1
c 11	00, 0	01, 0
b 10	00, 0	11, 0

$Y1 = \overline{y2} \cdot X$

$Y2 = y1 \cdot X$

$Z = \overline{y1} \cdot y2 \cdot X$

$D1 = Y1$

$D2 = Y2$

(d)

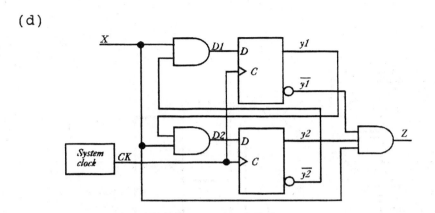

9-19. (a)

	NS		
PS	X = 0	X = 1	PS Z1 Z2
a	b	c	01
b	b	d	11
c	d	a	10
d	a	d	11

(b)

$$Ti = yi \oplus Yi$$

y1 y2\X			
Y1 Y2=	0	1	PS Z1 Z2
a 00	01	11	01
b 01	01	10	11
c 11	10	00	10
d 10	00	10	11

y1 y2\X		
T1 T2=	0	1
00	01	11
01	00	11
11	01	11
10	10	00

$$T1 = y1 \cdot \overline{y2} \cdot \overline{X} + \overline{y1} \cdot X + y2 \cdot X$$

$$T2 = \overline{y1 \cdot y2} + y1 \cdot y2 + \overline{y1} \cdot X$$

$$\overline{Z1} = \overline{y1 \cdot y2}$$

$$\overline{Z2} = y1 \cdot y2$$

(c)

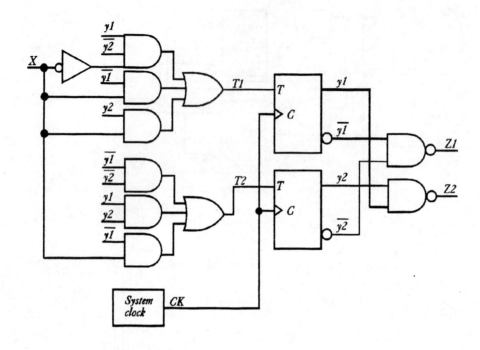

9-20. (a)

$$f_Z = \frac{1}{6} f_{CK}$$

$$\frac{1}{f_Z} = 6 \frac{1}{f_{CK}}$$

$$T_Z = 6 T_{CK}$$

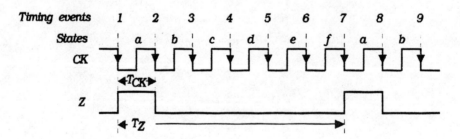

(b)

PS	NS	PS Z
a	b	1
b	c	0
c	d	0
d	e	0
e	f	0
f	a	0

(c)

$T_i = y_i \oplus Y_i$

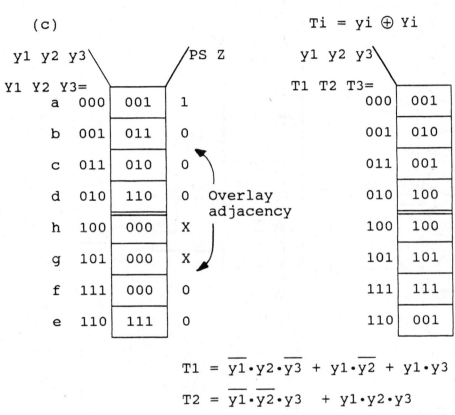

$$T1 = \overline{y1} \cdot y2 \cdot \overline{y3} + y1 \cdot \overline{y2} + y1 \cdot y3$$

$$T2 = \overline{y1} \cdot \overline{y2} \cdot y3 + y1 \cdot y2 \cdot y3$$

$$= (y1 \odot y2) \cdot y3$$

$$= (\overline{y1} \oplus y2) \cdot y3$$

$$T3 = \overline{y1} \cdot \overline{y2} \cdot \overline{y3} + y2 \cdot y3$$

$$+ \underline{y1 \cdot y3} + y1 \cdot y2$$

$$Z = \overline{y2 \cdot y3}$$

(d)

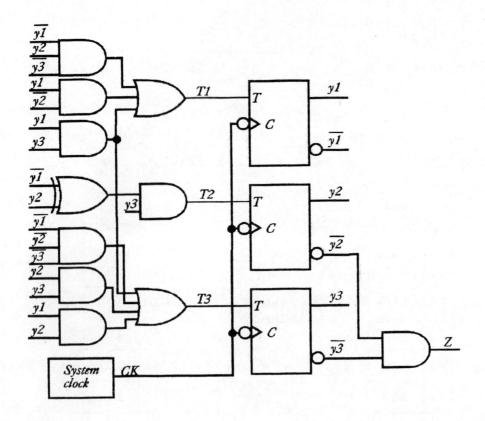

9-21. (a)
$T_i = y_i \oplus Y_i$

y1 y2\X		0	1
Y1 Y2,PS Z=			
a	00	01, 0	00, 0
b	01	10, 0	01, 1
d	11	00, 0	11, 1
c	10	11, 0	10, 0

y1 y2\X		0	1
T1 T2=			
00		01	00
01		11	00
11		11	00
10		01	00

(b)

$T1 = y2 \cdot \overline{X}$

$T2 = \overline{X}$

$Z = y2 \cdot X$

(c)

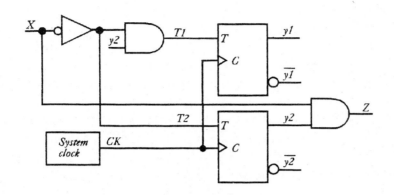

9-22.

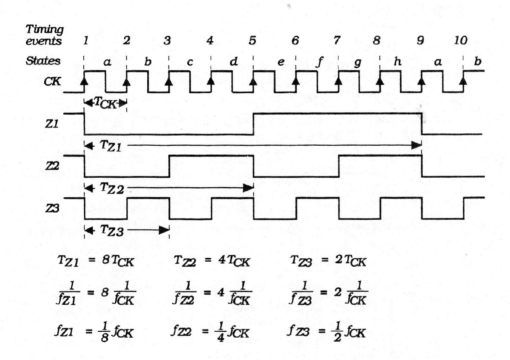

$$T_{Z1} = 8T_{CK} \qquad T_{Z2} = 4T_{CK} \qquad T_{Z3} = 2T_{CK}$$

$$\frac{1}{f_{Z1}} = 8\frac{1}{f_{CK}} \qquad \frac{1}{f_{Z2}} = 4\frac{1}{f_{CK}} \qquad \frac{1}{f_{Z3}} = 2\frac{1}{f_{CK}}$$

$$f_{Z1} = \frac{1}{8}f_{CK} \qquad f_{Z2} = \frac{1}{4}f_{CK} \qquad f_{Z3} = \frac{1}{2}f_{CK}$$

9-23. (a)

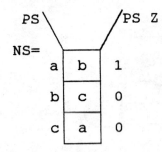

PS	NS=	PS Z
a	b	1
b	c	0
c	a	0

(b)

$y_1\,y_2$	$Y_1\,Y_2=$	PS Z
a 00	01	1
b 01	11	0
c 11	00	0
d 10	00	X

y_i	Y_i	J_i	K_i
0	0	0	X
0	1	1	X
1	0	X	1
1	1	X	0

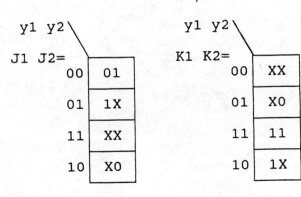

$y_1\,y_2$ — $J_1\,J_2=$

00	01
01	1X
11	XX
10	X0

$y_1\,y_2$ — $K_1\,K_2=$

00	XX
01	X0
11	11
10	1X

$J_1 = y_2$

$J_2 = \overline{y_1}$

$Z = \overline{y_2}$

$K_1 = 1$

$K_2 = y_1$

(c)

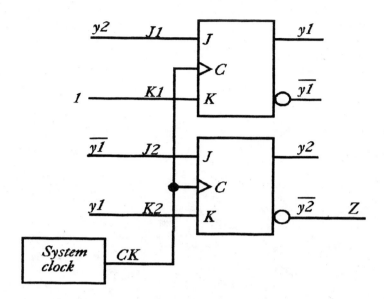

9-24. (a)

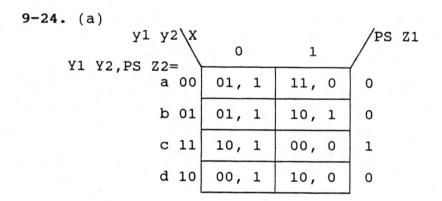

y1 y2\X	0	1	PS Z1
Y1 Y2,PS Z2=			
a 00	01, 1	11, 0	0
b 01	01, 1	10, 1	0
c 11	10, 1	00, 0	1
d 10	00, 1	10, 0	0

yi	Yi	Ji	Ki
0	0	0	X
0	1	1	X
1	0	X	1
1	1	X	0

y1 y2\X	0	1
J1 J2=		
00	01	11
01	0X	1X
11	XX	XX
10	X0	X0

y1 y2\X	0	1
K1 K2=		
00	XX	XX
01	X0	X1
11	01	11
10	1X	0X

$$J1 = X \qquad K1 = \overline{y2} \cdot \overline{X} + y2 \cdot X = y2 \odot X = \overline{y2} \oplus X$$

$$J2 = \overline{y1} \qquad K2 = y1 + X$$

$$Z1 = y1 \cdot y2$$

$$Z2 = \overline{X} + \overline{y1} \cdot y2$$

(c)

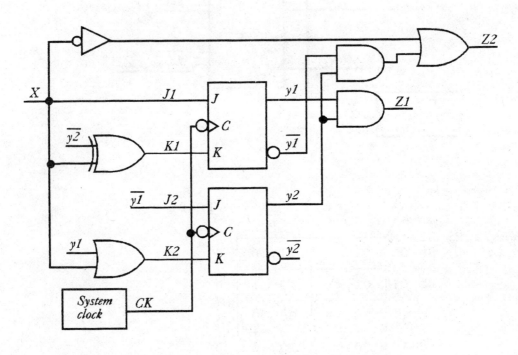

9-25. (a)

		0	1
NS,PS Z=			
a 00		b, 0	a, 1
b 01		b, 0	c, 0
c 11		a, 0	a, 0
d 10		a, 0	a, 0

y1 y2\X

(b)

y1 y2\X		
Y1 Y2,PS Z=	0	1
a 00	01, 0	00, 1
b 01	01, 0	11, 0
c 11	00, 0	00, 0
d 10	00, 0	00, 0

yi	Yi	Ji	Ki
0	0	0	X
0	1	1	X
1	0	X	1
1	1	X	0

y1 y2\X		
J1 J2=	0	1
00	01	00
01	0X	1X
11	XX	XX
10	X0	00

y1 y2\X		
K1 K2=	0	1
00	XX	XX
01	X0	X0
11	11	11
10	1X	XX

$J1 = y2 \cdot X$ $\qquad$ $K1 = 1$

$J2 = \overline{y1} \cdot \overline{X}$ $\qquad$ $K2 = y1$

$Z = \overline{y1} \cdot \overline{y2} \cdot X$

(c)

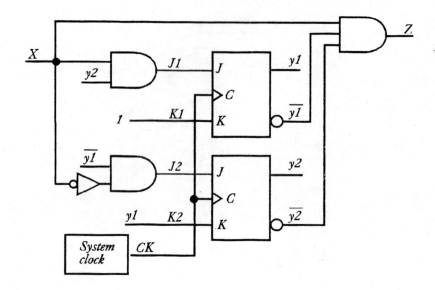

9-26. A controller is a circuit that is used to supervise (or control) the operation of other circuits.

Section 9-4   Synchronous Counter Design

9-27. Decoding glitches can occur at the outputs. Decoding glitches are false or momentary states, and they can cause false triggering of the device being driven by these outputs.

9-28. A synchronous counter has all its control inputs tied to the system clock while an asynchronous counter does not.

9-29. A counter must be self-correcting or self-starting to guarantee that proper operation of the circuit will always occur if the counter finds itself in any state.

9-30. 1. Analyze each state of the counter to verify that unknown or illegal states have a path back to the primary counting sequence.

   2. A D type flip-flop is generally better suited for illegal state recovery since its output goes to 0 (the reset condition) when the D input is 0. If the 0 state is one of the state assignments in the primary counting sequence and all the D flip-flop in the synchronous circuit have a 0 input then at the next clock timing event the circuit will recover to the primary counting sequence.

9-31. (a)   The state diagram for a synchronous counter that provides a BCD Moore output sequence with illegal state recovery is show as follows:

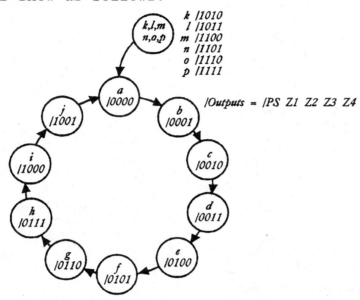

$$k \ /1010$$
$$l \ /1011$$
$$m \ /1100$$
$$n \ /1101$$
$$o \ /1110$$
$$p \ /1111$$

/Outputs = /PS Z1 Z2 Z3 Z4

For synchronous counters a state assignment of y1 = Z1, y2 = Z2, y3 = Z3, y4 = Z4, ··· is generally used. For this state assignment, no output combinational logic (also referred to a decoding logic) is required, and this is how the complexity of the circuit is reduced.

(b)

y1 y2\y3 y4

Y1 Y2 Y3 Y4=

	00	01	11	10
00	0001	0010	0100	0011
01	0101	0110	1000	0111
11	0000	0000	0000	0000
10	1001	0000	0000	0000

$$Y1 = \overline{y1} \cdot y2 \cdot y3 \cdot y4 + y1 \cdot \overline{y2} \cdot \overline{y3} \cdot \overline{y4}$$

$$Y2 = \overline{y1} \cdot \overline{y2} \cdot y3 \cdot y4 + \overline{y1} \cdot y2 \cdot \overline{y4} + \overline{y1} \cdot y2 \cdot \overline{y3}$$

$$Y3 = \overline{y1} \cdot \overline{y3} \cdot y4 + \overline{y1} \cdot y3 \cdot \overline{y4}$$

$$Y4 = \overline{y1} \cdot \overline{y4} + \overline{y2} \cdot \overline{y3} \cdot \overline{y4}$$

D1 = Y1, D2 = Y2, D3 = Y3, D4 = Y4

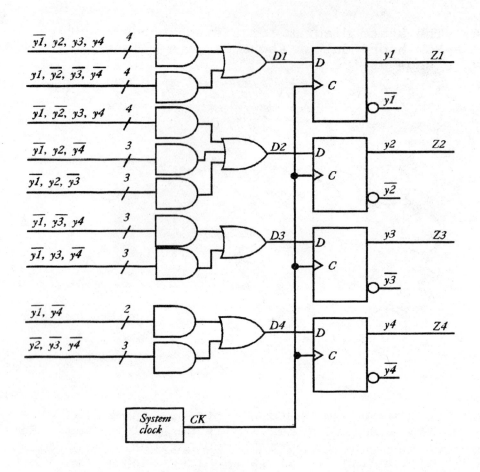

(c)

yi	Yi	Ji	Ki
0	0	0	X
0	1	1	X
1	0	X	1
1	1	X	0

J1 J2 J3 J4=

y1 y2\y3 y4

	00	01	11	10
00	0001	001X	01XX	00X1
01	0X01	0X1X	1XXX	0XX1
11	XX00	XX0X	XXXX	XXX0
10	X001	X00X	X0XX	X0X0

K1 K2 K3 K4=

y1 y2\y3 y4

	00	01	11	10
00	XXXX	XXX1	XX11	XX0X
01	X0XX	X0X1	X111	X00X
11	11XX	11X1	1111	111X
10	0XXX	1XX1	1X11	1X1X

$$J1 = y2 \cdot y3 \cdot y4 \qquad K1 = y2 + y3 + y4$$

$$J2 = \overline{y1} \cdot y3 \cdot y4 \qquad K2 = y1 + y3 \cdot y4$$

$$J3 = \overline{y1} \cdot y4 \qquad K3 = y1 + y4$$

298

$$J4 = \overline{y1} + \overline{y2} \cdot \overline{y3} \qquad\qquad K4 = 1$$

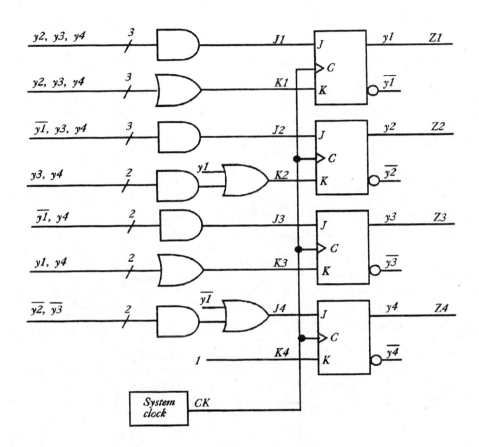

9-32. (a)

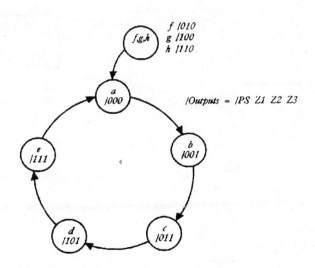

(b) Let the state assignment be y1 = Z1, y2 = Z2, y3 = Z3 for reduced circuit complexity, i.e., no output combinational logic requirement.

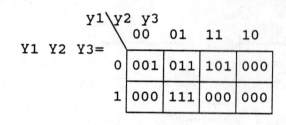

Y1 Y2 Y3 =

y1\y2 y3	00	01	11	10
0	001	011	101	000
1	000	111	000	000

$$Y1 = \overline{y1}\cdot y2 \cdot y3 + y1\cdot \overline{y2}\cdot y3$$

$$Y2 = \overline{y2}\cdot y3$$

$$Y3 = \overline{y1}\cdot \overline{y2} + \overline{y1}\cdot y3 + \overline{y2}\cdot y3$$

$$D1 = Y1, \quad D2 = Y2, \quad D3 = Y3$$

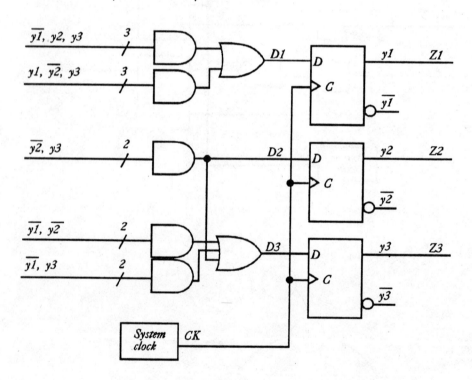

(c) $T_i = y_i \oplus Y_i$

T1 T2 T3 =

y1\y2 y3	00	01	11	10
0	001	010	110	010
1	100	010	111	110

300

$$T1 = y2 \cdot y3 + y1 \cdot \overline{y3}$$

$$T2 = y2 + y3$$

$$T3 = \overline{y1} \cdot \overline{y2} \cdot \overline{y3} + y1 \cdot y2 \cdot y3$$

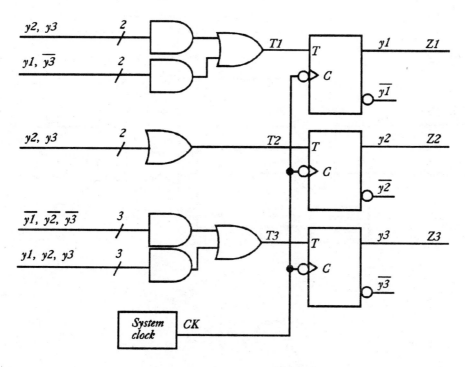

(d)

y_i	Y_i	J_i	K_i
0	0	0	X
0	1	1	X
1	0	X	1
1	1	X	0

J1 J2 J3=

y1\y2 y3	00	01	11	10
0	001	01X	1XX	0X0
1	X00	X1X	XXX	XX0

K1 K2 K3=

y1\y2 y3	00	01	11	10
0	XXX	XX0	X10	X1X
1	1XX	0X0	111	11X

$$J1 = y2 \cdot y3 \qquad\qquad K1 = y2 + \overline{y3}$$

$$J2 = y3 \qquad\qquad\qquad K2 = 1$$

$$J3 = \overline{y1} \cdot \overline{y2} \qquad\qquad K3 = y1 \cdot y2$$

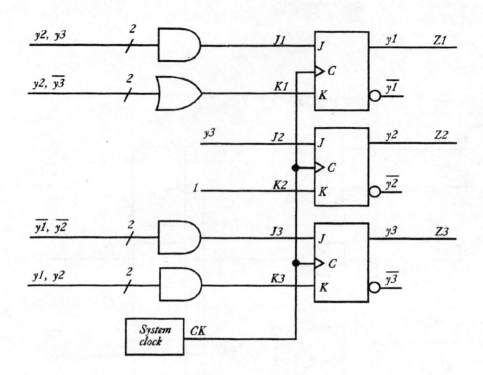

9-33. (a)

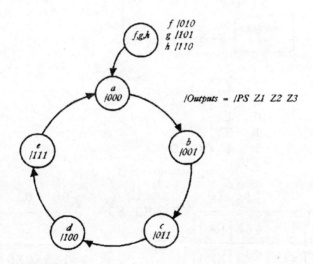

f /010
g /101
h /110

/Outputs = /PS Z1 Z2 Z3

Let the state assignment be y1 = Z1, y2 = Z2, y3 = Z3 for reduced circuit complexity, i.e., no output combinational logic requirement.

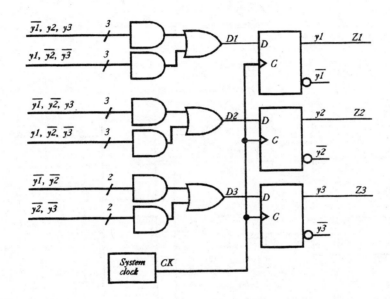

y1\y2 y3

Y1 Y2 Y3=		00	01	11	10
	0	001	011	100	000
	1	111	000	000	000

$$Y1 = \overline{y1} \cdot y2 \cdot y3 + y1 \cdot \overline{y2} \cdot \overline{y3}$$

$$Y2 = \overline{y1} \cdot \overline{y2} \cdot y3 + y1 \cdot \overline{y2} \cdot \overline{y3}$$

$$Y3 = \overline{y1} \cdot \overline{y2} + \overline{y2} \cdot \overline{y3}$$
$$D1 = Y1, \quad D2 = Y2, \quad D3 = Y3$$

(b)

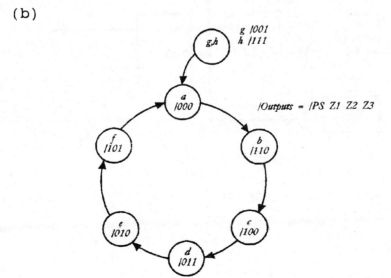

/Outputs = /PS Z1 Z2 Z3

Let the state assignment be y1 = Z1, y2 = Z2, y3 = Z3 for reduced circuit complexity, i.e., no output combinational logic requirement.

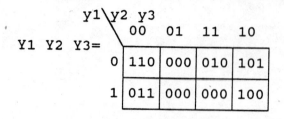

Y1 Y2 Y3=

y1\y2 y3	00	01	11	10
0	110	000	010	101
1	011	000	000	100

$$Y1 = \overline{y1} \cdot \overline{y3} + y2 \cdot \overline{y3}$$

$$Y2 = \overline{y2} \cdot \overline{y3} + \overline{y1} \cdot y2 \cdot y3$$

$$Y3 = \overline{y1} \cdot y2 \cdot \overline{y3} + y1 \cdot \overline{y2} \cdot \overline{y3}$$

$$D1 = Y1, \quad D2 = Y2, \quad D3 = Y3$$

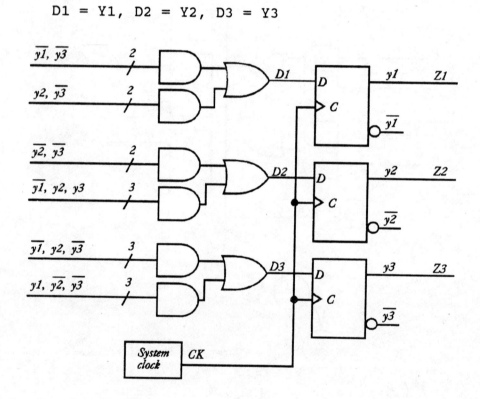

(c)

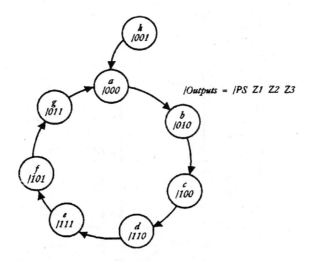

/Outputs = /PS Z1 Z2 Z3

Let the state assignment be y1 = Z1, y2 = Z2, y3 = Z3
for reduced circuit complexity, i.e., no output
combinational logic requirement.

Y1 Y2 Y3=

y1\y2 y3	00	01	11	10
0	010	000	000	100
1	110	011	101	111

$$Y1 = y2 \cdot \overline{y3} + y1 \cdot \overline{y3} + y1 \cdot y2$$

$$Y2 = \overline{y2} \cdot \overline{y3} + y1 \cdot \overline{y2} + \underline{y1 \cdot \overline{y3}}$$

$$Y3 = y1 \cdot y3 + \underline{y1 \cdot y2}$$

$$D1 = Y1, \quad D2 = Y2, \quad D3 = Y3$$

305

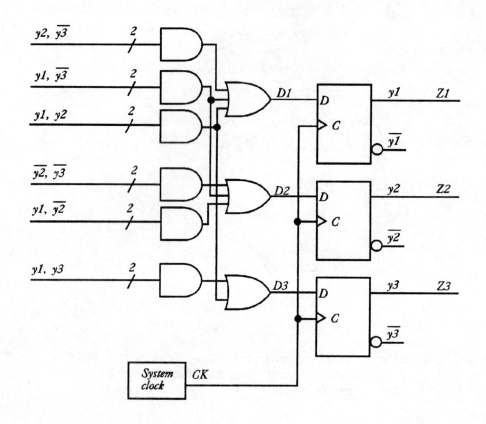

9-34. (a)

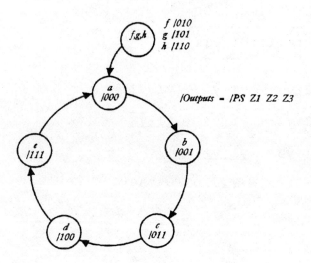

f /010
g /101
h /110

/Outputs = /PS Z1 Z2 Z3

Let the state assignment be y1 = Z1, y2 = Z2, y3 = Z3 for reduced circuit complexity, i.e., no output combinational logic requirement.

$Y1\ Y2\ Y3=$

y1\y2 y3	00	01	11	10
0	001	011	100	000
1	111	000	000	000

$$Ti = yi \oplus Yi$$

$T1\ T2\ T3=$

y1\y2 y3	00	01	11	10
0	001	010	111	010
1	011	101	111	110

$$T1 = y2 \cdot y3 + y1 \cdot y3 + y1 \cdot y2$$

$$T2 = \overline{y1} \cdot y3 + y1 \cdot \overline{y3} + y2 = (y1 \oplus y3) + y2$$

$$T3 = \overline{y2} \cdot \overline{y3} + y2 \cdot y3 + y1 \cdot \overline{y2}$$

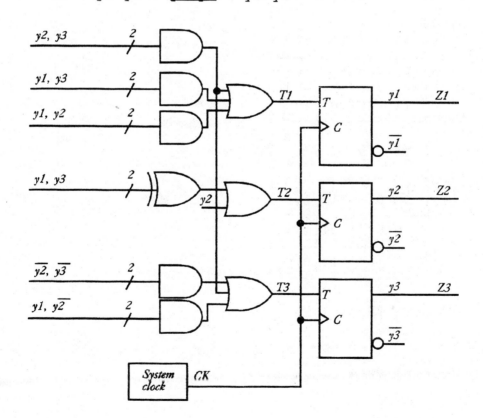

(b)

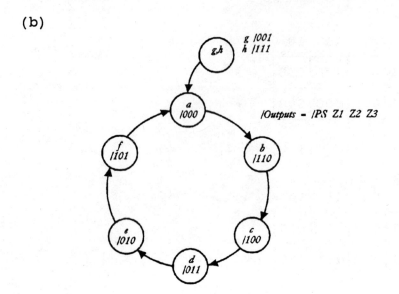

Let the state assignment be y1 = Z1, y2 = Z2, y3 = Z3 for reduced circuit complexity, i.e., no output combinational logic requirement.

Y1 Y2 Y3=

y1\y2 y3	00	01	11	10
0	110	000	010	101
1	011	000	000	100

$$T_i = y_i \oplus Y_i$$

T1 T2 T3=

y1\y2 y3	00	01	11	10
0	110	001	001	111
1	111	101	111	010

$$T1 = \overline{y1} \cdot \overline{y3} + y1 \cdot \overline{y2} + y1 \cdot y3$$

$$T2 = \overline{y3} + y1 \cdot y2$$

$$T3 = \overline{y1} \cdot y2 + \underline{y1 \cdot \overline{y2}} + y3$$

308

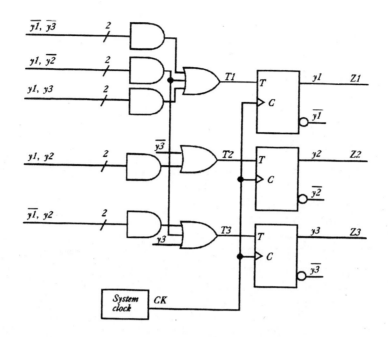

(c)

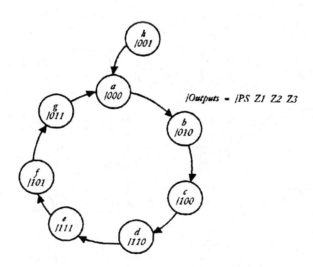

/Outputs = /PS Z1 Z2 Z3

Let the state assignment be y1 = Z1, y2 = Z2, y3 = Z3 for reduced circuit complexity, i.e., no output combinational logic requirement.

309

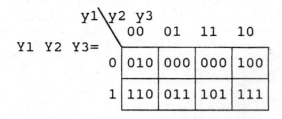

$Y_1\ Y_2\ Y_3 =$

y1\y2 y3	00	01	11	10
0	010	000	000	100
1	110	011	101	111

$$T_i = y_i \oplus Y_i$$

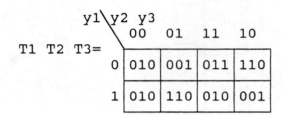

$T_1\ T_2\ T_3 =$

y1\y2 y3	00	01	11	10
0	010	001	011	110
1	010	110	010	001

$$T_1 = \overline{y_1} \cdot y_2 \cdot \overline{y_3} + y_1 \cdot \overline{y_2} \cdot y_3$$

$$T_2 = \overline{y_1} \cdot \overline{y_3} + y_1 \cdot \overline{y_2} + y_2 \cdot y_3$$

$$T_3 = \overline{y_1} \cdot y_3 + y_1 \cdot y_2 \cdot \overline{y_3}$$

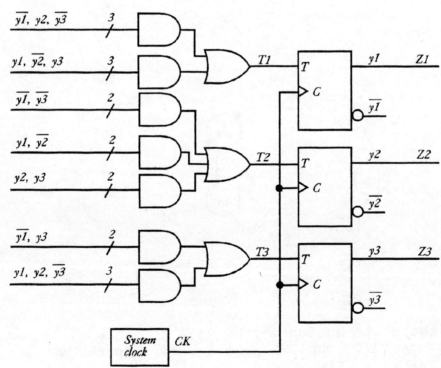

9-35. (a)

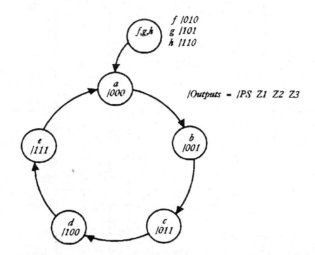

```
f /010
g /101
h /110
```

/Outputs = /PS Z1 Z2 Z3

Let the state assignment be y1 = Z1, y2 = Z2, y3 = Z3 for reduced circuit complexity, i.e., no output combinational logic requirement.

$Y1\ Y2\ Y3=$

y1\y2 y3

y1 \ y2 y3	00	01	11	10
0	001	011	100	000
1	111	000	000	000

yi	Yi	Ji	Ki
0	0	0	X
0	1	1	X
1	0	X	1
1	1	X	0

$J1\ J2\ J3=$

y1\y2 y3

y1 \ y2 y3	00	01	11	10
0	001	01X	1XX	0X0
1	X11	X0X	XXX	XX0

$K1\ K2\ K3=$

y1\y2 y3

y1 \ y2 y3	00	01	11	10
0	XXX	XX0	X11	X1X
1	0XX	1X1	111	11X

$J1 = y2 \cdot \overline{y3}$

$J2 = y1 \cdot \overline{y3} + \overline{y1} \cdot y3 = y1 \oplus y3$

$J3 = \overline{y2}$

$K1 = y2 + y3$

$K2 = 1$

$K3 = y1 + y2$

311

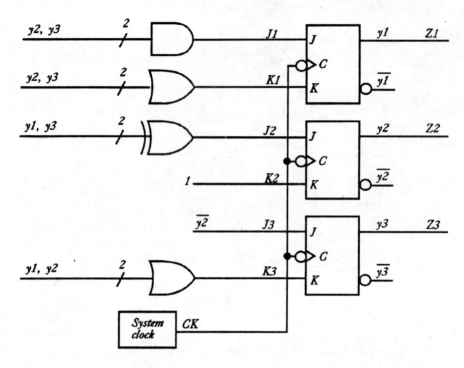

(b)

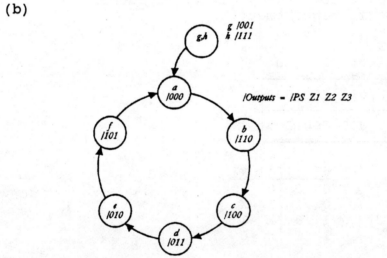

/Outputs = /PS Z1 Z2 Z3

Let the state assignment be y1 = Z1, y2 = Z2, y3 = Z3 for reduced circuit complexity, i.e., no output combinational logic requirement.

$Y1\ Y2\ Y3 =$

y1\y2 y3	00	01	11	10
0	110	000	010	101
1	011	000	000	100

312

```
yi Yi | Ji Ki
0  0  | 0  X
0  1  | 1  X
1  0  | X  1
1  1  | X  0
```

J1 J2 J3=

y1\y2 y3	00	01	11	10
0	110	00X	0XX	1X1
1	X11	X0X	XXX	XX0

K1 K2 K3=

y1\y2 y3	00	01	11	10
0	XXX	XX1	X01	X1X
1	1XX	1X1	111	01X

$J1 = \overline{y3}$

$J2 = \overline{y3}$

$J3 = y1 \cdot \overline{y2} + \overline{y1} \cdot y2 = y1 \oplus y2$

$K1 = \overline{y2} + y3$

$K2 = y1 + \overline{y3}$

$K3 = 1$

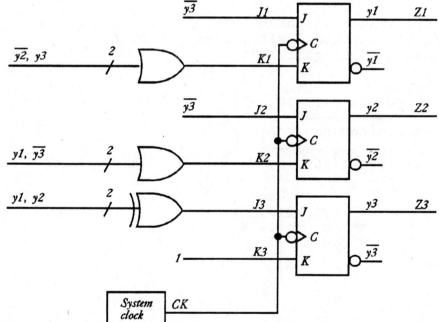

(c)

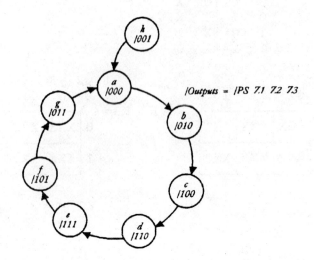

/Outputs = /PS Z1 Z2 Z3

Let the state assignment be y1 = Z1, y2 = Z2, y3 = Z3 for reduced circuit complexity, i.e., no output combinational logic requirement.

Y1 Y2 Y3=

y1 \ y2 y3	00	01	11	10
0	010	000	000	100
1	110	011	101	111

yi	Yi	Ji	Ki
0	0	0	X
0	1	1	X
1	0	X	1
1	1	X	0

J1 J2 J3=

y1 \ y2 y3	00	01	11	10
0	010	00X	0XX	1X0
1	X10	X1X	XXX	XX1

K1 K2 K3=

y1 \ y2 y3	00	01	11	10
0	XXX	XX1	X11	X1X
1	0XX	1X0	010	00X

$J1 = y2 \cdot \overline{y3}$

$J2 = y1 + \overline{y3}$

$J3 = y1 \cdot y2$

$K1 = \overline{y2} \cdot y3$

$K2 = \overline{y1} + y3$

$K3 = \overline{y1}$

314

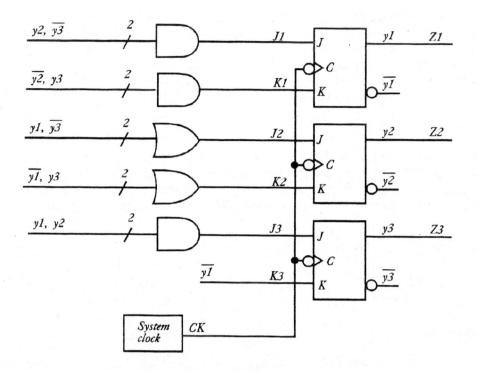

9-36. For 2 bits, n = 2 (number of primary states) leaving $2^n - n = 4 - 2 = 2$ (number of unused or illegal states); therefore, illegal state recovery circuity **is** required. For a synchronous 2-bit ring counter without illegal state recovery, the counting sequence follows the normal directed lines segments in the following simplified state diagram.

State = y1 y2

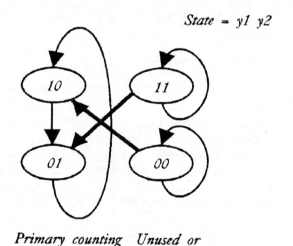

Primary counting   Unused or
sequence   illegal states

To be self-correcting the counter must follow the bold directed line segments for those states not in the primary counting sequence (remember Y2 = y1, i.e., y1 shifts right at the next clock timing event.) It can be seen from the state diagram that a self-correcting synchronous 2-bit ring counter has the following next state equation for Y1

$$Y1 = \overline{y1}.$$

The next state equation for Y1 can also be obtained from the composite Karnaugh map for the next state by reading the map for Y1 as follows. The composite Karnaugh map is obtained from the self-correcting state diagram.

```
          \y1 y2
           \  00 01 11 10
  Y1 Y2=    \
            ┌────┬────┬────┬────┐
            │ 10 │ 10 │ 01 │ 01 │
            └────┴────┴────┴────┘
```

therefore $Y1 = \overline{y1}$

The circuit for the synchronous 2-bit ring counter with illegal state recovery is shown as follow.

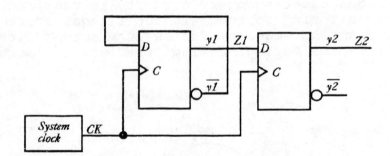

9-37. For 3 bits, n = 3 (number of primary states) leaving $2^n - n = 8 - 3 = 5$ (number of unused or illegal states); therefore, illegal state recovery circuity _is_ required. For a synchronous 3-bit ring counter without illegal state recovery, the counting sequence follows the normal directed lines segments in the following simplified state diagram.

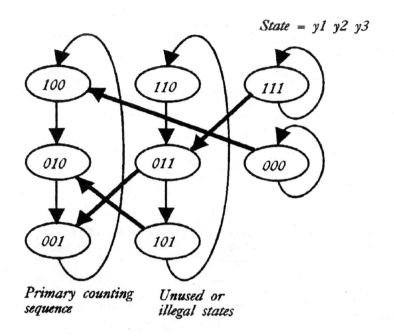

State = y1 y2 y3

*Primary counting
sequence*

*Unused or
illegal states*

To be self-correcting the counter should follow the bold directed line segments for those states not in the primary counting sequence (remember Y2 = y1, Y3 = y2 i.e., y1 y2 shifts right at the next clock timing event.) It is desirable to force the counter to return to the primary counting sequence as fast as possible. It can be seen from the state diagram that this self-correcting synchronous 3-bit ring counter has the following next state equation for Y1.

$$Y1 = \overline{y1} \cdot \overline{y2}$$

The next state equation for Y1 can also be obtained from the composite Karnaugh map for the next state by reading the map for Y1 as follows. The composite Karnaugh map is obtained from the self-correcting state diagram.

Y1 Y2 Y3=

y1\y2 y3	00	01	11	10
0	100	100	001	001
1	010	010	011	011

therefore $Y1 = \overline{y1} \cdot \overline{y2}$

The circuit for the synchronous 3-bit ring counter with illegal state recovery is shown as follows.

317

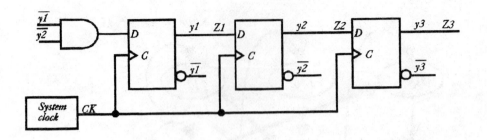

9-38. For 5 bits, n = 5 (number of primary states) leaving $2^n - n = 32 - 5 = 17$ (number of unused or illegal states); therefore, illegal state recovery circuity <u>is</u> required. For a synchronous 5-bit ring counter without illegal state recovery. From the solution for a synchronous 2-bit ring counter with illegal state recovery in Problem 9-36 we found

$$Y1 = \overline{\overline{y1}}$$

From the solution for a synchronous 3-bit ring counter with illegal state recovery in Problem 9-37 we found

$$Y1 = \overline{\overline{y1} \cdot \overline{y2}}$$

From the solution for a synchronous 4-bit ring counter with illegal state recovery as discussed in the text we said.

$$Y1 = \overline{\overline{y1} \cdot \overline{y2} \cdot \overline{y3}}$$

For a synchronous (n + 1)-bit ring counter with illegal state recovery it may be true that

$$Y1 = \overline{\overline{y1} \cdot \overline{y2} \cdot \overline{y3}} \cdots \overline{yn}$$

so for a 5-bit synchronous ring counter with illegal state recovery the next state equation for Y1 would be

$$Y1 = \overline{\overline{y1} \cdot \overline{y2} \cdot \overline{y3} \cdot \overline{y4}}$$

The circuit for the synchronous 5-bit ring counter with illegal state recovery is shown as follows:

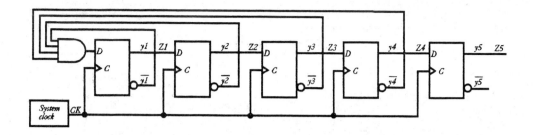

From the next state equations $Y1 = \overline{y1} \cdot \overline{y2} \cdot \overline{y3} \cdot \overline{y4}$, $Y2 = y1$, $Y3 = y2$, $Y4 = y3$, $Y5 = y4$ the composite Karnaugh maps for the next state can be filled in.

$y1=1, y2\ y3 \backslash y4\ y5$

	00	01	11	10
00	01000	01000	01001	01001
01	01010	01010	01011	01011
11	01110	01110	01111	01111
10	01100	01100	01101	01101

$y1=0, y2\ y3 \backslash y4\ y5$

Y1 Y2 Y3 Y4 Y5=

	00	01	11	10
00	10000	10000	00001	00001
01	00010	00010	00011	00011
11	00110	00110	00111	00111
10	00100	00100	00101	00101

Using the composite Karnaugh map we can draw the state diagram as follows. Notice that the Basic synchronous 5-bit ring counter state diagram uses normal directed line segments while the bold directed line segments provide the path followed by the machine as a result of the illegal state recovery circuitry. Notice that the machine does recover from all its unused or illegal states using the proposed next state for Y1.

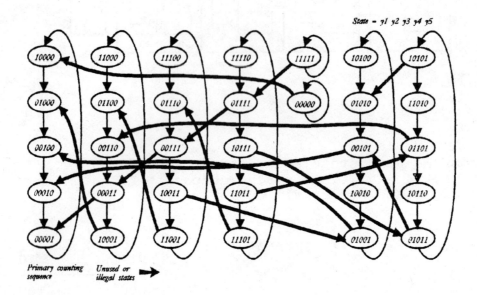

State = $y1$ $y2$ $y3$ $y4$ $y5$

Primary counting sequence

Unused or illegal states →

9-39. For 2 bits, 2n = 4 (number of primary states), leaving $2^n - 2n = 4 - 4 = 0$ (number of unused or illegal states); therefore, illegal state recovery circuity **is not** required.

The next state equations for a synchronous 2-bit twisted ring counter are written as follows:

$$Y1 = \overline{y2}, \quad Y2 = y1$$

From these equations we can obtain the following composite Karnaugh map for the next state.

Y1 Y2=

	y1 y2		
00	01	11	10
10	00	01	11

Using the map we can draw the state diagram of the 2-bit counter as follows.

320

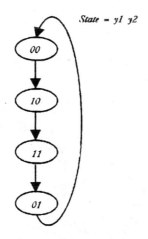

State = y1 y2

The circuit for the synchronous 2-bit twisted ring counter is shown as follows:

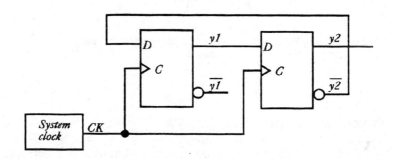

9-40. For 3 bits, 2n = 6 (number of primary states), leaving 2^n - 2n = 8 - 6 = 2 (number of unused or illegal states); therefore, illegal state recovery circuity **is** required.

Case 1   Case 2

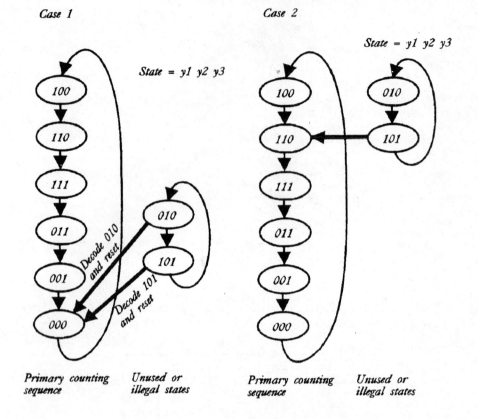

State = y1 y2 y3

Primary counting   Unused or   Primary counting   Unused or
sequence   illegal states   sequence   illegal states

In the state diagram illustrated above, normal
directed line segments show the operation of a basic
synchronous 3-bit twisted ring counter without illegal state
recovery while the bold directed line segments provide the
path followed by the illegal state recovery circuity. For
case 1 you can decode either 010 or 101 and use the decoded
output to reset the counter to get back to the primary
counting sequence. The design for case 1 is straight
forward since the decoded output is simply used to reset all
the flip-flops in the counter by connecting the output of
the decoder to the reset input of each flip-flop, assuming
of course that each flip-flop has a reset input. For case 2

you can decode 101 and use the decoded output ORed with $\overline{y3}$
to obtain Y1 to get back to the primary counting sequence as
discussed in the text. For the design for case 2 to be
self-correcting, the counter should follow the bold directed
line segment for state 101 back to the primary counting
sequence (remember Y2 = y1, Y3 = y2 i.e., y1 y2 shifts
right at the next clock timing event). For case 2 The
counter has the following next state equation for Y1.

$$Y1 = \overline{y3} + \text{illegal state recovery term}$$

$$= \overline{y3} + y1 \cdot \overline{y2} \cdot y3$$

$$= \overline{y3} + y1 \cdot \overline{y2}$$

The next state equation for Y1 can also be obtained from the composite Karnaugh map for the next state for case 2 by reading the map for Y1 as follows. The composite Karnaugh map is obtained from the self-correcting state diagram.

Y1 Y2 Y3=

y1\y2 y3	00	01	11	10
0	100	000	001	101
1	110	110	011	111

therefore $Y1 = \overline{y3} + y1 \cdot \overline{y2}$

The circuit for the synchronous 3-bit twisted ring counter with illegal state recovery is shown as follows.

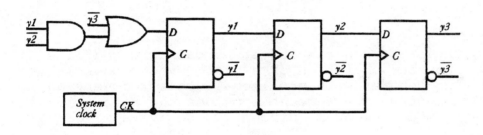

9-41. For 4 bits, 2n = 8 (number of primary states), leaving $2^n - 2n = 16 - 8 = 8$ (number of unused or illegal states); therefore, illegal state recovery circuity **is** required.

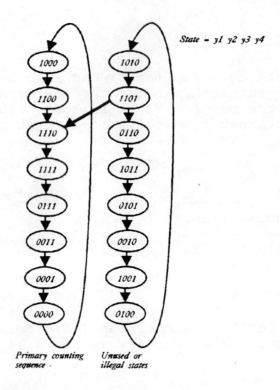

State = $y_1\ y_2\ y_3\ y_4$

1000	1010
1100	1101
1110	0110
1111	1011
0111	0101
0011	0010
0001	1001
0000	0100

Primary counting sequence   *Unused or illegal states*

In the state diagram above, normal directed line segments show the operation of a basic synchronous 4-bit twisted ring counter without illegal state recovery while the bold directed line segment provides the path followed by the illegal state recovery circuity. Notice how you can

decode 1101 and use the decoded output ORed with $\overline{y_4}$ to obtain Y1 to get back to the primary counting sequence which is similar to the discussion in the text for the synchronous 3-bit twisted ring counter (remember Y2 = y1, Y3 = y2, Y4 = y3, i.e., y1 y2 y3 shifts right at the next clock timing event). The counter has the following next state equation for Y1.

$$Y1 = \overline{y4} + \text{illegal state recovery term}$$

$$= \overline{y4} + y1 \cdot y2 \cdot \overline{y3} \cdot y4$$

$$= \overline{y4} + y1 \cdot y2 \cdot \overline{y3}$$

The next state equation for Y1 can also be obtained from the composite Karnaugh map for the next state by reading the map for Y1 as follows. The composite Karnaugh map is obtained from the self-correcting state diagram.

$$y1 \ y2 \backslash y3 \ y4$$

Y1 Y2 Y3 Y4=	00	01	11	10
00	1000	0000	0001	1001
01	1010	0010	0011	1011
11	1110	1110	0111	1111
10	1100	0100	0101	1101

Therefore $Y1 = \overline{y4} + y1 \cdot y2 \cdot \overline{y3}$

The circuit for the synchronous 4-bit twisted ring counter with illegal state recovery is shown as follows:

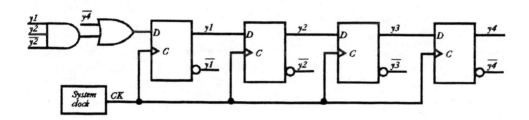

9-42. For 2 bits, $2^n - 1 = 3$ (number of primary states), leaving $2^n - (2^n - 1) = 4 - 3 = 1$ (number of unused or illegal states); therefore, illegal state recovery circuity **is** required.

The next state equations for a synchronous 2-bit maximum length shift counter with an Exclusive OR function are written as follows:

$$Y1 = y1 \oplus y2$$

$$Y2 = y1$$

From these equations we obtain the following composite Karnaugh map for the next state.

```
        y1 y2
          00  01  11  10
Y1 Y2=  ┌────┬────┬────┬────┐
        │ 00 │ 10 │ 01 │ 11 │
        └────┴────┴────┴────┘
```

Using the map we can draw the state diagram of the 2-bit
counter without illegal state recovery as follows.

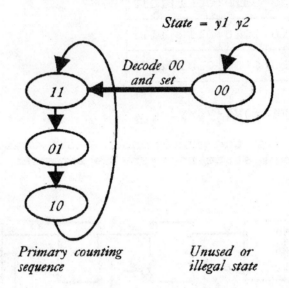

State = y1 y2

*Decode 00
and set*

*Primary counting
sequence*

*Unused or
illegal state*

     The path the counter follows without illegal state
recovery is shown using normal directed line segments.
The path provided by the illegal state recovery circuity is
shown in bold in the state diagram and is accomplished by
decoding state 00 and using the decoded output to either
synchronously or asynchronously set both flip-flops in the
shift register assuming of course each flip-flop has a set
input.

    The circuit diagram for the synchronous 2-bit maximum
length shift counter with an Exclusive OR function with
illegal state recovery circuitry using an asynchronous set
signal is shown below.

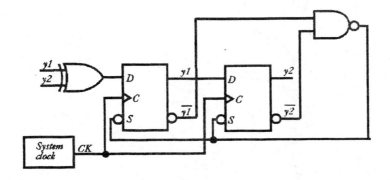

9-43. For 3 bits, $2^n - 1 = 7$ (number of primary states), leaving $2^n - (2^n - 1) = 8 - 7 = 1$ (number of unused or illegal states); therefore, illegal state recovery circuity **is** required.

The next state equations for a synchronous 3-bit maximum length shift counter with an Exclusive OR function are written as follows:

$$Y1 = y2 \oplus y3$$

$$Y2 = y1$$

$$Y3 = y2$$

From these equations we obtain the following composite Karnaugh map for the next state.

y1\y2 y3	00	01	11	10
0	000	100	001	101
1	010	110	011	111

Y1 Y2 Y3=

Using the map we can draw the state diagram of the 3-bit counter without illegal state recovery as follows.

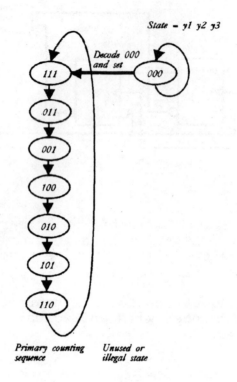

State – y1 y2 y3

Decode 000
and set

111

011

001

100

010

101

110

000

Primary counting
sequence

Unused or
illegal state

The path the counter follows without illegal state recovery is shown using normal directed line segments. The path provided by the illegal state recovery circuitry is shown in bold in the state diagram and is accomplished by decoding state 000 and using the decoded output to either synchronously or asynchronously set all flip-flops in the shift register assuming of course each flip-flop has a set input.

The circuit diagram for the synchronous 3-bit maximum length shift counter with an Exclusive OR function with illegal state recovery circuitry using an asynchronous set signal is shown below.

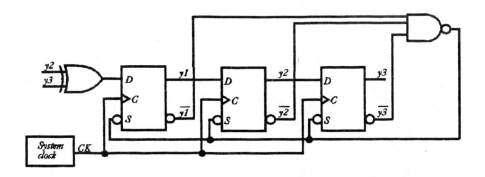

9-44. For 4 bits, $2^n - 1 = 15$ (number of primary states), leaving $2^n - (2^n - 1) = 16 - 15 = 1$ (number of unused or illegal states); therefore, illegal state recovery circuity **is** required.

The next state equations for a synchronous 4-bit maximum length shift counter with an Exclusive OR function are written as follows:

$$Y1 = y3 \oplus y4$$

$$Y2 = y1$$

$$Y3 = y2$$

$$Y4 = y3$$

From these equations we obtain the following composite Karnaugh map for the next state.

y1 y2 \ y3 y4	00	01	11	10
Y1 Y2 Y3 Y4 = 00	0000	1000	0001	1001
01	0010	1010	0011	1011
11	0110	1110	0111	1111
10	0100	1100	0101	1101

Using the map we can draw the state diagram of the 4-bit counter without illegal state recovery as follows.

329

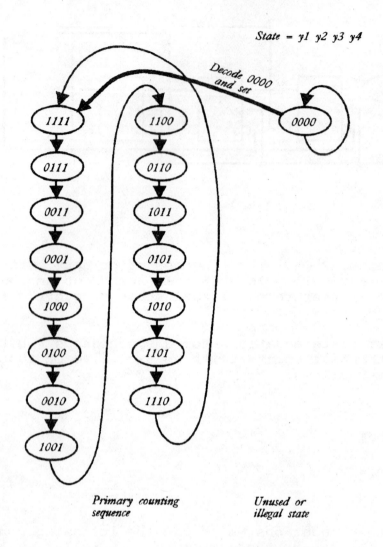

State = $y1$ $y2$ $y3$ $y4$

Decode 0000
and set

Primary counting
sequence

Unused or
illegal state

    The path the counter follows without illegal state
recovery is shown using normal directed line segments.
The path provided by the illegal state recovery circuity is
shown in bold in the state diagram and is accomplished by
decoding state 0000 and using the decoded output to either
synchronously or asynchronously set all flip-flops in the
shift register assuming of course each flip-flop has a set
input.

    The circuit diagram for the synchronous 4-bit maximum
length shift counter with an Exclusive OR function with
illegal state recovery circuitry using an asynchronous set
signal is shown below.

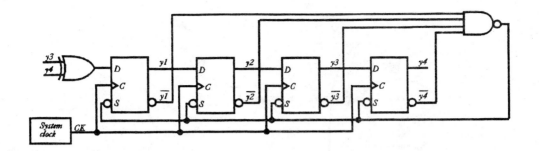

9-45. For 2 bits, $2^n - 1 = 3$ (number of primary states), leaving $2^n - (2^n - 1) = 4 - 3 = 1$ (number of unused or illegal states); therefore, illegal state recovery circuitry **is** required.

The next state equations for a synchronous 2-bit maximum length shift counter with an Exclusive NOR function are written as follows:

$$Y1 = \overline{y1 \oplus y2} = y1 \odot y2$$

$$Y2 = y1$$

From these equations we obtain the following composite Karnaugh map for the next state.

Y1 Y2= \ y1 y2	00	01	11	10
0	10	00	11	01

Using the map we can draw the state diagram of the 2-bit counter without illegal state recovery as follows.

331

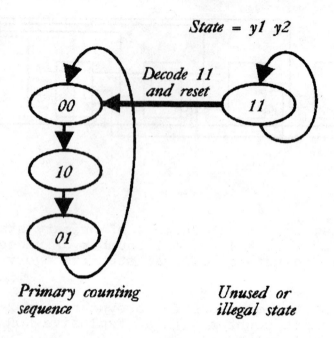

State = y1 y2

Decode 11
and reset

00

10

01

11

Primary counting
sequence

Unused or
illegal state

The path the counter follows without illegal state
recovery is shown using normal directed line segments.
The path provided by the illegal state recovery circuity is
shown in bold in the state diagram and is accomplished by
decoding state 11 and using the decoded output to either
synchronously or asynchronously reset both flip-flops in the
shift register assuming of course each flip-flop has a reset
input.

The circuit diagram for the synchronous 2-bit maximum
length shift counter with an Exclusive NOR function with
illegal state recovery circuitry using a synchronous reset
signal is shown below.

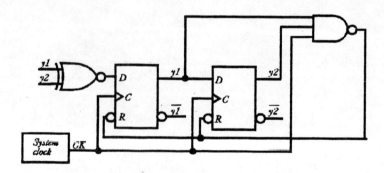

9-46. For 3 bits, $2^n - 1 = 7$ (number of primary states), leaving $2^n - (2^n - 1) = 8 - 7 = 1$ (number of unused or illegal states); therefore, illegal state recovery circuitry **is** required.

The next state equations for a synchronous 3-bit maximum length shift counter with an Exclusive NOR function are written as follows:

$$Y1 = \overline{y2 \oplus y3} = y2 \odot y3$$

$$Y2 = y1$$

$$Y3 = y2$$

From these equations we obtain the following composite Karnaugh map for the next state.

Y1 Y2 Y3=

y1 \ y2 y3	00	01	11	10
0	100	000	101	001
1	110	010	111	011

Using the map we can draw the state diagram of the 3-bit counter without illegal state recovery as follows.

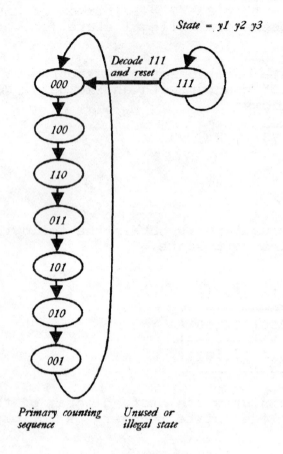

State = y1 y2 y3

Decode 111 and reset

Primary counting sequence

Unused or illegal state

    The path the counter follows without illegal state recovery is shown using normal directed line segments. The path provided by the illegal state recovery circuity is shown in bold in the state diagram and is accomplished by decoding state 111 and using the decoded output to either synchronously or asynchronously reset all flip-flops in the shift register assuming of course each flip-flop has a reset input.

    The circuit diagram for the synchronous 3-bit maximum length shift counter with an Exclusive NOR function with illegal state recovery circuitry using a synchronous reset signal is shown below.

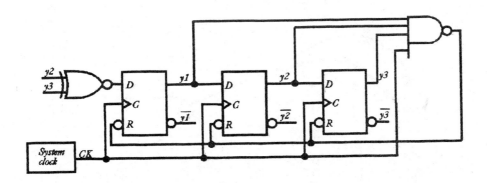

9-47. For 4 bits, $2^n - 1 = 15$ (number of primary states), leaving $2^n - (2^n - 1) = 16 - 15 = 1$ (number of unused or illegal states); therefore, illegal state recovery circuitry **is** required.

The next state equations for a synchronous 4-bit maximum length shift counter with an Exclusive NOR function are written as follows:

$$Y1 = \overline{y3 \oplus y4} = y3 \odot y4$$

$$Y2 = y1$$

$$Y3 = y2$$

$$Y4 = y3$$

From these equations we obtain the following composite Karnaugh map for the next state.

y1 y2\y3 y4

Y1 Y2 Y3 Y4=	00	01	11	10
00	1000	0000	1001	0001
01	1010	0010	1011	0011
11	1110	0110	1111	0111
10	1100	0100	1101	0101

Using the map we can draw the state diagram of the 4-bit counter without illegal state recovery as follows.

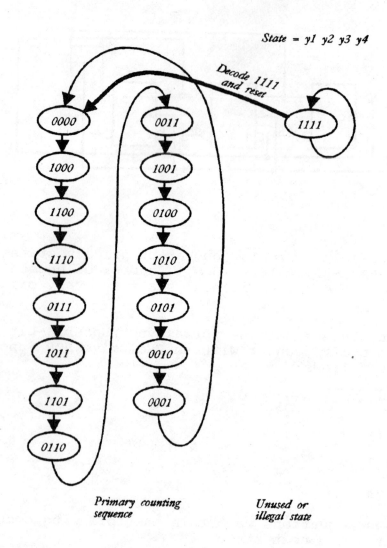

State = $y1$ $y2$ $y3$ $y4$

Decode 1111 and reset

Primary counting sequence

Unused or illegal state

    The path the counter follows without illegal state recovery is shown using normal directed line segments. The path provided by the illegal state recovery circuity is shown in bold in the state diagram and is accomplished by decoding state 1111 and using the decoded output to either synchronously or asynchronously reset all flip-flops in the shift register assuming of course each flip-flop has a reset input.

    The circuit diagram for the synchronous 4-bit maximum length shift counter with an Exclusive NOR function with illegal state recovery circuitry using a synchronous reset signal is shown below.

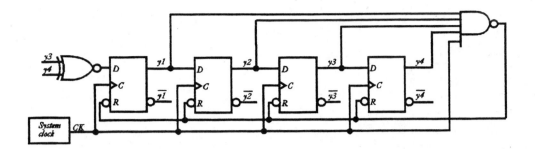

Section 9-5   Designing Synchronous State Machines Using Programmable Devices

9-48. (a) FA, FB, and FC in the following design file represent the output F for part (a), (b), and (c) respectively for problem 3-17.

```
"Design file
TITLE                   Combinational logic p317;
COMMENT                 Logic Equation entry format;

FUNCTION p317_le;

INPUT X, Y, Z;
OUTPUT FA, FB, FC;

 "Logic equations can be entered in practically any form as
 "illustrated below, but the results provided by PLDesigner
 "V1.2 in the documentation file is always the SOP expression
 "for the 1s of each function F.

FA = /X*/Y*Z + X*/Y*Z + X*Y;

FB = /X*(/Y*/Z + Y*/Z) + X*(/Y*Z + Y*Z);

/FC = /X + /X*Z + /X*Y + X*Y*Z;

END p317_le;
```

The following solution provided in the documentation file was generated using PLDesigner V1.2 with the equation reduction level set to Espresso via the CONFIGURATION tool. Espresso is a trademark of the University of California at Berkeley.

EQUATIONS FOR FUNCTION P317_LE

INPUT SIGNALS:
        HIGH_TRUE   X
        HIGH_TRUE   Y
        HIGH_TRUE   Z
OUTPUT SIGNALS:
        HIGH_TRUE   FA
        HIGH_TRUE   FB
        HIGH_TRUE   FC

REDUCED EQUATIONS:

FA.EQN   = Y*X + /Y*Z ;

FB.EQN   = /X*/Z + X*Z ;

FC.EQN   = X*/Z + X*/Y ;

        (b) FA, FB, and FC in the following design file
represent the output F for part (a), (b), and (c)
respectively for problem 3-18.

"Design file
TITLE   Combinational logic p318;
COMMENT   Logic Equation entry format;

FUNCTION p318_le;

INPUT X, Y, Z;
OUTPUT FA, FB, FC;

 "Logic equations can be entered in practically any form as
 "illustrated below, but the results provided by PLDesigner
 "V1.2 in the documentation file is always the SOP expression
 "for the 1s of each function F.

 "To use the software to obtain POS expressions we can
 "simply replace 1s with 0s and 0s with 1s in the function
 "by complementing the signal name. Each result provided
 "by PLDesigner V1.2 in the documentation file will now be
 "the SOP form of F for what it thinks are the 1s of the
 "function but what we know are now the 0s of the function.
 "To obtain the correct result we must now recomplement
 "each signal name in the documentation file to obtain
 "the actual SOP form of F for the 0s of the function.

 "Complementing both sides of the equation for the SOP form
 "of F for the 0s of the function and applying DeMorgan's
 "theorem results in the POS form of F for the 0s of the
 "function. This statement is the same as using the

```
"following expression represented by Equation 1-6 in the
"text
"                        _____
"POS form of F = SOP form of F̄

 "Complementing FA results in
/FA = /X*/Y*Z + X*/Y*Z + X*Y;

 "Complementing FB results in
/FB = /X*(/Y*/Z + Y*/Z) + X*(/Y*Z + Y*Z);
 "
                ___
 "Complementing FC results in
FC = /X + /X*Z + /X*Y + X*Y*Z;

END p318_le;
```

The following solution provided in the documentation
file was generated using PLDesigner V1.2 with the equation
reduction level set to Espresso via the CONFIGURATION tool.
Espresso is a trademark of the University of California at
Berkeley.

```
EQUATIONS FOR FUNCTION P318_LE

INPUT SIGNALS:
        HIGH_TRUE   X
        HIGH_TRUE   Y
        HIGH_TRUE   Z
OUTPUT SIGNALS:
        HIGH_TRUE   FA
        HIGH_TRUE   FB
        HIGH_TRUE   FC

REDUCED EQUATIONS:

FA.EQN          = /Y*/Z + /X*Y ;

FB.EQN          = X*/Z + /X*Z ;

FC.EQN          = Y*Z + /X ;
```

Using Equation 1-6 we obtain the following POS expressions.

```
            FA = (Y + Z)*(X + /Y)
            FB = (/X + Z)*(X + /Z)
            FC = (/Y + /Z)*X
```

(c) FA, FB, FC, FD, and FE in the following design
file represent the output F for part (a), (b), (c), (d),
and (e) respectively for problem 3-19.

```
"Design file
TITLE               Combinational logic p319;
COMMENT             Logic Equation entry format;
```

```
FUNCTION p319_le;

INPUT X, Y, Z;
OUTPUT FA, FB, FC, FD, FE;

 "Logic equations can be entered in practically any form as
 "illustrated below, but the results provided by PLDesigner
 "V1.2 in the documentation file is always the SOP expression
 "for the 1s of each function F.

FA = X*/Y*Z + X*/Y*/Z + X*Y*/Z;

FB = X*(Y*Z + Y*/Z) + X*(/Y*Z + Y*Z);

/FC = X + /X*Z + /X*Y + /X*Y*Z;

FD = (X + Y)*(X + /Y + Z);

/FE = (X + /Z)*(/Y + /Z)*(/X + /Y)*(X + /Y + Z) ;

END p319_le;
```

The following solution provided in the documentation
file was generated using PLDesigner V1.2 with the equation
reduction level set to Espresso via the CONFIGURATION tool.
Espresso is a trademark of the University of California at
Berkeley.

```
EQUATIONS FOR FUNCTION P319_LE

INPUT SIGNALS:
      HIGH_TRUE   X
      HIGH_TRUE   Y
      HIGH_TRUE   Z
OUTPUT SIGNALS:
      HIGH_TRUE   FA
      HIGH_TRUE   FB
      HIGH_TRUE   FC
      HIGH_TRUE   FD
      HIGH_TRUE   FE

REDUCED EQUATIONS:

FA.EQN            = X*/Z + X*/Y ;

FB.EQN            = X*Z + X*Y ;

FC.EQN            = /X*/Y*/Z ;

FD.EQN            = Y*Z + X ;

FE.EQN            = Y + /X*Z ;
```

9-49. (a) FA, FB, and FC in the following design file
represent the output F for part (a), (b), and (c)
respectively for problem 3-23.

```
"Design file
TITLE                   Combinational logic p323;
COMMENT                 Logic Equation entry format;

FUNCTION p323_le;

INPUT W, X, Y, Z;
OUTPUT FA, FB, FC;

  "Logic equations can be entered in practically any form as
  "illustrated below, but the results provided by PLDesigner
  "V1.2 in the documentation file is always the SOP expression
  "for the 1s of each function F.  PLDesigner V1.2 requires
  "that we write each Boolean equation in term of the
  "independent variables. To save typing we elected to
  "write each Boolean equation using the SOP expression
  "that requires the fewest number of product terms.

  "To use the software to obtain POS expressions we can
  "simply replace 1s with 0s and 0s with 1s in the function
  "by complementing the signal name.  Each result provided
  "by PLDesigner V1.2 in the documentation file will now be
  "the SOP form of F for what it thinks are the 1s of the
  "function but what we know are now the 0s of the function.
  "To obtain the correct result we must now recomplement
  "each signal name in the documentation file to obtain
  "the actual SOP form of F for the 0s of the function.

  "Writing the SOP expression for the 1s of FA and then
  "complementing FA results in
/FA = /W*/X*Y*/Z + /W*/X*Y*Z + /W*X*Y*/Z + /W*X*Y*Z

    + W*/X*/Y*/Z + W*/X*/Y*Z + W*X*/Y*/Z + W*X*/Y*Z;

  "Writing the SOP expression for the 0s of FB and then
  "
  "complementing FB results in
FB = /W*/X*/Y*Z + /W*/X*Y*/Z + W*/X*/Y*/Z + W*/X*/Y*Z

    + W*/X*Y*/Z;

  "Writing the SOP expression for the 0s of FC and then
  "
  "complementing FC results in
FC = /W*/X*/Y*/Z + /W*/X*Y*/Z + /W*X*/Y*Z + /W*X*Y*Z

    + W*/X*/Y*/Z + W*/X*Y*/Z;

END p323_le;
```

The following solution provided in the documentation file was generated using PLDesigner V1.2 with the equation reduction level set to Espresso via the CONFIGURATION tool. Espresso is a trademark of the University of California at Berkeley.

EQUATIONS FOR FUNCTION P323_LE

```
INPUT SIGNALS:
        HIGH_TRUE   W
        HIGH_TRUE   X
        HIGH_TRUE   Y
        HIGH_TRUE   Z
OUTPUT SIGNALS:
        HIGH_TRUE   FA
        HIGH_TRUE   FB
        HIGH_TRUE   FC
```

REDUCED EQUATIONS:

```
FA.EQN          = /W*/Y + W*Y ;

FB.EQN          = /X*/Y*W + /X*Y*/Z + /X*/Y*Z ;

FC.EQN          = X*Z*/W + /X*/Z ;
```

By recomplementing each signal name we obtain the SOP expressions for the 0s of each function.

```
        /FA = /W*/Y + W*Y

        /FB = /X*/Y*W + /X*Y*/Z + /X*/Y*Z

        /FC = X*Z*/W + /X*/Z
```

(b) FA, FB, and FC in the following design file represent the output F for part (a), (b), and (c) respectively for problem 3-26.

```
"Design file
TITLE               Combinational logic p326;
COMMENT             Logic Equation entry format;

FUNCTION p326_le;

INPUT W, X, Y, Z;
OUTPUT FA, FB, FC;

 "Logic equations can be entered in practically any form as
 "illustrated below, but the results provided by PLDesigner
 "V1.2 in the documentation file is always the SOP expression
 "for the 1s of each function F.
```

"To use the software to obtain POS expressions we can
"simply replace 1s with 0s and 0s with 1s in the function
"by complementing the signal name. Each result provided
"by PLDesigner V1.2 in the documentation file will now be
"the SOP form of F for what it thinks are the 1s of the
"function but what we know are now the 0s of the function.
"To obtain the correct result we must now recomplement
"each signal name in the documentation file to obtain
"the actual SOP form of F for the 0s of the function.

"Writing the SOP expression for FA and then
"complementing FA results in
/FA = /W*Y*/Z + X*Y*/Z + W*X + W*Z + X*/Y*/Z;

"Writing the SOP expression for FB and then
"complementing FB results in
/FB = Y*(/Z + W*/X*Z) + /Y*/Z + /W*X;

"
"Writing the SOP expression for $\overline{\overline{FC}}$ and then
"
"complementing $\overline{\overline{FC}}$ results in
FC = /W*Z + W*/X*/Z + X*Z + /W*/Z;

END p326_le;

        The following solution provided in the documentation
file was generated using PLDesigner V1.2 with the equation
reduction level set to Espresso via the CONFIGURATION tool.
Espresso is a trademark of the University of California at
Berkeley.

EQUATIONS FOR FUNCTION P326_LE

INPUT SIGNALS:
        HIGH_TRUE   W
        HIGH_TRUE   X
        HIGH_TRUE   Y
        HIGH_TRUE   Z
OUTPUT SIGNALS:
        HIGH_TRUE   FA
        HIGH_TRUE   FB
        HIGH_TRUE   FC

REDUCED EQUATIONS:

FA.EQN   = /W*/X*/Y + W*/X*/Z + /W*Z ;

FB.EQN   = /Y*Z*/X + /W*Z*/X + W*Z*X ;

FC.EQN   = /X*/Z + X*Z + /W ;

By recomplementing each signal name we obtain the SOP
expression for /F for each function as follows..

343

```
            /FA = /W*/X*/Y + W*/X*/Z + /W*Z ;
            /FB = /Y*Z*/X + /W*Z*/X + W*Z*X ;
            /FC = /X*/Z + X*Z + /W ;
```

      (c) FA, FB, and FC in the following design file
represent the output F for part (a), (b), and (c)
respectively for problem 3-27.

```
"Design file
TITLE               Combinational logic p327;
COMMENT             Logic Equation entry format;

FUNCTION p327_le;

INPUT V, W, X, Y, Z;
OUTPUT FA, FB, FC;

 "Logic equations can be entered in practically any form as
 "illustrated below, but the results provided by PLDesigner
 "V1.2 in the documentation file is always the SOP expression
 "for the 1s of each function F.  PLDesigner V1.2 requires
 "that we write each Boolean equation in term of the
 "independent variables. To save typing we elected to
 "write each Boolean equation using the SOP expression
 "that requires the fewest number of product terms.

 "Writing the SOP expression for the 0s of FA results in
/FA = /V*/W*/X*Y*/Z + /V*/W*/X*Y*Z + /V*/W*X*Y*Z + /V*W*/X*Y*/Z

    + /V*W*/X*Y*Z + V*/W*/X*Y*Z + V*/W*X*Y*/Z + V*/W*X*Y*Z

    + V*W*/X*Y*Z + V*W*X*Y*/Z;

 "Writing the SOP expression for the 1s of FB results in
FB = /V*/W*/X*Y*Z + /V*/W*X*/Y*Z + /V*W*/X*/Y*/Z + /V*W*/X*/Y*Z

    + /V*W*/X*Y*/Z + /V*W*/X*Y*Z + /V*W*X*/Y*Z + V*/W*/X*/Y*Z

    + V*/W*/X*Y*Z + V*W*/X*/Y*/Z + V*W*/X*/Y*Z + V*W*/X*Y*/Z

    + V*W*/X*Y*Z;

 "Writing the SOP expression for the 0s of FC results in
/FC = /V*/W*/X*Y*/Z + /V*/W*/X*Y*Z + /V*/W*X*Y*/Z + /V*/W*X*Y*Z

    + /V*W*X*/Y*/Z + /V*W*X*/Y*Z + /V*W*X*Y*/Z + /V*W*X*Y*Z

    + V*/W*/X*Y*/Z + V*/W*/X*Y*Z + V*/W*X*Y*/Z + V*/W*X*Y*Z

    + V*W*/X*/Y*Z + V*W*/X*Y*Z + V*W*X*/Y*/Z + V*W*X*/Y*Z;

END p327_le;
```

The following solution provided in the documentation file was generated using PLDesigner V1.2 with the equation reduction level set to Espresso via the CONFIGURATION tool. Espresso is a trademark of the University of California at Berkeley.

EQUATIONS FOR FUNCTION P327_LE

INPUT SIGNALS:
```
        HIGH_TRUE   V
        HIGH_TRUE   W
        HIGH_TRUE   X
        HIGH_TRUE   Y
        HIGH_TRUE   Z
```
OUTPUT SIGNALS:
```
        HIGH_TRUE   FA
        HIGH_TRUE   FB
        HIGH_TRUE   FC
```

REDUCED EQUATIONS:

FA.EQN   = V*/X*/Z + X*Z*W + /V*X*/Z + /Y ;

FB.EQN   = X*/Y*Z*/V + /X*Z*V + /X*Y*Z + W*/X ;

FC.EQN   = W*/X*/Z + V*W*X*Y + /V*W*/X + /W*/Y ;

(d) In the following design file for problem 3-42, FA and CFA represent the outputs for part (a) while FB and CFB represent the outputs for part (b).

```
"Design file
TITLE               Combinational logic p342;
COMMENT             Truth Table entry format;

FUNCTION p342_tt;

INPUT A, B, C, D, E;
OUTPUT FA, CFA, FB, CFB; "CFA and CFB represent
                        "FA and FB complemented
```

```
TRUTH_TABLE

A,B,C,D,E::FA,CFA,FB,CFB;

0,0,0,0,0::X,X,1,0;
0,0,0,0,1::1,0,1,0;
0,0,0,1,0::X,X,X,X;
0,0,0,1,1::1,0,1,0;
0,0,1,0,0::0,1,0,1;
0,0,1,0,1::1,0,0,1;
0,0,1,1,0::1,0,0,1;
0,0,1,1,1::0,1,0,1;
0,1,0,0,0::X,X,X,X;
0,1,0,0,1::1,0,1,0;
0,1,0,1,0::X,X,1,0;
0,1,0,1,1::0,1,1,0;
0,1,1,0,0::X,X,1,0;
0,1,1,0,1::1,0,1,0;
0,1,1,1,0::1,0,0,1;
0,1,1,1,1::X,X,0,1;
1,0,0,0,0::0,1,1,0;
1,0,0,0,1::0,1,1,0;
1,0,0,1,0::X,X,1,0;
1,0,0,1,1::1,0,1,0;
1,0,1,0,0::0,1,1,0;
1,0,1,0,1::0,1,1,0;
1,0,1,1,0::1,0,0,1;
1,0,1,1,1::0,1,0,1;
1,1,0,0,0::X,X,1,0;
1,1,0,0,1::0,1,1,0;
1,1,0,1,0::X,X,1,0;
1,1,0,1,1::0,1,1,0;
1,1,1,0,0::0,1,1,0;
1,1,1,0,1::0,1,1,0;
1,1,1,1,0::1,0,X,X;
1,1,1,1,1::0,1,X,X;

END;"Truth table
```

END p342_tt;

The following solution provided in the documentation
file was generated using PLDesigner V1.2 with the equation
reduction level set to Espresso via the CONFIGURATION tool.
Espresso is a trademark of the University of California at
Berkeley.

EQUATIONS FOR FUNCTION P342_TT

INPUT SIGNALS:
```
        HIGH_TRUE   A
        HIGH_TRUE   B
        HIGH_TRUE   C
        HIGH_TRUE   D
```

```
            HIGH_TRUE   E
OUTPUT SIGNALS:
            HIGH_TRUE   FA
            HIGH_TRUE   CFA
            HIGH_TRUE   FB
            HIGH_TRUE   CFB

REDUCED EQUATIONS:

FA.EQN          = /C*D*E*/B + /D*E*/A + C*D*/E ;
CFA.EQN         = A*E*B + /C*E*B*D + C*/E*/B*/D + C*E*/B*D +
                  A*C*/D + A*/B*/D ;
FB.EQN          = /C*E + B*C*/D + A*/D + B*/C*D + A*/C +
                  /B*/C*/D ;
CFB.EQN         = /A*C*D + /B*C*D + /A*/B*C ;
```

Since PLDesigner V1.2 treats don't cares as 0s these SOP expressions are not minimum expressions. Simply complement both sides of CFA and CFB to obtain the POS expressions for FA and FB as follows.

```
    FA = (/A + /E + /B)*(C + /E + /B + /D)*(/C + E + B + D)
       *(/C + /E + B + /D)*(/A + /C + D)*(/A + B + D)
    FB = (A + /C + /D)*(B + /C + /D)*(A + B + /C)
```

9-50. (a) F1, F2, and F3 in the following design file represent the output F for part (a), (b), and (c) respectively for problem 4-18.

```
"Design file
TITLE                   Combinational logic p418;
COMMENT                 Truth Table entry format;

FUNCTION p418_tt;

INPUT A, B, C;
OUTPUT F1, F2, F3;

    TRUTH_TABLE

    A,B,C::F1,F2,F3;
    0,0,0::1,1,1;
    0,0,1::0,1,0;
    0,1,0::1,0,1;
    0,1,1::1,1,0;
    1,0,0::1,1,1;
    1,0,1::0,0,0;
    1,1,0::1,0,0;
    1,1,1::0,0,X;
    END;"Truth table

END p418_tt;
```

The following solution provided in the documentation file was generated using PLDesigner V1.2 with the equation

347

reduction level set to Espresso via the CONFIGURATION tool.
Espresso is a trademark of the University of California at
Berkeley.

EQUATIONS FOR FUNCTION P418_TT

INPUT SIGNALS:
        HIGH_TRUE   A
        HIGH_TRUE   B
        HIGH_TRUE   C
OUTPUT SIGNALS:
        HIGH_TRUE   F1
        HIGH_TRUE   F2
        HIGH_TRUE   F3

REDUCED EQUATIONS:

F1.EQN   = /A*B + /C ;

F2.EQN   = /B*/C + /A*C ;

F3.EQN   = /C*/B + /A*/C ;

        (b) The following design file is for problem 4-20.

"Design file
TITLE   Combinational logic p420;
COMMENT   Truth Table entry format;

FUNCTION p420_tt;

INPUT I3, I2, I1, I0;
OUTPUT F3, F2, F1, F0;

    TRUTH_TABLE

    I3,I2,I1,I0::F3,F2,F1,F0;

    0,0,0,0::0,0,1,1;
    0,0,0,1::0,1,0,0;
    0,0,1,0::0,1,0,1;
    0,0,1,1::0,1,1,0;
    0,1,0,0::0,1,1,1;
    0,1,0,1::1,0,0,0;
    0,1,1,0::1,0,0,1;
    0,1,1,1::1,0,1,0;
    1,0,0,0::1,0,1,1;
    1,0,0,1::1,1,0,0;
    1,0,1,0::X,X,X,X;
    1,0,1,1::X,X,X,X;
    1,1,0,0::X,X,X,X;
    1,1,0,1::X,X,X,X;
    1,1,1,0::X,X,X,X;
    1,1,1,1::X,X,X,X;

```
        END;"Truth table

END p420_tt;
```

        The following solution provided in the documentation file was generated using PLDesigner V1.2 with the equation reduction level set to Espresso via the CONFIGURATION tool. Espresso is a trademark of the University of California at Berkeley.

```
EQUATIONS FOR FUNCTION P420_TT

INPUT SIGNALS:
        HIGH_TRUE    I3
        HIGH_TRUE    I2
        HIGH_TRUE    I1
        HIGH_TRUE    I0
OUTPUT SIGNALS:
        HIGH_TRUE    F3
        HIGH_TRUE    F2
        HIGH_TRUE    F1
        HIGH_TRUE    F0

REDUCED EQUATIONS:

F3.EQN          = I3*/I2*/I1 + /I3*I2*I1 + /I3*I2*I0 ;

F2.EQN          = /I3*I2*/I0*/I1 + /I2*I0*/I1 + /I3*/I2*I1 ;

F1.EQN          = /I3*I1*I0 + /I1*/I0*/I2 + /I3*/I1*/I0 ;

F0.EQN          = /I0*/I2*/I1 + /I3*/I0 ;
```

Since PLDesigner V1.2 treats don't cares as 0s these SOP expressions are not minimum expressions.

9-51. The following design file is for the full-comparator truth table in Example 5-7.

```
"Design file
TITLE                   Combinational logic ex57;
COMMENT                 Truth Table entry format;

FUNCTION ex57_tt;

INPUT ILES1, IEQU1, IGRE1, A1, B1;
OUTPUT OLES1, OEQU1, OGRE1;
```

```
TRUTH_TABLE

ILES1,IEQU1,IGRE1,A1,B1::OLES1,OEQU1,OGRE1;

0,0,0,0,0::X,X,X;
0,0,0,0,1::X,X,X;
0,0,0,1,0::X,X,X;
0,0,0,1,1::X,X,X;
0,0,1,0,0::0,0,1;
0,0,1,0,1::1,0,0;
0,0,1,1,0::0,0,1;
0,0,1,1,1::0,0,1;
0,1,0,0,0::0,1,0;
0,1,0,0,1::1,0,0;
0,1,0,1,0::0,0,1;
0,1,0,1,1::0,1,0;
0,1,1,0,0::X,X,X;
0,1,1,0,1::X,X,X;
0,1,1,1,0::X,X,X;
0,1,1,1,1::X,X,X;
1,0,0,0,0::1,0,0;
1,0,0,0,1::1,0,0;
1,0,0,1,0::0,0,1;
1,0,0,1,1::1,0,0;
1,0,1,0,0::X,X,X;
1,0,1,0,1::X,X,X;
1,0,1,1,0::X,X,X;
1,0,1,1,1::X,X,X;
1,1,0,0,0::X,X,X;
1,1,0,0,1::X,X,X;
1,1,0,1,0::X,X,X;
1,1,0,1,1::X,X,X;
1,1,1,0,0::X,X,X;
1,1,1,0,1::X,X,X;
1,1,1,1,0::X,X,X;
1,1,1,1,1::X,X,X;

        END;"Truth table
END ex57_tt;
```

The following solution provided in the documentation file was generated using PLDesigner V1.2 with the equation reduction level set to Espresso via the CONFIGURATION tool. Espresso is a trademark of the University of California at Berkeley.

EQUATIONS FOR FUNCTION EX57_TT

```
INPUT SIGNALS:
        HIGH_TRUE    ILES1
        HIGH_TRUE    IEQU1
        HIGH_TRUE    IGRE1
        HIGH_TRUE    A1
        HIGH_TRUE    B1
```

OUTPUT SIGNALS:
```
        HIGH_TRUE   OLES1
        HIGH_TRUE   OEQU1
        HIGH_TRUE   OGRE1
```

REDUCED EQUATIONS:

```
OLES1.EQN          = /ILES1*IEQU1*/IGRE1*/A1*B1 + /ILES1*/IEQU1*
                   IGRE1*/A1*B1 + ILES1*/IEQU1*/IGRE1*B1 +
                   ILES1*/IEQU1*/IGRE1*/A1 ;

OEQU1.EQN          = /ILES1*IEQU1*/IGRE1*/A1*/B1 +
                   /ILES1*IEQU1*/IGRE1*A1*B1 ;

OGRE1.EQN          = ILES1*/IEQU1*/IGRE1*/B1*A1 + /ILES1*IEQU1*
                   /IGRE1*/B1*A1 + /ILES1*/IEQU1*IGRE1*A1 +
                   /ILES1*/IEQU1*IGRE1*/B1 ;
```

Since PLDesigner V1.2 treats don't cares as 0s these
SOP expressions are not minimum expressions.

9-52. (a) The following design file is for the seven segment
outputs in the truth table in Fig. E5-17a
for a common cathode display.

```
"Design file
TITLE                 Combinational logic ccdis;
COMMENT               Truth Table entry format;

FUNCTION ccdis_tt;

INPUT D, C, B, A;
OUTPUT OA, OB, OC, OD, OE, OF, OG;

    TRUTH_TABLE

    D,C,B,A::OA,OB,OC,OD,OE,OF,OG;

    0,0,0,0::1,1,1,1,1,1,0;
    0,0,0,1::0,1,1,0,0,0,0;
    0,0,1,0::1,1,0,1,1,0,1;
    0,0,1,1::1,1,1,1,0,0,1;
    0,1,0,0::0,1,1,0,0,1,1;
    0,1,0,1::1,0,1,1,0,1,1;
    0,1,1,0::1,0,1,1,1,1,1;
    0,1,1,1::1,1,1,0,0,0,0;
    1,0,0,0::1,1,1,1,1,1,1;
    1,0,0,1::1,1,1,1,0,1,1;
    1,0,1,0::1,1,1,0,1,1,1;
    1,0,1,1::0,0,1,1,1,1,1;
    1,1,0,0::1,0,0,1,1,1,0;
    1,1,0,1::0,1,1,1,1,0,1;
    1,1,1,0::1,0,0,1,1,1,1;
    1,1,1,1::1,0,0,0,1,1,1;
```

```
        END;"Truth table

END ccdis_tt;

        The following solution provided in the documentation
file was generated using PLDesigner V1.2 with the equation
reduction level set to Espresso via the CONFIGURATION tool.
Espresso is a trademark of the University of California at
Berkeley.

EQUATIONS FOR FUNCTION CCDIS_TT

INPUT SIGNALS:
        HIGH_TRUE    D
        HIGH_TRUE    C
        HIGH_TRUE    B
        HIGH_TRUE    A
OUTPUT SIGNALS:
        HIGH_TRUE    OA
        HIGH_TRUE    OB
        HIGH_TRUE    OC
        HIGH_TRUE    OD
        HIGH_TRUE    OE
        HIGH_TRUE    OF
        HIGH_TRUE    OG

REDUCED EQUATIONS:

OA.EQN            = /D*C*A + D*/A + D*/C*/B + C*B + /D*B +
                   /C*/A ;

OB.EQN            = /D*/C + /D*B*A + /D*/B*/A + /C*/A + D*/B*A ;

OC.EQN            = /C*/B + /C*A + /B*A + D*/C + /D*C ;

OD.EQN            = /D*/C*/A + B*C*/A + B*/C*A + D*/B + /B*C*A ;

OE.EQN            = B*/A + D*B + /C*/A + D*C ;

OF.EQN            = D*B + C*/D*/B + /A*/B + /C*D + C*/A ;

OG.EQN            = B*/A + D*A + /D*C*/B + /C*B + D*/C ;

        (b) The following design file is for the seven segment
outputs in the truth table in Fig. E5-17a for a common anode
display.

"Design file
TITLE                Combinational logic cadis;
COMMENT              Truth Table entry format;

FUNCTION cadis_tt;
```

```
INPUT D, C, B, A;
low_true OUTPUT OA, OB, OC, OD, OE, OF, OG;

    TRUTH_TABLE

    D,C,B,A::OA,OB,OC,OD,OE,OF,OG;

    0,0,0,0::1,1,1,1,1,1,0;
    0,0,0,1::0,1,1,0,0,0,0;
    0,0,1,0::1,1,0,1,1,0,1;
    0,0,1,1::1,1,1,1,0,0,1;
    0,1,0,0::0,1,1,0,0,1,1;
    0,1,0,1::1,0,1,1,0,1,1;
    0,1,1,0::1,0,1,1,1,1,1;
    0,1,1,1::1,1,1,0,0,0,0;
    1,0,0,0::1,1,1,1,1,1,1;
    1,0,0,1::1,1,1,1,0,1,1;
    1,0,1,0::1,1,1,0,1,1,1;
    1,0,1,1::0,0,1,1,1,1,1;
    1,1,0,0::1,0,0,1,1,1,0;
    1,1,0,1::0,1,1,1,1,0,1;
    1,1,1,0::1,0,0,1,1,1,1;
    1,1,1,1::1,0,0,0,1,1,1;

    END;"Truth table

END cadis_tt;
```

The following solution provided in the documentation file was generated using PLDesigner V1.2 with the equation reduction level set to Espresso via the CONFIGURATION tool. Espresso is a trademark of the University of California at Berkeley.

EQUATIONS FOR FUNCTION CADIS_TT

```
INPUT SIGNALS:
        HIGH_TRUE    D
        HIGH_TRUE    C
        HIGH_TRUE    B
        HIGH_TRUE    A
OUTPUT SIGNALS:
        LOW_TRUE     OA
        LOW_TRUE     OB
        LOW_TRUE     OC
        LOW_TRUE     OD
        LOW_TRUE     OE
        LOW_TRUE     OF
        LOW_TRUE     OG
```

REDUCED EQUATIONS:

```
OA.EQN              = /D*C*A + D*/A + D*/C*/B + C*B + /D*B +
                    /C*/A ;
```

```
OB.EQN           = /D*/C + /D*B*A + /D*/B*/A + /C*/A + D*/B*A ;

OC.EQN           = /C*/B + /C*A + /B*A + D*/C + /D*C ;

OD.EQN           = /D*/C*/A + B*C*/A + B*/C*A + D*/B + /B*C*A ;

OE.EQN           = B*/A + D*B + /C*/A + D*C ;

OF.EQN           = D*B + C*/D*/B + /A*/B + /C*D + C*/A ;

OG.EQN           = B*/A + D*A + /D*C*/B + /C*B + D*/C ;
```

9-53. (a) The following design file is for the synchronous state machine represented by the ASM chart in Fig. E9-1a. The next state and output equations in the design file are written using the 'set or hold' method and the logic equation entry method respectively.

```
"Design file
TITLE              State Machine fe91a;
COMMENT            Logic Equation entry format;

FUNCTION fe91a_le;

INPUT CK, X1, X2, X3;
OUTPUT y CLOCKED_BY CK;
OUTPUT Z;

    Y = /y*X1*X2*/X3       "set
      + y*/x3;             "hold
    Z = y;

END fe91a_le;
```

The following solution provided in the documentation file was generated using PLDesigner V1.2 with the equation reduction level set to Espresso via the CONFIGURATION tool. Espresso is a trademark of the University of California at Berkeley.

```
EQUATIONS FOR FUNCTION FE91A_LE

INPUT SIGNALS:
        HIGH_TRUE   CK
        HIGH_TRUE   X1
        HIGH_TRUE   X2
        HIGH_TRUE   X3
OUTPUT SIGNALS:
        HIGH_TRUE   Y
        HIGH_TRUE   Z
```

REDUCED EQUATIONS:

```
Y.CLK            = CK ;
 .D              = X1*X2*/X3 + /X3*Y ;

Z.EQN.           = Y ;
```

   (b) The following design file is for the synchronous state machine represented by the ASM chart in Fig. E9-1a. The design file is written using the state machine entry method.

```
"Design file
TITLE            State Machine fe91a;
COMMENT          State Machine entry format;

FUNCTION fe91a_sm;

INPUT CK, X1, X2, X3;
OUTPUT y CLOCKED_BY CK;
OUTPUT Z;

    STATE_MACHINE fe91a;
    CLOCKED_BY CK;
    STATE_BITS[y];
        STATE a[0]: Z = 0; IF X1*X2*/X3 = 1 THEN GOTO b;
                           ELSE GOTO a;
        STATE b[1]: Z = 1; IF X3 = 0 THEN GOTO b;
                           ELSE GOTO a;
    END fe91a;"State machine

END fe91a_sm;
```

   The following solution provided in the documentation file was generated using PLDesigner V1.2 with the equation reduction level set to Espresso via the CONFIGURATION tool. Espresso is a trademark of the University of California at Berkeley.

EQUATIONS FOR FUNCTION FE91A_SM

```
INPUT SIGNALS:
        HIGH_TRUE   CK
        HIGH_TRUE   X1
        HIGH_TRUE   X2
        HIGH_TRUE   X3
OUTPUT SIGNALS:
        HIGH_TRUE   Y
        HIGH_TRUE   Z
```

REDUCED EQUATIONS:

```
Y.CLK            = CK ;
 .D              = /X3*Y + X1*X2*/X3 ;
```

```
Z.EQN                = Y ;
```

9-54. (a) The following design file is for the synchronous
state machine represented by the composite Karnaugh map in
Fig. E9-2b. The design file is written using the truth table
entry method.

```
"Design file
TITLE                State Machine fe92b;
COMMENT              Truth Table or State Table entry format;

FUNCTION fe92b_tt;

INPUT CK, X;
OUTPUT y1, y2 CLOCKED_BY CK;
OUTPUT Z;

     TRUTH_TABLE
     y1,y2,X::Y1,Y2,Z;

       0,0,0::0,0,0;
       0,0,1::0,1,0;
       0,1,0::1,1,0;
       0,1,1::0,1,0;
       1,1,0::0,0,0;
       1,1,1::0,1,1;
       1,0,0::0,0,0;
       1,0,1::0,1,0;

     END;"Truth table

END fe92b_tt;
```

The following solution provided in the documentation
file was generated using PLDesigner V1.2 with the equation
reduction level set to Espresso via the CONFIGURATION tool.
Espresso is a trademark of the University of California at
Berkeley.

```
EQUATIONS FOR FUNCTION FE92B_TT

INPUT SIGNALS:
        HIGH_TRUE    CK
        HIGH_TRUE    X
OUTPUT SIGNALS:
        HIGH_TRUE    Y1
        HIGH_TRUE    Y2
        HIGH_TRUE    Z
```

REDUCED EQUATIONS:

```
Y1.CLK              = CK ;
  .D                = /X*/Y1*Y2 ;

Y2.CLK              = CK ;
  .D                = /Y1*Y2 + X ;

Z.EQN               = X*Y1*Y2 ;
```

(b) The following design file is for the synchronous state machine represented by the composite Karnaugh map in Fig. E9-2b. The design file is written using the state machine entry method.

```
"Design file
TITLE               State Machine fe92b;
COMMENT             State Machine entry format;

FUNCTION fe92b_sm;

INPUT CK, X;
OUTPUT y1, y2 CLOCKED_BY CK;
OUTPUT Z;

    STATE_MACHINE fe92b;
    CLOCKED_BY CK;
    STATE_BITS[y1,y2];
        STATE a[0]: IF X = 0 THEN BEGIN Z = 0; GOTO a; END
                             ELSE BEGIN Z = 0; GOTO b; END;
        STATE b[1]: IF X = 0 THEN BEGIN Z = 0; GOTO c; END
                             ELSE BEGIN Z = 0; GOTO b; END;
        STATE c[3]: IF X = 0 THEN BEGIN Z = 0; GOTO a; END
                             ELSE BEGIN Z = 1; GOTO b; END;
        STATE d[2]: IF X = 0 THEN BEGIN Z = 0; GOTO a; END
                             ELSE BEGIN Z = 0; GOTO b; END;
    END fe92b;"State machine

END fe92b_sm;
```

The following solution provided in the documentation file was generated using PLDesigner V1.2 with the equation reduction level set to Espresso via the CONFIGURATION tool. Espresso is a trademark of the University of California at Berkeley.

EQUATIONS FOR FUNCTION FE92B_SM

```
INPUT SIGNALS:
        HIGH_TRUE   CK
        HIGH_TRUE   X
OUTPUT SIGNALS:
        HIGH_TRUE   Y1
        HIGH_TRUE   Y2
```

```
            HIGH_TRUE   Z

REDUCED EQUATIONS:

Y1.CLK              = CK ;
   .D               = /X*/Y1*Y2 ;

Y2.CLK              = CK ;
   .D               = X + /Y1*Y2 ;

Z.EQN               = X*Y1*Y2 ;
```

9-55. (a) The following design file is for the synchronous state machine represented by the composite Karnaugh map in Fig. E9-3b. The design file is written using the truth table entry method.

```
"Design file
TITLE               State Machine fe93b;
COMMENT             Truth Table or State Table entry format;

FUNCTION fe93b_tt;

INPUT CK, X;
OUTPUT y1, y2 CLOCKED_BY CK;
OUTPUT Z;

     TRUTH_TABLE
     y1,y2,X::Y1,Y2,Z;

     0,0,0::0,0,0;
     0,0,1::0,1,0;
     0,1,0::0,0,0;
     0,1,1::1,1,0;
     1,1,0::1,0,0;
     1,1,1::1,1,0;
     1,0,0::0,0,1;
     1,0,1::0,1,1;

     END;"Truth table

END fe93b_tt;
```

The following solution provided in the documentation file was generated using PLDesigner V1.2 with the equation reduction level set to Espresso via the CONFIGURATION tool. Espresso is a trademark of the University of California at Berkeley.

EQUATIONS FOR FUNCTION FE93B_TT

INPUT SIGNALS:

```
        HIGH_TRUE   CK
```

```
              HIGH_TRUE   X
OUTPUT SIGNALS:
              HIGH_TRUE   Y1
              HIGH_TRUE   Y2
              HIGH_TRUE   Z

REDUCED EQUATIONS:

Y1.CLK              = CK ;
  .D                = Y2*Y1 + X*Y2 ;

Y2.CLK              = CK ;
  .D                = X ;

Z.EQN               = Y1*/Y2 ;
```

(b) The following design file is for the synchronous state machine represented by the composite Karnaugh map in Fig. E9-3b. The design file is written using the state machine entry method.

```
"Design file
TITLE               State Machine fe93b;
COMMENT             State Machine entry format;

FUNCTION fe93b_sm;

INPUT CK, X;
OUTPUT y1, y2 CLOCKED_BY CK;
OUTPUT Z;

    STATE_MACHINE fe93b;
    CLOCKED_BY CK;
    STATE_BITS[y1,y2];
        STATE a[0]: Z = 0; IF X = 0 THEN GOTO a;
                                   ELSE GOTO b;
        STATE b[1]: Z = 0; IF X = 0 THEN GOTO a;
                                   ELSE GOTO c;
        STATE c[3]: Z = 0; IF X = 0 THEN GOTO d;
                                   ELSE GOTO c;
        STATE d[2]: Z = 1; IF X = 0 THEN GOTO a;
                                   ELSE GOTO b;
    END fe93b;"State machine

END fe93b_sm;
```

The following solution provided in the documentation file was generated using PLDesigner V1.2 with the equation reduction level set to Espresso via the CONFIGURATION tool. Espresso is a trademark of the University of California at Berkeley.

EQUATIONS FOR FUNCTION FE93B_SM

INPUT SIGNALS:
        HIGH_TRUE   CK
        HIGH_TRUE   X
OUTPUT SIGNALS:
        HIGH_TRUE   Y1
        HIGH_TRUE   Y2
        HIGH_TRUE   Z

REDUCED EQUATIONS:

```
Y1.CLK          = CK ;
  .D            = Y2*Y1 + X*Y2 ;

Y2.CLK          = CK ;
  .D            = X ;

Z.EQN           = Y1*/Y2 ;
```

9-56. The following design file is for the synchronous state machine represented by the state diagram shown in Fig. 9-15. The design file is written using the state machine entry method.

```
"Design file
TITLE               State Machine f915;
COMMENT             State Machine entry format;

FUNCTION f915_sm;

INPUT CK, X;
OUTPUT y1, y2, y3, y4 CLOCKED_BY CK;
OUTPUT Z1, Z2, Z3, Z4;

    STATE_MACHINE SM;
    CLOCKED_BY CK;
    STATE_BITS[y1..y4];
        STATE a[0]: [Z1..Z4] = 0; IF X = 1 THEN GOTO b;
                                  ELSE GOTO j;
        STATE b[1]: [Z1..Z4] = 1; IF X = 1 THEN GOTO c;
                                  ELSE GOTO a;
        STATE c[2]: [Z1..Z4] = 2; IF X = 1 THEN GOTO d;
                                  ELSE GOTO b;
        STATE d[3]: [Z1..Z4] = 3; IF X = 1 THEN GOTO e;
                                  ELSE GOTO c;
        STATE e[4]: [Z1..Z4] = 4; IF X = 1 THEN GOTO f;
                                  ELSE GOTO d;
        STATE f[5]: [Z1..Z4] = 5; IF X = 1 THEN GOTO g;
                                  ELSE GOTO e;
        STATE g[6]: [Z1..Z4] = 6; IF X = 1 THEN GOTO h;
                                  ELSE GOTO f;
        STATE h[7]: [Z1..Z4] = 7; IF X = 1 THEN GOTO i;
                                  ELSE GOTO g;
```

```
         STATE i[8]:  [Z1..Z4] = 8;  IF X = 1 THEN GOTO j;
                                          ELSE GOTO h;
         STATE j[9]:  [Z1..Z4] = 9;  IF X = 1 THEN GOTO a;
                                          ELSE GOTO i;
         STATE k[10]: [Z1..Z4] = 10; IF X = 0 THEN GOTO a;
         STATE l[11]: [Z1..Z4] = 11; IF X = 0 THEN GOTO a;
         STATE m[12]: [Z1..Z4] = 12; IF X = 0 THEN GOTO a;
         STATE n[13]: [Z1..Z4] = 13; IF X = 0 THEN GOTO a;
         STATE o[14]: [Z1..Z4] = 14; IF X = 0 THEN GOTO a;
         STATE p[15]: [Z1..Z4] = 15; IF X = 0 THEN GOTO a;

      END SM;"State machine

END f915_sm;
```

The following solution provided in the documentation file was generated using PLDesigner V1.2 with the equation reduction level set to Espresso via the CONFIGURATION tool. Espresso is a trademark of the University of California at Berkeley.

```
EQUATIONS FOR FUNCTION F915_SM

INPUT SIGNALS:
        HIGH_TRUE   CK
        HIGH_TRUE   X
OUTPUT SIGNALS:
        HIGH_TRUE   Y1
        HIGH_TRUE   Y2
        HIGH_TRUE   Y3
        HIGH_TRUE   Y4
        HIGH_TRUE   Z1
        HIGH_TRUE   Z2
        HIGH_TRUE   Z3
        HIGH_TRUE   Z4

REDUCED EQUATIONS:

Y1.CLK           = CK ;
  .D             = /X*/Y1*/Y4*/Y3*/Y2 + /X*Y1*Y4*/Y3*/Y2 +
        X*Y4*Y3*Y2 + X*Y1*Y2 + X*Y1*Y3 + X*Y1*/Y4 ;

Y2.CLK           = CK ;
  .D             = Y2*Y3*/Y4*/Y1 + /X*Y2*Y4*/Y1 + X*Y2*/Y3 +
        /X*/Y2*/Y3*/Y4*Y1 + X*/Y2*Y3*Y4*/Y1 + X*Y2*Y1 ;

Y3.CLK           = CK ;
  .D             = /X*/Y3*/Y4*Y1*/Y2 + /X*/Y3*/Y4*/Y1*Y2 +
        X*/Y3*Y4*/Y1 + /X*Y3*Y4*/Y1 + X*Y3*Y1 + X*Y3*/Y4 ;

Y4.CLK           = CK ;
  .D             = X*Y1*Y4*Y2 + X*Y1*Y3*Y4 + /Y3*/Y4*/Y2 +
        /Y1*/Y4 ;
```

361

```
Z1.EQN              = Y1 ;
Z2.EQN              = Y2 ;
Z3.EQN              = Y3 ;
Z4.EQN              = Y4 ;
```

9-57. The following design file is for the synchronous state machine represented by the state diagram in Fig. 9-18 after modifying the diagram so that unused or illegal state h goes to state b to provide illegal state recovery. The design file is written using the state machine entry method.

```
"Design file
TITLE               State Machine f918 with illegal state
recovery;
COMMENT             State Machine entry format;

FUNCTION f918_sm;

INPUT CK;
OUTPUT y1, y2, y3 CLOCKED_BY CK;
OUTPUT Z;

    STATE_MACHINE f918;
    CLOCKED_BY CK;
    STATE_BITS[y1,y2,y3];
        STATE a[4]: Z = 0; GOTO b;
        STATE b[6]: Z = 0; GOTO c;
        STATE c[7]: Z = 0; GOTO d;
        STATE d[3]: Z = 0; GOTO e;
        STATE e[1]: Z = 1; GOTO f;
        STATE f[0]: Z = 0; GOTO a;
        STATE g[2]: Z = 0; GOTO h;
        STATE h[5]: Z = 0; GOTO b;

    END f918;"State machine

END f918_sm;
```

The following solution provided in the documentation file was generated using PLDesigner V1.2 with the equation reduction level set to Espresso via the CONFIGURATION tool. Espresso is a trademark of the University of California at Berkeley.

```
EQUATIONS FOR FUNCTION F918_SM

INPUT SIGNALS:
      HIGH_TRUE   CK
OUTPUT SIGNALS:
      HIGH_TRUE   Y1
      HIGH_TRUE   Y2
      HIGH_TRUE   Y3
      HIGH_TRUE   Z
```

REDUCED EQUATIONS:

```
Y1.CLK          = CK ;
   .D           = Y3*Y1*/Y2 + /Y3 ;

Y2.CLK          = CK ;
   .D           = Y1 ;

Y3.CLK          = CK ;
   .D           = Y2 ;

Z.EQN           = /Y1*/Y2*Y3 ;
```

You may notice that we entered 0s for the don't cares for Z. This is the same as not entering anything for Z for states g and h. PLDesigner V1.2 treats the absence of a variable as a don't care and then sets each don't care to 0 thus providing the same result. If the don't care for Z in state h (101) is teated as a 1 instead of a 0 then Z.EQN = /y2*y3. Also, notice that the Espresso equation reduction level did not obtain a minimum equation for Y1.D. The minimum equation for Y1.D can be easily obtained as Y1.D = y1*/y2 + /y3 using the Simplification Theorem T9a. If we were going to put these equation into a PAL the fact that the equations were not reduced to a minimum would make little difference when using most modern PALs such as the 16RP8A series illustrated in the text in Appendix C.

9-58. The following design file is for the synchronous state machine represented by the timing diagram in Fig. E9-4e. The design file is written using the waveform entry method with the following state assignments y1 y2 = 00 (state d), 01 (state a), 10 (state c), and 11 (state b).

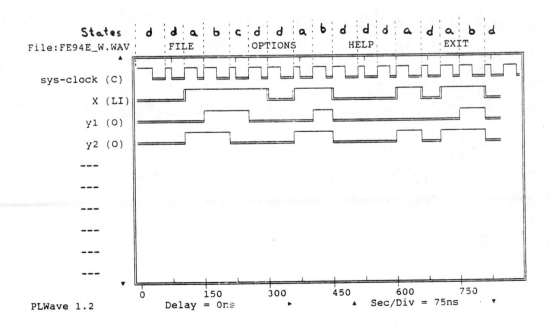

The following design file is for the synchronous state machine represented by the timing diagram in Fig. E9-4e. The design file is written just for output Z and uses the logic equation entry method.

```
"Design file
TITLE               Combinational logic fe94e;
COMMENT             Logic Equation entry format;

FUNCTION fe94e_le;

INPUT X, y1, y2;
OUTPUT Z;

Z = /y1*y2*X + y1*y2*X;

END fe94e_le;
```

The design file for output Z must be compiled prior to compiling the waveform file. When the file fe94e_le is compiled a file called fe94e_le.fb is generated. To incorporate the output file as part of the waveform design file, select OPTIONS in the waveform file fe94e_w and add it to the list of FBINCLUDE files. The waveform file is then compiled and automatically selects and incorporates all FBINCLUDE files in its list into the final waveform documentation file fe94e_w.DOC generated at compilation time.

The following solution provided in the documentation file was generated using PLDesigner V1.2 with the equation reduction level set to Espresso via the CONFIGURATION tool. Espresso is a trademark of the University of California at Berkeley.

```
EQUATIONS FOR FUNCTION FE94E_W

INPUT SIGNALS:
     HIGH_TRUE   X
OUTPUT SIGNALS:
     HIGH_TRUE   y1
     HIGH_TRUE   y2

REDUCED EQUATIONS:

y1.CLK          = sys_clock ;
  .D            = X*y2 ;

y2.CLK          = sys_clock ;
  .D            = X*/y1 ;
```

EQUATIONS FOR FUNCTION FE94E_LE

INPUT SIGNALS:
        HIGH_TRUE   X
        HIGH_TRUE   Y1
        HIGH_TRUE   Y2
OUTPUT SIGNALS:
        HIGH_TRUE   Z

REDUCED EQUATIONS:

Z.EQN   = X*Y2 ;

9-59. (a) The following design file is for the synchronous state machine represented by the timing diagram in Fig. P9-11. The design file is written using the waveform entry method with the following state assignments y1 y2 = 00 (state a), 01 (state b), and 10 (state c).

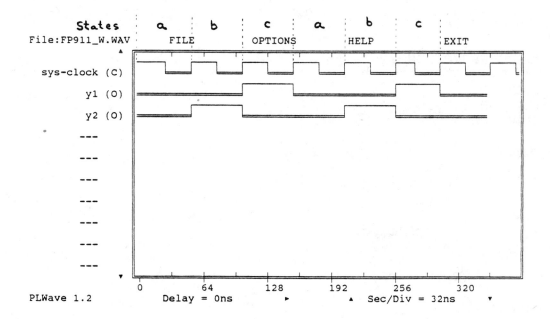

The following design file is for the synchronous state machine represented by the timing diagram in Fig. P9-11. The design file is written just for output Z and uses the logic equation entry method.

```
"Design file
TITLE                Combinational logic fp911;
COMMENT              Logic Equation entry format;

FUNCTION fp911_le;

INPUT y1, y2;
```

OUTPUT Z;

Z = /y1*/y2;

END fp911_le;

The design file for output Z must be compiled prior to compiling the waveform file. When the file fp911_le is compiled a file called fp911_le.fb is generated. To incorporate the output file as part of the waveform design file, select OPTIONS in the waveform file fp911_W and add it to the list of FBINCLUDE files. The waveform file is then compiled and automatically selects and incorporates all FBINCLUDE files in its list into the final waveform documentation file fp911_w.DOC generated at compilation time.

The following solution provided in the documentation file was generated using PLDesigner V1.2 with the equation reduction level set to Espresso via the CONFIGURATION tool. Espresso is a trademark of the University of California at Berkeley.

EQUATIONS FOR FUNCTION FP911_W

OUTPUT SIGNALS:
         HIGH_TRUE   y1
         HIGH_TRUE   y2

REDUCED EQUATIONS:

y1.CLK   = sys_clock ;
  .D   = y2 ;

y2.CLK   = sys_clock ;
  .D   = /y1*/y2 ;

EQUATIONS FOR FUNCTION FP911_LE

INPUT SIGNALS:
         HIGH_TRUE   Y1
         HIGH_TRUE   Y2
OUTPUT SIGNALS:
         HIGH_TRUE   Z

REDUCED EQUATIONS:

Z.EQN   = /Y1*/Y2 ;

(b) To check for illegal state recovery all we have to do is write the next state equations and substitute into these equations the unused or illegal state condition 11 (state d) to see if the next state remains in the illegal state (then there is no illegal state recovery) or the next

366

state returns to the primary counting sequence which is one of the other three states (then there is illegal state recovery).

$$Y1.D = y2$$

$$Y2.D = /y1*/y2$$

for state d, y1 y2 = 11

$$Y1.D = y2 = 1$$

$$Y2.D = /y1*/y2 = /1*/1 = 0$$

so at the next clock timing event
the circuit goes to state
y1 y2 = 10 or state c

therefore, the circuit can recover if it
finds itself in state d.

Section 10-2   Asynchronous Design Fundamentals

10-1. The two advantages of asynchronous sequential circuits compared to synchronous sequential circuits are asynchronous sequential circuits have 1. no clock requirement, and 2. operate faster.

10-2. (a) For synchronous sequential circuits only the clock signal changing from 0 to 1 (or 1 to 0) causes a state transition.

(b) For asynchronous (Fundamental mode) sequential circuits an input signal changing from 0 to 1 or from 1 to 0 causes a state transition.

(c) For asynchronous (Pulse mode) sequential circuits an input signal positive pulse (or negative pulse) causes a state transition.

10-3.   The input signal restrictions for a fundamental mode asynchronous circuit can be stated as follows.
   Rule 1: Only one input signal is allowed to change at one time
   Rule 2: Before the next input signal is allowed to change, the circuit must be given time to reach a new stable state.

10-4. No, the inputs X1 X2 are not allowed to change from 01 to 10 simultaneously. This violates Rule 1 for the input signal restrictions for a fundamental mode asynchronous circuit which says **Only one input signal is allowed to change at one time**. If this rule is violated the circuit is not being operated in fundamental mode and will probably not function as desired.

10-5. (a) The speed of an asynchronous circuit is determined by the longest delay path from the input to the output of the circuit.

(b) the speed of a synchronous circuit is determined by the period of the system clock, where T = 1/f and f = system clock frequency.

10-6. The two type of problem that must be considered when designing asynchrnonous sequential circuit that do not have to be considered when designing synchronous sequential circuit are 1. hazards or spurious signals, and 2. critical races.

10-7.

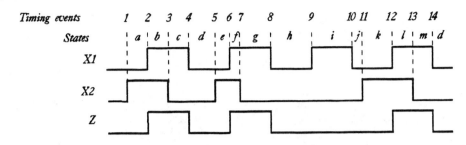

10-8.

State	X1	X2	Z	
a	0	1	0	
b	1	1	1	
c	1	0	1	←
d	0	0	0	
e	0	1	0	
f	1	1	1	
g	1	0	1	←
h	0	0	0	
i	1	0	0	← i ≠ c, g, or m
j	0	0	0	
k	0	1	0	
l	1	1	1	
m	1	0	1	←
d	0	0	0	

From the timing table we see states c, g, and m occur
when X1 X2 = 10 with an output Z = 1; however, state i
occurs when X1 X2 = 10 with an output Z = 0. The output Z
is not just dependent on the inputs X1 and X2, but also on
the previous state of the circuit. The circuit required to
generate the Z output must therefore be some form of
sequential circuit with feedback and memory capability.

10-9.

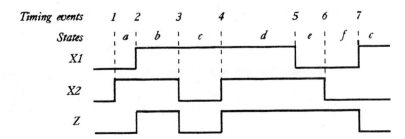

10-10.

State	X1	X2	Z
a	0	1	0
b	1	1	1
c	1	0	0
d	1	1	1
e	0	1	1
f	0	0	1
c	1	0	0

← a ≠ e

← (for e)

From the timing table we see that state a occurs when X1 X2 = 01 with an output Z = 0; however, state e occurs when X1 X2 = 01 with an output Z = 1. The output Z is not just dependent on the inputs X1 and X2, but also on the previous state of the circuit. The circuit required to generate the Z output must therefore be some form of sequential circuit with feedback and memory capability.

10-11. First we must identify the timing events and the states in the timing diagram in Fig. P10-7 as shown below.

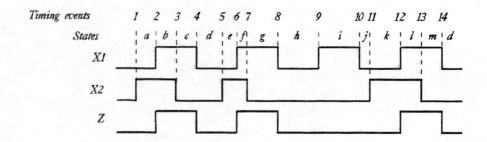

(a) Transferring the information from the timing diagram shown above we obtain the following primitive flow map.

PS \ X1 X2	00	01	11	10	PS Z
NS=					
a	–	<u>a</u>	b	–	0
b	–	–	<u>b</u>	c	1
c	d	–	–	<u>c</u>	1
d	<u>d</u>	e	–	–	0
e	–	<u>e</u>	f	–	0
f	–	–	<u>f</u>	g	1
g	h	–	–	<u>g</u>	1
h	<u>h</u>	–	–	i	0
i	j	–	–	<u>i</u>	0
j	<u>j</u>	k	–	–	0
k	–	<u>k</u>	l	–	0
l	–	–	<u>l</u>	m	1
m	d	–	–	<u>m</u>	1

371

(b) The following implication chart (or table) is filled in using the primitive flow map in (a)

	a	b	c	d	e	f	g	h	i	j	k	l
b	X											
c	X	✓										
d	✓	X	X									
e	b=f ✓	X	X	✓								
f	X	c=g ✓	c=g ✓	X	X							
g	X	c=g ✓	d=h ✓	X	X	✓						
h	✓	X	X	✓	✓	X	X					
i	✓	X	X	d=j ✓	✓	X	X	h=j ✓				
j	a=k ✓	X	X	e=k ✓	e=k ✓	X	X	✓	✓			
k	b=l ✓	X	X	e=k ✓	f=l ✓	X	X	✓	✓	✓		
l	X	c=m ✓	c=m ✓	X	X	g=m ✓	g=m ✓	X	X	X	X	
m	X	c=m ✓	✓	X	X	g=m ✓	d=h ✓	X	X	X	X	✓

The following list represents pairs of equivalent or compatible states which can be merged.

a,d a,e b,f a,h a,i a,j a,k b,l b,c c,g b,g c,m

b,m c,f d,h c,l d,e d,i d,j e,k d,k e,h e,i e,j

f,l f,g g,m f,m g,l h,i h,j h,k i,j i,k j,k l,m

(c)

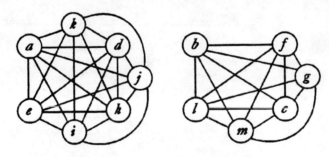

The merger diagram has two groups of strongly connected states. The two state are obtained by merging the following groups of states.

a,d,e,h,i,j,k b,c,f,g,l,m

(d) Merging states in the primitive flow map into the first and second rows of a reduced flow map results in

PS\X1 X2 NS=	00	01	11	10	PS Z
a,d,e,h,i,j,k	d̲	a̲	b	i̲	0
b,c,f,g,l,m	d	-	b̲	c̲	1

Changing to a single state name for each row results in

PS\X1 X2 NS=	00	01	11	10	PS Z
n (a,d,e,h,i,j,k)	n̲	n̲	o	n̲	0
o (b,c,f,g,l,m)	n	-	o̲	o̲	1

(e)

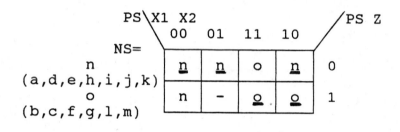

(f)

y\X1 X2 Y=	00	01	11	10	PS Z
n 0	0̲	0̲	1	0̲	0
o 1	0	-	1̲	1̲	1

There cannot be a critical race due to the state assignment since it contains only one state variable. Also since there is only one state variable, there are no illegal state to recover from.

(g) Only one next state equation is required which is given below and it is logic hazard-free without adding any extra product terms. One Moore output equation is also required.

373

$$Y = X1 \cdot X2 + y \cdot X1$$

$$Z = y$$

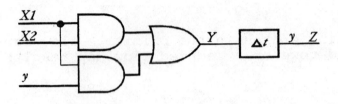

(h) Since there is only one next state equation only one S-R NOR latch is required.

y	Y	S	R
0	0	0	X
0	1	1	0
1	0	0	1
1	1	X	0

X = don't care

y \ X1 X2	00	01	11	10
S R= 0	0X	0X	10	0X
1	01	XX	X0	X0

$$S = X1 \cdot X2$$

$$R = \overline{X1}$$

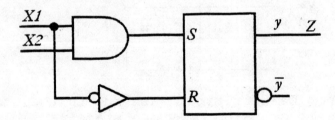

(i) $\sim S = \overline{X1 \cdot X2}$

         $\sim R = \overline{\overline{X1}} = X1$

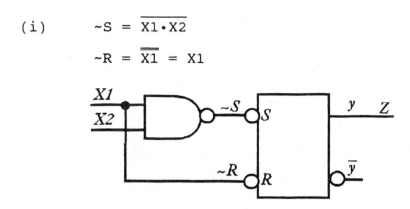

10-12. The following transition diagram shows state transitions for all possible asynchronous machines with four stable states.

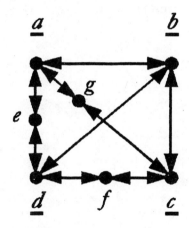

A universal race-free state assignment for any asynchronous fundamental mode machine with four stable states is shown in the following state assignment code map.

```
         y1\y2 y3
              00   01   11   10
Assignment=
           0 | a  | b  | c  | g |
           1 | e  | d  | f  |    |
```

10-13. First we must identify the timing events and the states in the timing diagram in Fig. P10-9 as shown below.

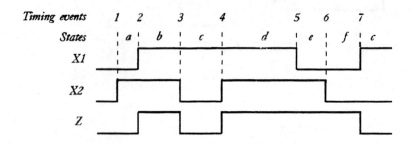

(a) Transferring the information from the timing diagram shown above we obtain the following primitive flow map.

```
PS\X1 X2                           PS Z
       00    01    11    10
NS=
   a |  -  |  a  |  b  |  -  |      0
   b |  -  |  -  |  b  |  c  |      1
   c |  -  |  -  |  d  |  c  |      0
   d |  -  |  e  |  d  |  -  |      1
   e |  f  |  e  |  -  |  -  |      1
   f |  f  |  -  |  -  |  -  |      1
```

(b) The following implication chart (or table) is filled in using the primitive flow map in (a)

	a	b	c	d	e
b	X				
c	b=d ✓	X			
d	X	✓	X		
e	X	✓	X	✓	
f	X	✓	X	✓	✓

The following list represents pairs of equivalent or compatible states which can be merged.

a,c b,d b,e b,f d,e d,f e,f

(c)

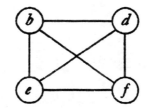

The merger diagram has two groups of strongly connected states. The two state are obtained by merging the following groups of states.

a,c b,d,e,f

(d) Merging states in the primitive flow map into the first and second rows of a reduced flow map results in

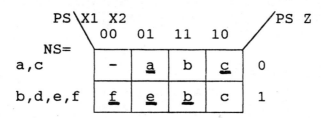

Changing to a single state name for each row results in

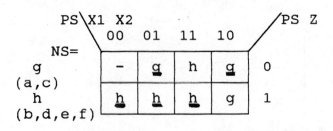

NS=	PS\X1 X2				PS Z
	00	01	11	10	
g (a,c)	–	g	h	g	0
h (b,d,e,f)	h	h	h	g	1

(e)

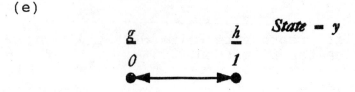

(f)

Y=	y\X1 X2				PS Z
	00	01	11	10	
g 0	–	0	1	0	0
h 1	1	1	1	0	1

There cannot be a critical race due to the state assignment since it contains only one state variable. Also since there is only one state variable, there are no illegal state to recover from.

(g) Only one next state equation is required and an extra product term is added to make the equation logic hazard-free. One Moore output equation is also required.

$$Y = y \cdot \overline{X1} + X1 \cdot X2 + y \cdot X2$$
(The y·X2 product term insures a logic hazard-free equation.)
$$Z = y$$

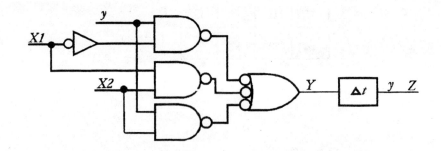

378

(h) Since there is only one next state equation only one S-R NOR latch is required.

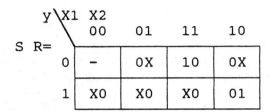

y	Y	S	R
0	0	0	X
0	1	1	0
1	0	0	1
1	1	X	0

$$X = \text{don't care}$$

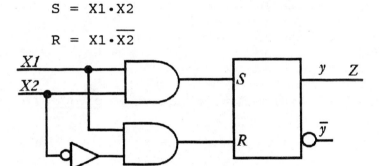

S R= $y\backslash$X1 X2	00	01	11	10
0	–	0X	10	0X
1	X0	X0	X0	01

$$S = X1 \cdot X2$$

$$R = X1 \cdot \overline{X2}$$

In this case the version of the circuit using the S-R NOR latch circuit is somewhat simpler than the gate level circuit. In fact, this is the design for a Gated D latch circuit where C = X1 and D = X2.

(i) $\sim S = \overline{X1 \cdot X2}$

     $\sim R = \overline{X1 \cdot \overline{X2}}$

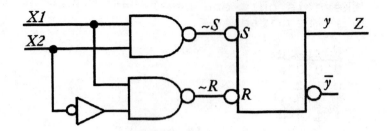

10-14. (a) The composite Karnaugh map for the next state and external outputs for Fig. P10-14 is shown as follow.

Y,PS SUM=

y\A B	00	01	11	10	PS CARRYOUT
m 0	0,0	0,1	1,0	0,1	0
n 1	0,0	0,1	1,0	0,1	1

(b) The logic hazard-free equations for the circuit are

$$Y = A \cdot B$$

$$SUM = \overline{A} \cdot B + A \cdot \overline{B} = A \oplus B$$

$$CARRYOUT = y$$

(c) Using the logic hazard-free equations we obtain the following gate level circuit.

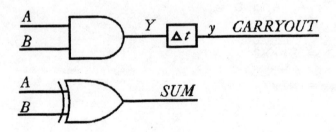

(d) The composite Karnaugh map for the S and R excitation inputs can be written as

y	Y	S	R
0	0	0	X
0	1	1	0
1	0	0	1
1	1	X	0

X = don't care

$$y \backslash A \; B$$

S R=	00	01	11	10
0	0X	0X	10	0X
1	01	01	X0	01

(e) Using the composite Karnaugh map for the S and R excitation inputs we can write the following S and R excitation input equations and draw the circuit as shown below.

$$S = A \cdot B$$

$$R = \overline{A} + \overline{B}$$

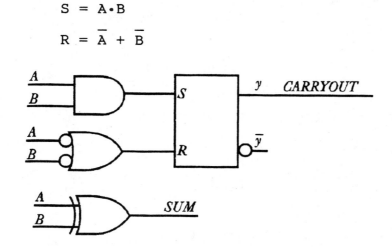

In this case the gate level circuit is somewhat simpler than the version of the circuit using a S-R NOR latch.

(f) Complementing the equations for the S-R NOR latch and using these equations we can draw the circuit using a S-R NAND latch as shown below.

$$\sim S = \overline{A \cdot B}$$

$$\sim R = \overline{\overline{A} + \overline{B}}$$

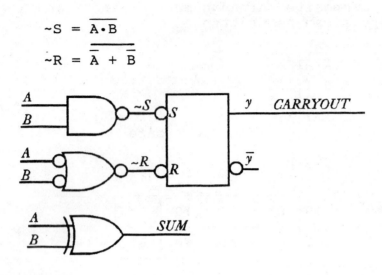

10-15. (a) The composite Karnaugh map for the next state and external outputs for Fig. P10-15 is shown as follow. This map is shown without regards for a race-free assignment or illegal state recovery.

y1 y2\A B	00	01	11	10	PS A=B A>B A<B
Y1 Y2=					
i 00	<u>00</u>	11	<u>00</u>	01	100
j 01	00	11	00	<u>01</u>	010
k 11	00	<u>11</u>	00	01	001
l 10	-	-	-	-	-

- = don't care

In the following map we modified the above composite Karnaugh map to go through transitory state l from state k in two places to get to state i to insure a race-free state assignment. We also fixed the second and fourth columns of state l such that the circuit changes to one of the stable states in the other rows of the map to insure illegal state recovery. Illegal state recovery was not mentioned in the problem on purpose to get each students to think and remember that all digital circuits should be designed to allow illegal state recovery.

```
y1 y2\A B                              /PS  A=B A>B A<B
            00     01     11     10   /
Y1 Y2=
     i 00 | 00  |  11  |  00  |  01  |  100
         |‾‾   |      | ‾‾   |      |
     j 01 | 00  |  11  |  00  |  01  |  010
         |     |      |      | ‾‾   |
     k 11 | 10  |  11  |  10  |  01  |  001
         |     | ‾‾   |      |      |
     l 10 | 00  |  11  |  00  |  01  |   -
```

$$- = \text{don't care}$$

(b) The logic hazard-free equations for the circuit are

$$Y1 = \overline{A} \cdot B + y1 \cdot y2 \cdot \overline{A} + y1 \cdot y2 \cdot B$$

$$Y2 = \overline{A} \cdot B + A \cdot \overline{B} = A \oplus B$$

$$A=B = \overline{y2}$$

$$A>B = \overline{y1} \cdot y2$$

$$A<B = y1$$

(c) Using the logic hazard-free equations we obtain the following gate level circuit.

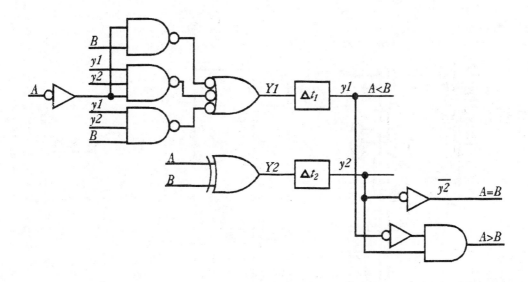

(d) Composite Karnaugh maps for the S and R excitation inputs can be written as

383

y	Y	S	R
0	0	0	X
0	1	1	0
1	0	0	1
1	1	X	0

X = don't care

y1 y2\A B

S1 S2=

	00	01	11	10
00	00	11	00	01
01	00	1X	00	0X
11	X0	XX	X0	0X
10	00	X1	00	01

y1 y2\A B

R1 R2=

	00	01	11	10
00	XX	00	XX	X0
01	X1	00	X1	X0
11	01	00	01	10
10	1X	00	1X	10

(e) Using the composite Karnaugh maps for the S and R excitation inputs we can write the following S and R excitation input equations and draw the circuit as shown below.

$$S1 = \overline{A} \cdot B$$

$$R1 = \overline{y2} \cdot \overline{B} + \overline{y2} \cdot A + A \cdot \overline{B}$$

$$S2 = \overline{A} \cdot B + A \cdot \overline{B}$$
$$\quad = A \oplus B$$

$$R2 = \overline{A} \cdot \overline{B} + A \cdot B$$
$$\quad = A \odot B$$
$$\quad = \overline{A} \oplus B$$

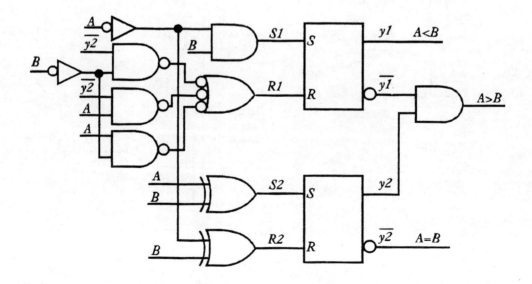

(f) Complementing the equations for the S-R NOR latches and using these equations we can draw the circuit using S-R NAND latches as shown below.

$$\sim\!S1 = \overline{\overline{A}\cdot B}$$

$$\sim\!R1 = \overline{\overline{y2\cdot\overline{B}} + \overline{y2\cdot A} + A\cdot\overline{B}}$$

$$\sim\!S2 = \overline{A \oplus B}$$

$$\sim\!R2 = \overline{\overline{A} \oplus B}$$

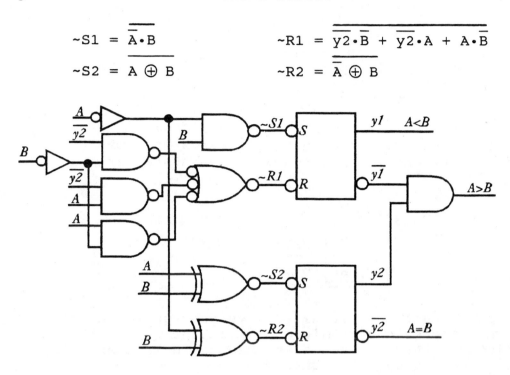

10-16. Yes. The state assignment in Fig. 10-12 is a race-free state assignment. Since only one bit changes between each stable state transition in the assignment, i. e., adjacent codes have been chosen such that there binary values differ by only one bit thus preventing the possibility of a critical race.

10-17. An essential hazard occurs in an asynchronous sequential circuit as a result of certain specified next state sequences. Sequences occur as a result of transitory states. Essential hazards are due to unequal signal delay paths in the circuit. To observe from a Karnaugh map if an essential hazard can occur simply test the circuit by changing a single input variable one time (change the single input variable from 0 to 1 or 1 to 0) then change the same input variable three times consecutively (from 0 to 1 to 0 to 1 or 1 to 0 to 1 to 0). If the circuit always ends up in the same state then the essential hazard cannot occur otherwise it can occur.

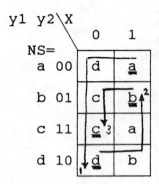

y1 y2\X

NS=

	0	1
a 00	d	<u>a</u>
b 01	c	<u>b</u>
c 11	<u>c</u>	a
d 10	<u>d</u>	b

No. An essential hazard cannot occur in a flow map that contains an external input and a single state variable.

y\X

NS=

	0	1
a 0	<u>a</u>	b
b 1	a	<u>b</u>

Notice that the circuit always ends up in the same state when there is a single change in the external input variable X or three consecutive changes in X; therefore, an essential hazard cannot occur.

10-18. Yes. The circuit diagram shown in Fig. 10-13 is logic hazard-free. This is true because the next state equation for Y1 and Y2 that are used to generate the asynchronous circuit were obtained using the following two step procedure provided in the text on page 416 and repeated below. (1) Obtain a minimum covering of the 1s (0s) of the function, and (2) add a product term to cover each occurrence of adjacent 1s (0s) that are not already contained in the same p-subcube (r-subcube) in the minimum covering of the function.

10-19.   y1 y2\X

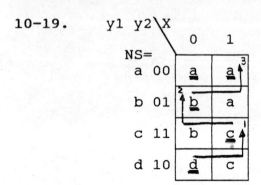

NS=

	0	1
a 00	<u>a</u>	<u>a</u>
b 01	<u>b</u>	a
c 11	b	<u>c</u>
d 10	<u>d</u>	c

Begin in stable state d and observe X changing from 0 to 1 to 0 to 1 (three consecutive changes) and the circuit can end up in a different state (state a) compared to ending up in state c for only one change in X. Also start in stable state c and notice that the circuit can end up in state a not b.

```
y1 y2\X1 X2
         00    01    11    10
Y1 Y2=
     m 00 |  m  |  m  |  r  |  r  |
     n 01 |2 n  |  n  |  n  |  s  |
     r 11 |  n  |  s  |  r  |  r  |
     s 10 |  s  |  s  |  n  |3 s  |
```

Begin in stable state m and observe X1 changing from 0 to 1 to 0 to 1 (three consecutive changes) and the circuit can end up in a different state (state s) compared to ending up in state r for only one change in X1.

Also not shown by arrows, begin in stable state m in the second column and observe X1 changing from 0 to 1 to 0 to 1 (three consecutive changes) and the circuit can end up in a different state (state n) compared to ending up in state r for only one change in X1. Also begin in stable state r in the third column and observe the circuit can end up in state n not s due to X1 changing three consecutive times compared to only one time.

10-20. To fix an essential hazard we must first determine that such a hazard can exist. If it is true it can exist by applying Unger's test it is then necessary to insure that the feedback delays for the state variables in the asynchronous circuit are adjusted such that the essential hazard does not occur. This generally involves increasing and/or decreasing certain feedback delay paths in the circuit so that the essential hazard does not occur.

10-21. The four important items a designer must carefully consider when designing fundamental mode asynchronous circuits in order that they operate properly, without hanging up and generating spurious signals, is to insure the circuits have

1. illegal state recovery
2. a race-free or critical race-free state assignment
3. logic hazard-free equations, and
4. essential hazard removal.

Only the first item, illegal state recovery, is important when designing synchronous circuits. The last three items are not a requirement when designing synchronous sequential circuits generally making the design of synchronous sequential circuits much easier.

10-22. There are four restrictions that must be placed on the manner in which the input signals are applied to pulse mode asynchronous circuit. These are listed as follows.

> Rule 1: Only one input signal pulse is allowed to occur at one time.
>
> Rule 2: Before the next input pulse is allowed to occur, the circuit must be given time to reach to reach a new stable state via a single state change.
>
> Rule 3: Each applied input pulse has a minimum pulse width that is determined by the time it takes to change the slowest flip-flop in the circuit to a new stable state.
>
> Rule 4: The maximum pulse width of an applied input pulse must be sufficiently narrow so that it is no longer present when the new present state output signals become available.

10-23. The term mutually exclusive when applied to a pulse mode asynchronous circuit means that the input signal pulses must not occur at the same time.

10-24. If the pulse width is not sufficiently narrow in a pulse mode circuit and a pulse is still present when the new present state output signals becomes available, then the input pulse can cause more than one state change (this is not allowed in a pulse mode design). The use of a double-rank circuit removes the maximum pulse width requirement by not allowing a new present state to occur until after the input pulse is gone.

10-25. (a) A state diagram for the circuit as be drawn as follows.

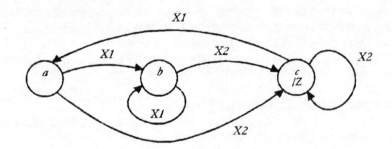

(b) From the state diagram we obtain the following flow map.

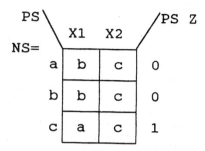

PS＼ ／PS Z

NS=

	X1	X2	
a	b	c	0
b	b	c	0
c	a	c	1

(c) Using the flow map we obtain the implication table as follows.

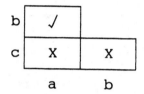

b	✓	
c	X	X
	a	b

a,b can be merged

Yes, the flow map can be reduced since states a and b are identical, i.e., they lead to the same next state and output condition. This can be seen by inspection of the flow map, but the implication table confirms it.

Merging rows a and b in the flow map in (b) results in the following reduced flow map.

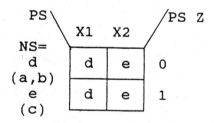

PS＼ ／PS Z

NS=

	X1	X2	
d (a,b)	d	e	0
e (c)	d	e	1

We can also draw the state diagram as follows.

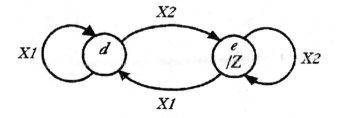

If you started with this state diagram, it obviously
will not reduce and its implication table shows that it will
not reduce.

e

(d) Choosing y as the state variable and the state
assignment as state a (0) and state b (1) we obtain the
composite Karnaugh map for the next state and external
outputs as follows.

$Y=$

y \	X1	X2	PS Z
d 0	0	1	0
e 1	0	1	1

(e) Using the composite Karnaugh map for the next state
and external outputs and the T excitation equation
$Ti = yi \oplus Yi$, we can obtain the composite Karnaugh map for
the T excitation inputs as follows.

$$Ti = yi \oplus Yi$$

$T=$

y \	X1	X2
0	0	1
1	1	0

(f) The equations for the pulse mode circuit that uses
positive edge triggered flip-flops for the master rank and D
flip-flops for the slave rank can be written as follows.

$$T = y \cdot X1 + \bar{y} \cdot X2$$

$$Z = y$$

(g) The double-rank implementation for the pulse mode
circuit is shown below.

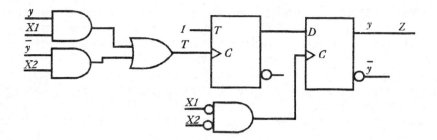

10-26. (a) A state diagram for the circuit as be drawn as follows.

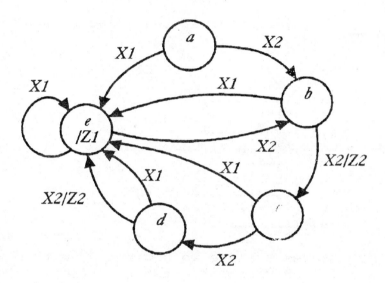

(b) From the state diagram we obtain the following flow map.

NS,PS Z2=

PS \	X1	X2	PS Z1
a	e	b	0
b	e	c,Z2	0
c	e	d	0
d	e	e,Z2	0
e	e	b	1

391

For five states it will require three state variables for this design, that is three flip-flops, if the number of states cannot be reduced. Notice that state a is not in the primary counting sequence and can be treated in the same manner as an unused or illegal state since it can only be arrived at by forcing the circuit into state a by perhaps using the set and/or reset inputs of the flip-flops, as a coincidence at power up, or as a result of a power glitch that may cause the flip-flops to go to state a. If we allow the Moore output to be 1, that is Z1 = 1, in state a, then state a and state e are identical since they both lead to the same next state for X1 (both go to state e) and X2 (both go to state b) and provide the same Moore and Mealy outputs, namely Moore output Z1 = 1 and Mealy output Z2 = 0.

Merging our now identical states a and e results in the following flow map.

NS,PS Z2= PS	X1	X2	PS Z1
f (a,e)	f	g	1
g (b)	f	h,Z2	0
h (c)	f	i	0
i (d)	f	f,Z2	0

By merging states a and e we now end up with only two state variable which only requires two flip-flops. We can also draw the state diagram in the following simpler form.

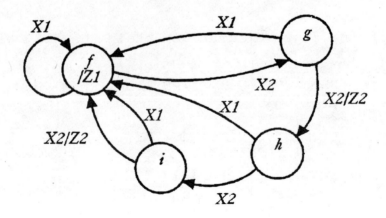

(c) Using the flow map we obtain the following implication table. This is the implication table before changing the Moore output to a 1 in state a and merging states a and e. This implication table tells us if there was any possible reduction prior to making the change indicated above.

```
b |  X
  |------
c | b=d  X |  X
  |----------------
d |  X   | c=e  X |  X
  |--------------------------
e |  X   |  X     |  X  |  X
  |------------------------------
      a       b      c     d
```

Notice that the implication table shows that no reduction was possible since there are no mergeable states.

The implication table after changing the Moore output to 1 in state a and merging states a and e is shown below.

```
g |  X
  |------
h |  X  |  X
  |------------
i |  X  | f=h  X |  X
  |------------------------
      f      g       h
```

This implication table shows that no further reduction is possible since there are no mergeable states.

(d) Choosing y1 and y2 as the state variables and the state assignments as state f (00), state g (01), state h (11), and state i (10) we obtain the composite Karnaugh map for the next state and external outputs as follows.

y1 y2 Y1 Y2,PS Z2=	X1	X2	PS Z1
f 00	00,0	01,0	1
g 01	00,0	11,1	0
h 11	00,0	10,0	0
i 10	00,0	00,1	0

393

(e) Using the composite Karnaugh map for the next state and external outputs and the T excitation equation $T_i = y_i \oplus Y_i$, we can obtain the composite Karnaugh map for the T excitation inputs as follows.

$$T_i = y_i \oplus Y_i$$

y1 y2 T1 T2=	X1	X2
f 00	00	01
g 01	01	10
h 11	11	01
i 10	10	10

(f) The equations for the pulse mode circuit that uses positive edge triggered flip-flops for the master rank and D flip-flops for the slave rank can be written as follows.

$$T_1 = y_1 \cdot X_1 + \overline{y_1} \cdot y_2 \cdot X_2 + y_1 \cdot \overline{y_2} \cdot X_2$$

$$T_2 = y_2 \cdot X_1 + \overline{y_1} \cdot \overline{y_2} \cdot X_2 + y_1 \cdot y_2 \cdot X_2$$

$$Z_1 = \overline{y_1} \cdot \overline{y_2}$$

$$Z_2 = \overline{y_1} \cdot y_2 \cdot X_2 + y_1 \cdot \overline{y_2} \cdot X_2$$

(g) The double-rank implementation for the pulse mode circuit is shown below.

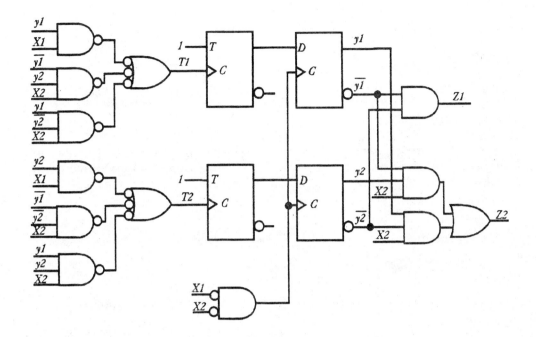

10-27. After assigning state variable y to the q output of the D flip-flop for the slave rank in Fig. P10-27 we can write the following equations.

$$T = \overline{y} \cdot X1 + X2 \cdot \overline{X3}$$

$$= \overline{y} \cdot X1 + X2$$

$$Z1 = y$$

$$Z2 = y \cdot X1$$

Recall that $\overline{X3}$ does not convey any information in a pulse mode circuit.

We can now find the next state output equation for the circuit by substituting the T excitation equation into the bistable equation for the T type bistable.

$$Q = \overline{T} \cdot q + T \cdot \overline{q}$$

so,

$$Y = \overline{T} \cdot y + T \cdot \overline{y}$$

and,

$$Y = \overline{(\overline{y} \cdot X1 + X2)} \cdot y + (\overline{y} \cdot X1 + X2) \cdot y$$

$$= ((\overline{\overline{y} \cdot X1}) \cdot \overline{X2}) \cdot y + \overline{y} \cdot X1 + \overline{y} \cdot X2$$

$$= (y + \overline{X1}) \cdot \overline{X2} \cdot y + \overline{y} \cdot X1 + \overline{y} \cdot X2$$

$$= y \cdot \overline{X2} + y \cdot \overline{X1} \cdot \overline{X2} + \overline{y} \cdot X1 + \overline{y} \cdot X2$$

$$= \overline{y} \cdot X1 + y \cdot X2$$

Recall that $\overline{X1}$ and $\overline{X2}$ do not convey any information and y alone cannot cause a circuit to change states in a pulse mode circuit.

Using the next state output equation and the Moore and Mealy output equations we can obtain the composite Karnaugh map for the next state and external outputs as follows.

Y,PS Z2=	y	X1	X2	PS Z1
a	0	1,0	0,0	0
b	1	0,1	1,0	1

Assigning 0 to state a and 1 to state b as shown we can draw the flow map as

NS,PS Z2=	PS	X1	X2	PS Z1
	a	b	a	0
	b	a,Z2	b	1

Using the flow map we can draw the state diagram for the circuit as follows.

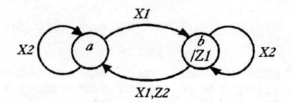

396

10-28. From the state diagram in Fig. P10-28 we can write the following flow map.

NS,PS Z2=	X1	X2	PS Z1
a	b,Z2	a	0
b	d	a	1
c	b	a	0
d	c	a	0

Testing the circuit for state reduction we can generate the following implication table.

b	X		
c	X	X	
d	X	X	b=c X
	a	b	c

No state reduction is possible.

Choosing $y1$ and $y2$ as the state variables and the state assignment state a (00), state b (01), state c (11), and state d (10) we can obtain the composite Karnaugh map for the next state and external outputs as follows.

Y1 Y2,PS Z2=	X1	X2	PS Z1
a 00	01,1	00,0	0
b 01	10,0	00,0	1
c 11	01,0	00,0	0
d 10	11,0	00,0	0

Using the composite Karnaugh map for the next state and external outputs and the T excitation equation $Ti = yi \oplus Yi$, we can obtain the composite Karnaugh map for the T excitation inputs as follows.

$$T_i = y_i \oplus Y_i$$

y1 y2 T1 T2=	X1	X2
a 00	01	00
b 01	11	01
c 11	10	11
d 10	01	10

The equations for the pulse mode circuit that uses positive edge triggered flip-flops for the master rank and D flip-flops for the slave rank can be written as follows.

$$T1 = y2 \cdot X1 + y1 \cdot X2$$

$$T2 = \overline{y1} \cdot X1 + \overline{y2} \cdot X1 + y2 \cdot X2$$

$$Z1 = \overline{y1} \cdot y2$$

$$Z2 = \overline{y1} \cdot \overline{y2} \cdot X1$$

The double-rank pulse mode circuit for Fig. P10-28 is shown below.

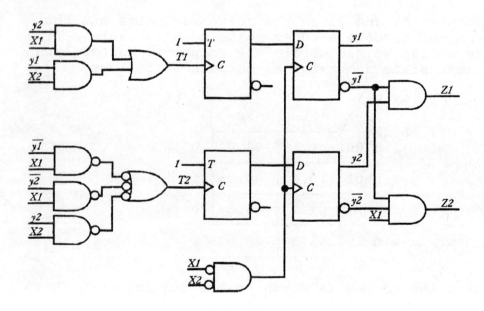

10-29. From the state diagram in Fig. P10-29 we can write the following flow map.

NS,PS Z1= PS	X1	X2	X3	PS Z2
a	b	-	a	0
b	c	a	b	1
c	a,Z1	a	c	0

Testing the circuit for state reduction we can generate the following implication table.

b	X	
c	X	X
	a	b

No state reduction is possible.

At this point we need to answer the questions. For proper circuit operation does the design require:

(a) Illegal state recovery? Yes, since the circuit can hang up in the illegal state if it is not directed back to the primary counting sequence.

(b) A race-free state assignment? No, only asynchronous fundamental mode circuits require a race-free state assignment. For asynchronous pulse mode circuits there are no transitory states, all states are stable. This make the design easier since any state assignment can be chosen.

(c) Logic hazard-free equations? Generally yes. Asynchronous fundamental mode and asynchronous pulse mode gate level circuits require logic hazard-free equations. NOR and NAND latches and memory devices in general are designed to be logic hazard-free and thus only require there inputs be logic hazard-free to make the circuit they are used in logic hazard-free.

(d) Essential hazard removal? No, only asynchronous fundamental mode circuits require essential hazard removal because they have transitory states, since there are no transitory states in asynchronous pulse mode circuits essential hazards cannot occur in pulse mode circuits.

Choosing y_1 and y_2 as the state variables and the state assignment state a (00), state b (01), state c (11), and unused or illegal state d (10) we can obtain the

composite Karnaugh map for the next state and external
outputs as follows.

y1 y2 Y1 Y2,PS Z1=	X1	X2	X3	PS Z2
a 00	01,0	-,-	00,0	0
b 01	11,0	00,0	01,0	1
c 11	00,1	00,0	11,0	0
d 10	00,0	00,0	00,0	0

To provide illegal state recovery we chose state d to
return to state a for all input pulses and both the Moore
and Mealy outputs to be 0 in illegal state d.

Using the composite Karnaugh map for the next state
and external outputs and the T excitation equation
$T_i = y_i \oplus Y_i$, we can obtain the composite Karnaugh map for
the T excitation inputs as follows.

$$T_i = y_i \oplus Y_i$$

y1 y2 T1 T2=	X1	X2	X3
a 00	01	-	00
b 01	10	01	00
c 11	11	11	00
d 10	10	10	10

The equations for the pulse mode circuit that uses
positive edge triggered flip-flops for the master rank and D
flip-flops for the slave rank can be written as follows.

$$T1 = y2 \cdot X1 + y1 \cdot X1 + y1 \cdot X2 + y1 \cdot \overline{y2} \cdot X3$$

$$T2 = \overline{y1} \cdot \overline{y2} \cdot X1 + y1 \cdot y2 \cdot X1 + y2 \cdot X2$$

$$Z1 = y1 \cdot y2 \cdot X1$$

$$Z2 = \overline{y1} \cdot y2$$

The double-rank pulse mode circuit for Fig. P10-29 is shown below.

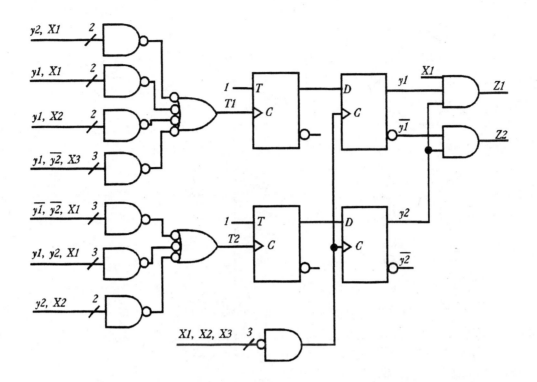

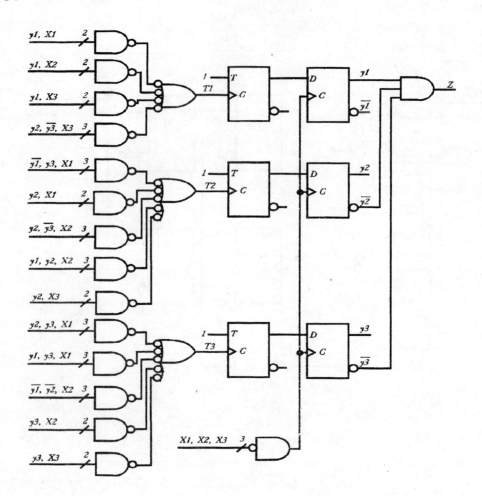

PART
II

Computer Solutions
Using PLDesigner

USING THE STUDENT VERSION OF PLDesigner V1.5

by Richard S. Sandige

The University of Wyoming

Electrical Engineering Department

Copyright 1990

In support of the text *Modern Digital Design*
by Richard S. Sandige, McGraw Hill, 1990

6/29/90

About the Student Version of PLDesigner V1.5

The student version is derived from the software package PLDesigner. The student version allows the user to enter a logic description in a high level language that includes truth tables, state machines, and Boolean equation constructs. The software generates reduced Boolean equations based on the design entry and allows the user to functionally simulate the digital design.

The software can be use in the following ways:

· To generate reduced Boolean equations from a high level language. These Boolean equations can then be implemented in any desired technology, including SSI and MSI devices, or PLDs.

· To simulate Boolean equations at a functional level. This allows the user to explore the functionality of a set of Boolean equations.

· To obtain a design file for PLDesigner by Minc Incorporated. PLDesigner will take the design description provided in the design file and implement it in PLDs without requiring the user to be familiar with the devices.

1-1 INTRODUCTION

Software design synthesis tools do not have to be presented in a beginning course in digital design; however, since these tools are being used daily in industry, the author feels strongly that they should be. The purpose of this document is to present a number of different design examples using the Student Version of PLDesigner V1.5 (a programmable logic device design synthesis tool). **This software tool can be used without knowing anything about PLDs.**

NOTE THAT THE "DESIGN PHASE" OF PLDesigner CAN BE EFFECTIVELY USED AS A <u>TEACHING TOOL</u> TO DESIGN AND PERFORM FUNCTIONAL SIMULATIONS FOR COMBINATIONAL AND SEQUENTIAL LOGIC FUNCTIONS *even if those functions may never be put in a PLD.*

The system flow chart for PLDesigner is shown in Fig. 1. The Student Version of PLDesigner V1.5 contains the following features: (1) High Level Language Entry (including logic equation, truth table, and state machine entry

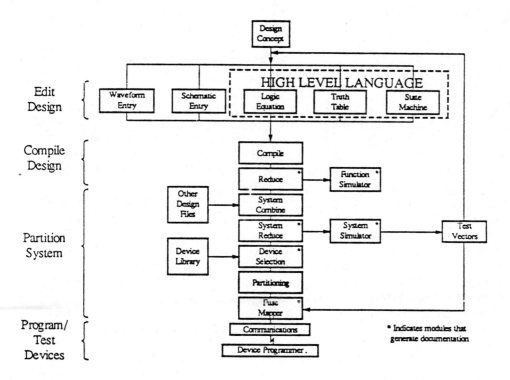

Figure 1.
PLDesigner System Flow Chart (Courtesy Minc Incorporated)

methods), (2) Simulation and Test Vector Generation for expected outputs for specified input combinations; and (3) Simulation Output for viewing simulation results. These features compose the "design phase" of PLDesigner and provide an excellent tool for learning to design and analyze both **COMBINATIONAL** and **SEQUENTIAL** digital logic circuits. In the PLDesigner System Flow Chart the design phase consists of the Edit Design and Compile Design sections.

To teach the "design phase" of PLDesigner only requires the user to obtain a copy of the student version. A *design file* is first created for the design description using a text editor. Compiling the *design file* results in the following documentation: (1) reduced equations for the design, and (2) simulation outputs in truth table format for the design (if a simulation section is specified). The *design file* created with the student version can be used, if desired, with any commercial version of PLDesigner. Commercial versions of PLDesigner provide not only the "design phase" but also the "device-selection phase" which provides the software support to choose devices and program them using a device programmer. If an instructor wants to teach the entire design process from the "design phase", "device-selection phase" all the way to device programming, one commercial version of PLDesigner and a device programmer will be required. The "device-selection phase" and device programming are rather routine after the "design phase" is completed. Refer to Chapter 9, Section 9-5 in the text *Modern Digital Design* by Richard S. Sandige, McGraw Hill, 1990 for more information about PLDesigner.

With the availability of the Student Version, each student may now obtain his or her own copy of the "design phase" of PLDesigner. Prior to compiling the *design file* the CONFIGURATION tool is used to select either the Sum of Products or the Espresso equation reduction level. The *design file* is then compiled using the COMPILE DESIGN tool and the Boolean equations (reduced if specified) and the signal list for the design are placed in the Documentation [.DOC] file. These Boolean equations can then be implemented in whatever technology one desires including SSI and MSI devices, or PLDs. A set of Boolean equations can also be investigated using test vectors at a functional level and the results observed via a simulation output in truth table format. The test vectors are simply a specification of the forcing values for the input signals and the expected values for the output signals for a design. Using a simulation section, specified test vectors generated for the design can either be checked against the output provided by the functional simulator, or the output of the functional simulator can be viewed for requested input combinations (i.e., the truth table for the function can be displayed).

2

An instructor may elect to follow the traditional sequence of topics provided in the text *Modern Digital Design* with or without discussing programmable logic device design synthesis tools such as PALASM and PLDesigner. It is the recommendation of this author and teacher that introducing students to one of the industry's most advanced software design package for PLDs, PLDesigner, will help make these students better prepared after graduation. Just how and when this tool is introduced is left to the discretion of the instructor to provide maximum flexibility.

1-2 A SAMPLE LABORATORY

One example of a classroom tested PLDesigner laboratory exercise follows.

University of Wyoming

Department of Electrical Engineering

EE 439   Lab #X

Learning PLDesigner

Bring to lab your own 5¼" 360 Kbyte disk (2S/2D) or 1.2 Mbyte (2S/HD) disk to save your work files. You must use your own floppy disk to save your work files so you do not add to the DOS lab's hard disk.

1. To start PLDesigner type cd \minc_ed to change directories to the \minc_ed directory then type pld then press enter.

2. Investigate PLDesigner's Main Menu (i.e., using the arrow key to select each tool and the F1 key to get help).

3. Select Current file and type a:\ followed by a valid DOS file name (example a:\test). Press enter then select Edit Design. Select Equation, State Machine, or Truth Table and press enter.
   (a) Create a small test file with the Minc Editor.
   (b) Leave the Minc Editor using Exit&Save to verify your test file is being saved.
   (c) Return to your test file and verify it is intact.

4. Print your source file ([.SRC] file) by leaving the Minc Editor and exiting PLDesigner. From the DOS prompt type print a: <your test file name>. Obtain a printed copy of your test file. Restart PLDesigner.

5. Learn the basic features of the Minc Editor and how to use it.

6. Use the Minc Editor to create a language based logic design using the logic equation entry format. Under Current file in the main menu type a:\T_LE (where T_LE is the file name chosen for this example). Select any [.SRC] entry in the Edit Design tool. Type the following *design file* which uses logic equation entry format.

```
"Design file
TITLE               Test T;
COMMENT             Logic Equation entry format;
ENGINEER            type your name here;

FUNCTION T_LE;

INPUT X, Y, Z;
OUTPUT F1, F2;
    F1 = X*Y + X*(/X + Z) + X*(/Z + X*Y) + X*/Y*Z;

    F2 = /(/(X*/Z)*/(Y*Z)*/(/X*Y));
END T_LE; "Function
```

7. Save the *design file* in 6. after you type it in, and then verify you can access it after you save it. Select the Sum of Products equation reduction level using the CONFIGURATION tool then exit the tool and use the COMPILE DESIGN tool to compile the *design file*. If no errors are detected the Documentation [.DOC] file accessed via the EDIT DESIGN tool will contain the equations for the design in SOP form. If an error is detected the COMPILE DESIGN tool indicates an error by line number. Repeat the process until you successfully get a *design file* that will compile. Observe in the Documentation [.DOC] file the form of the resulting logic equations. Use Boolean algebra and write F1 and F2 in SOP form and verify that your SOP forms agree with the results obtained by the PLDesigner program.

8. Change to the Espresso equation reduction level and recompile the design. Use Boolean algebra and write F1 and F2 in a reduced SOP form and verify that your reduced SOP forms agrees with the results obtained by the PLDesigner program.

9. Create your own *design file* to allow the PLDesigner software to reduce the left side of the expression of each of the Theorems T1, T2a through T9a on page 27 in *Modern Digital Design* using the Espresso equation reduction level. Compile your design then print out the result in the Documentation [.DOC] file and check which expressions the PLDesigner software was able to reduce to a minimum literal count or a constant.

10. Truth Table entry format can also be used in a *design file* as illustrated below using the NAND truth table for

4

inputs X and Y and output F. The following truth table
format must be placed after the OUTPUT line and before the
END line in the example T_LE shown above in place of the F1
and F2 logic equations.

    TRUTH_TABLE

        X,Y::F;
        0,0::1;
        0,1::1;
        1,0::1;
        1,1::0;

    END; "Truth table

    Make a copy of *design file* T_LE using File, WriteTo
T_TT.SRC in the Minc Editor. Change the current file in the
main menu to T_TT. Modify the T_TT file by removing input Z
and modifying the output to contain only F. Change T_LE to
T_TT at each entry in the *design file* and also change the
COMMENT line to reflect the entry method being used for the
design. Save the modified *design file* T_TT to your disk
then compile the T_TT design. After successfully compiling
the design print a copy of the Documentation [.DOC] file for
T_TT and verify the equation obtained for the NAND truth
table is correct.

11. Create a *design file* using logic equation entry for the
following Boolean function. Create a second *design file*
using truth table entry for the same Boolean function. Use
the Espresso equation reduction level and compile each
design file and print out the results in the Documentation
[.DOC] file.   Compare the results obtain for each design.
The reduced equations obtained for each design should be the
same, if they are not find any errors and correct them.
Obtain a print out of the correct Documentation [.DOC] file
for each *design file*.

$$F(A,B,C) = \Sigma m(0,4,5,6,7)$$

12. Your report consists of an explanation of how to use the
PLDesigner program for both logic equation entry and truth
table entry with a print out of each of the [.SRC] and
compiled [.DOC] files for each design.

    ================ END OF LAB EXERCISE ================

1-3 Design Examples

    The Student Version of PLDesigner V1.5 was used for the
following design examples. The examples are taken from the
text *Modern Digital Design* on the page numbers indicated.
The simulation section in the *design file* is optional. When
it is used it shows either (1) sets of test vectors to allow

design verification (which can be helpful in locating problems in a design description), or (2) sets of selected input combinations to allow a design output from the simulator to be displayed. After the *design file* is compiled using the COMPILE DESIGN tool the Documentation [.DOC] file or the Sim Table [.SIM] file provides the simulation results.

Design Example 1
Page 22 in *Modern Digital Design*

Obtain the *design file* for the circuit in Figure E1-2 on page 22 using the logic equation entry method. For simulation use a table format (truth table) and list all possible binary values 000 through 111 in the table for the input variables. List an S for the output value for the output variable so that the simulator will know that it is suppose to supply the value. Before running the COMPILE DESIGN tool on the *design file* choose the CONFIGURATION tool and select the Sum of Products equation reduction level. Obtain the equation and simulation results contained in the Documentation [.DOC] file for this configuration. Then choose the CONFIGURATION tool and select the Espresso equation reduction level and rerun the COMPILE DESIGN tool on the *design file* to obtain the equation and simulation results for this configuration.

Solution

```
'Design file
TITLE                    Circuit Analysis CKTA;
COMMENT                  Logic Equation entry format;

FUNCTION CKTA_LE;

INPUT X, Y, Z;
OUTPUT F;

    F = Z*((X*Y) + /X);

    SIMULATION "Simulation section is optional
    STEP 25ns;

        X, Y, Z :: F;
        0, 0, 0 :: S;
        0, 0, 1 :: S;
        0, 1, 0 :: S;
        0, 1, 1 :: S;
        1, 0, 0 :: S;
        1, 0, 1 :: S;
        1, 1, 0 :: S;
```

```
            MESSAGE ('All inputs are 1');
            X, Y, Z :: F;
            1, 1, 1 :: S;

        END; "Simulation
    END CKTA_LE; "Function
```

The Documentation [.DOC] file contains the following
sum of products equation. The simulation results are also
shown below. The Sim Table [.SIM] file contains just the
simulation results. Notice that the equation F.EQN is simply
generated in SOP form with all the signals specified in the
design file (no reduction applied). Also notice that the
simulation results are displayed in truth table format.

```
    EQUATIONS FOR FUNCTION CKTA_LE

    INPUT SIGNALS:
            HIGH_TRUE   X
            HIGH_TRUE   Y
            HIGH_TRUE   Z
    OUTPUT SIGNALS:
            HIGH_TRUE   F

    REDUCED EQUATIONS:

    F.EQN               = Y*X*Z
                        + /X*Z ;   "(2 terms)

    SIMULATION OF CKTA_LE  -  Report Level 2

    Time ns   X Y Z F      Messages
    ------------------------------------------------
    0         0 0 0 0
    25        0 0 0 0
    50        0 0 1 1
    75        0 1 0 0
    100       0 1 1 1
    125       1 0 0 0
    150       1 0 1 0
    175       1 1 0 0
    200       1 1 1 1      All inputs are 1
```

If test values were provided for the output in the
simulation sections of the *design file* instead of the letter
S the test values would be compared with the values
generated for the original design and any anomalies would be
indicated by the letters U or E. A trailing U in the truth
table simulation output would indicate where an unstable
state occurs while a trailing E would indicate where a
discrepancy occurs between the simulation design and the
original design. The word Expected would appear in an
output to show where the simulation design differs from the
original design on the line following the trailing E.   The

7

column titled Messages in the simulation output can be
filled as shown above by the user using the Keyword MESSAGE
to insert messages for specified input combinations.

The *design file* for Example 1 could have been written
using intermediate variables for each output as shown below.
With the same simulation section as shown in the *design file*
for Example 1, the simulation truth table would now include
each of the outputs F1, F2, F3 and F and thus represent a
complete description for each logic line in the original
circuit.

```
INPUT X, Y, Z;
OUTPUT F1, F2, F3, F;

        F1 = X*Y;
        F2 = /X;
        F3 = F1 + F2;
        F  = Z*F3;
```

The Documentation [.DOC] file contains the following
reduced equation using the Espresso equation reduction
level. Notice that the equation F.EQN has been reduced as a
result of selecting the Espresso equation reduction level
via the CONFIGURATION tool prior to running the COMPILE
DESIGN tool on the *design file*. The simulation results are
the same for either reduction level and are not repeated.

```
EQUATIONS FOR FUNCTION CKTA

INPUT SIGNALS:
        HIGH_TRUE   X
        HIGH_TRUE   Y
        HIGH_TRUE   Z
OUTPUT SIGNALS:
        HIGH_TRUE   F

REDUCED EQUATIONS:

F.EQN               = Y*Z
                    + /X*Z ;   "(2 terms)
```

Design Example 2
Page 38 in *Modern Digital Design*

Obtain the *design file* for the Boolean function
$F(X,Y,Z) = \Sigma m(1,2,4)$ represented by the truth table in Fig.
1-16 using the truth table entry method. For simulation use
the test language constructs SET and CLOCKF to specify the
input combinations for the input variables and let the
outputs be supplied by the simulator. Choose the
CONFIGURATION tool and select the Sum Of Products equation
reduction level then compile the *design file* by running the

8

COMPILE DESIGN tool. Obtain the equation and simulation
results contained in the Documentation [.DOC] file.

Solution

```
"Design file
TITLE                    Standard sum of products form SSOP;
COMMENT                  Truth Table entry format;

FUNCTION SSOP_TT;

INPUT X, Y, Z;
OUTPUT F;

        TRUTH_TABLE

            X,Y,Z::F;
            0,0,0::0;
            0,0,1::1;
            0,1,0::1;
            0,1,1::0;
            1,0,0::1;
            1,0,1::0;
            1,1,0::0;
            1,1,1::0;

        END; "Truth table

        SIMULATION "Simulation section is optional
        STEP 25ns;

            MESSAGE ('All outputs supplied by simulator');
            SET X = 0, Y = 0, Z = 0, F = .S.;
            CLOCKF;
            SET                Z = 1;
            CLOCKF;
            SET        Y = 1, Z = 0;
            CLOCKF;
            SET                Z = 1;
            CLOCKF;
            SET X = 1, Y = 0, Z = 0;
            CLOCKF;
            SET                Z = 1;
            CLOCKF;
            SET        Y = 1, Z = 0;
            CLOCKF;
            SET                Z = 1;
            CLOCKF;

        END; "Simulation
END SSOP_TT; "Function
```

An alternate test language construct using the Keywords
SET and CLOCKF can also be used in the Simulation section as

indicated above. This also allows you to mix table format
and test language constructs as desired. To tell the
simulator to supply the output value we must use .S. for the
test language construct or simply S for the table format.
One advantage of using the test language constructs is
because it allows us to write either test vectors or input
combinations in a more concise manner. This is true since
you only need to list signal values with the SET statement
that change at a specified time unit. An unlisted signal
retains its previous value for each succeeding time unit.

     The sum of products equation provided in the
documentation file is shown below. The primary benefit of
the Sum Of Products equation reduction level is to allow
circuits to be designed logic hazard free by adding the
appropriate product terms (see Chapter 7, Section 7-9-2,
Logic Hazards, in *Modern Digital Design*). The simulation
results are also shown below in truth table format.

     EQUATIONS FOR FUNCTION SSOP_TT

     INPUT SIGNALS:
          HIGH_TRUE X
          HIGH_TRUE Y
          HIGH_TRUE Z
     OUTPUT SIGNALS:
          HIGH_TRUE F

     REDUCED EQUATIONS:

     F.EQN   = /Z*/Y*X
                    + /Z*Y*/X
                    + Z*/Y*/X ; "(3 terms)

     SIMULATION OF SSOP_TT - Report Level 2

Time ns	X	Y	Z	F	Messages
0	0	0	0	0	
25	0	0	0	0	All outputs supplied by simulator
50	0	0	1	1	
75	0	1	0	1	
100	0	1	1	0	
125	1	0	0	1	
150	1	0	1	0	
175	1	1	0	0	
200	1	1	1	0	

Design Example 3
Page 43 in *Modern Digital Design*

     Obtain the *design file* for the Boolean function
$F(X,Y,Z) = \Sigma M(0,3,5,6,7)$ shown on page 43 using the Logic

10

Equation entry method. For simulation use the table format
(truth table) to specify the input combinations and let the
output be provided by the simulator. Choose the
CONFIGURATION tool and select the Espresso equation
reduction level then compile the *design file* by running the
COMPILE DESIGN tool. Obtain the reduced equation and
simulation results contained in the Documentation [.DOC]
file.

Solution

```
"Design file
TITLE                   Standard product of sums SPOS;
COMMENT                 Logic Equation entry format;

FUNCTION SPOS_LE;

INPUT X, Y, Z;
OUTPUT F;

    F = (X + Y + Z)*(X + /Y + /Z)*(/X + Y + /Z)
      *(/X + /Y + Z)*(/X + /Y + /X);

    SIMULATION "Simulation section is optional
    STEP 25ns;

        MESSAGE ('All outputs supplied by simulator');
        X, Y, Z :: F;
        0, 0, 0 :: S;
        0, 0, 1 :: S;
        0, 1, 0 :: S;
        0, 1, 1 :: S;
        1, 0, 0 :: S;
        1, 0, 1 :: S;
        1, 1, 0 :: S;
        1, 1, 1 :: S;

    END; "Simulation
END SPOS_LE; "Function
```

The reduced equation provided in the documentation file
is shown below. The simulation results are also shown below
in truth table format.

EQUATIONS FOR FUNCTION SPOS_LE

```
INPUT SIGNALS:
        HIGH_TRUE   X
        HIGH_TRUE   Y
        HIGH_TRUE   Z
OUTPUT SIGNALS:
        HIGH_TRUE   F
```

REDUCED EQUATIONS:

```
F.EQN              = Z*/Y*/X
                   + /Z*Y*/X
                   + /Z*/Y*X ;   "(3 terms)
```

SIMULATION OF SPOS_LE - Report Level 2

Time ns	X	Y	Z	F	Messages
0	0	0	0	0	
25	0	0	0	0	All outputs supplied by simulator
50	0	0	1	1	
75	0	1	0	1	
100	0	1	1	0	
125	1	0	0	1	
150	1	0	1	0	
175	1	1	0	0	
200	1	1	1	0	

Design Example 4
Page 114 in *Modern Digital Design*

Obtain the *design file* for the Boolean function
$F(X,Y,Z) = \Sigma m(0,2,3,5,6,7)$ represented by the Karnaugh map
in Fig. 3-7. Use the Logic Equation entry method and write
the Boolean function in terms of its 0s, i.e., write the
standard sum of products for the 0s of the function, and
let the software do any reduction. For simulation use the
table format (truth table) to specify test vectors for the
Boolean function to check your equation entry, i.e., run a
functional simulation using test vectors to compare the
output against your original design. Choose the
CONFIGURATION tool and select the Espresso equation
reduction level then compile the *design file* by running the
COMPILE DESIGN tool. Obtain the reduced equation and the
simulation results contained in the Documentation [.DOC]
file.

Solution

```
"Design file
TITLE                Karnaugh map one KMAP1;
COMMENT              Logic Equation entry format;

FUNCTION KMAP1_LE;

INPUT X, Y, Z;

OUTPUT F;

    /F = /X*/Y*Z + X*/Y*/Z;   "standard SOP for the 0s
```

12

```
        SIMULATION "Simulation section is optional
        STEP 25ns;

            X, Y, Z :: F;
            0, 0, 0 :: 1;
            MESSAGE ('F is 0 for these inputs');
            X, Y, Z :: F;
            0, 0, 1 :: 0;
            0, 1, 0 :: 1;
            0, 1, 1 :: 1;
            MESSAGE ('F is 0 for these inputs');
            X, Y, Z :: F;
            1, 0, 0 :: 0;
            1, 0, 1 :: 1;
            1, 1, 0 :: 1;
            1, 1, 1 :: 1;

        END; "Simulation
    END KMAP1_LE; "Function
```

The reduced equation provided in the documentation file
is shown below. The simulation results are also shown below
in truth table format. The simulation ran successfully, i.
e. no errors were listed in the simulation output.

EQUATIONS FOR FUNCTION KMAP1_LE

INPUT SIGNALS:
        HIGH_TRUE   X
        HIGH_TRUE   Y
        HIGH_TRUE   Z
OUTPUT SIGNALS:
        HIGH_TRUE   F

REDUCED EQUATIONS:

F.EQN   = /Z*/X
                + Z*X
                + Y ;   "(3 terms)

SIMULATION OF KMAP1_LE - Report Level 2

Time ns	X	Y	Z	F	Messages
0	0	0	0	1	
25	0	0	0	1	
50	0	0	1	0	F is 0 for these inputs
75	0	1	0	1	
100	0	1	1	1	
125	1	0	0	0	F is 0 for these inputs
150	1	0	1	1	
175	1	1	0	1	
200	1	1	1	1	

Design Example 5
Page 130 in *Modern Digital Design*

Obtain the *design file* for the Boolean functions
F1(X,Y,Z) = Σm(0,1,2,4,6) and F2(X,Y,Z) = Σm(1,2,6)
represented by the Karnaugh maps in Fig. 3-15. Use the Logic
Equation entry method and write the Boolean functions in
terms of their 1s, i.e., write the standard sum of products
for the 1s of the functions, and let the software do any
reduction. For simulation use the table format (truth
table) to specify test vectors for the Boolean function to
check your equation entry, i.e., run a functional simulation
using test vectors to compare the output against your
original design. Choose the CONFIGURATION tool and select
the Espresso equation reduction level then compile the
design file by running the COMPILE DESIGN tool. Obtain the
reduced equations and the simulation results contained in
the Documentation [.DOC] file.

Solution

```
"Design file
TITLE                 Karnaugh maps KMAPS;
COMMENT               Logic Equation entry format;

FUNCTION KMAPS_LE;

INPUT X, Y, Z;
OUTPUT F1, F2;

    F1 = /X*/Y*/Z + /X*/Y*Z + /X*Y*/Z + X*/Y*/Z + X*Y*/Z;
    F2 = /X*/Y*Z + /X*Y*/Z + X*Y*/Z;

    SIMULATION "Simulation section is optional
    STEP 25ns;

        X, Y, Z :: F1, F2;
        0, 0, 0 :: 1 , 0 ;
        0, 0, 1 :: 1 , 1 ;
        0, 1, 0 :: 1 , 1 ;
        0, 1, 1 :: 0 , 0 ;
        1, 0, 0 :: 1 , 0 ;
        1, 0, 1 :: 0 , 0 ;
        1, 1, 0 :: 1 , 1 ;
        1, 1, 1 :: 0 , 0 ;

    END; "Simulation
  END KMAPS_LE; "Function
```

The reduced equations provided in the documentation
file are shown below. The simulation results are also shown
below in truth table format. The simulation ran
successfully, i. e. no errors were listed in the simulation
output.

14

EQUATIONS FOR FUNCTION KMAPS_LE

INPUT SIGNALS:
        HIGH_TRUE X
        HIGH_TRUE Y
        HIGH_TRUE Z
OUTPUT SIGNALS:
        HIGH_TRUE F1
        HIGH_TRUE F2

REDUCED EQUATIONS:

F1.EQN   = /Y*/X
                    + /Z ; "(2 terms)

F2.EQN   = Z*/Y*/X
                    + /Z*Y ; "(2 terms)

SIMULATION OF KMAPS_LE -   Report Level 2

| | | | | F | F | |
Time ns	X	Y	Z	1	2	Messages
0	0	0	0	1	0	
25	0	0	0	1	0	
50	0	0	1	1	1	
75	0	1	0	1	1	
100	0	1	1	0	0	
125	1	0	0	1	0	
150	1	0	1	0	0	
175	1	1	0	1	1	
200	1	1	1	0	0	

It can be observed that PLDesigner performs independent function minimization rather than multiple function minimization. PLDesigner was designed to use independent function minimization because this is better suited for PALs, the architecture predominantly used for PLDs.

Design Example 6
Page 134 in *Modern Digital Design*

Obtain the *design file* for the Boolean functions $F1(X,Y,Z) = \Sigma m(0,1,2) + d(6)$ and $F2(X,Y,Z) = \Sigma m(0,4) + d(2,5,6,7)$ represented by the truth table in Fig. 3-17 using the Truth Table entry method. For simulation use the table format (truth table) to specify test vectors for the Boolean function to check your equation entry, i.e., run a functional simulation using test vectors to compare the output against your original design. Choose the CONFIGURATION tool and select the Espresso equation reduction level then compile the *design file* by running the COMPILE DESIGN tool. Obtain the reduced equations and

15

simulation results contained in the Documentation [.DOC]
file.

Solution

```
  "Design file
  TITLE                    Don't cares DCS;
  COMMENT                  Truth Table entry format;

  FUNCTION DCS_TT;

  INPUT X, Y, Z;
  OUTPUT F1, F2;

        TRUTH_TABLE

            X, Y, Z :: F1, F2;
            0, 0, 0 :: 1 , 1 ;
            0, 0, 1 :: 1 , 0 ;
            0, 1, 0 :: 1 , X ;
            0, 1, 1 :: 0 , 0 ;
            1, 0, 0 :: 0 , 1 ;
            1, 0, 1 :: 0 , X ;
            1, 1, 0 :: X , X ;
            1, 1, 1 :: 0 , X ;

        END; "Truth table

        SIMULATION "Simulation section is optional
        STEP 25ns;

            X, Y, Z :: F1, F2;
            0, 0, 0 :: 1 , 1 ;
            0, 0, 1 :: 1 , 0 ;
            0, 1, 0 :: 1 , X ;
            0, 1, 1 :: 0 , 0 ;
            1, 0, 0 :: 0 , 1 ;
            1, 0, 1 :: 0 , X ;
            1, 1, 0 :: X , X ;
            1, 1, 1 :: 0 , X ;

        END; "Simulation
  END DCS_TT; "Function
```

The reduced equations provided in the documentation
file are shown below. The simulation results are also shown
below in truth table format. The simulation ran
successfully since there are no errors in the simulation
output truth table. It can be observed that the Espresso
equation reduction level is fast but it doesn't always
provide minimum SOP solutions, especially when there are
don't care outputs available. From this example, all the
don't care output values (Xs in the truth table in Fig. 3-

16

17) were treated as 0s when the Boolean functions F1 and F2
were minimized.

EQUATIONS FOR FUNCTION DCS_TT

INPUT SIGNALS:
        HIGH_TRUE   X
        HIGH_TRUE   Y
        HIGH_TRUE   Z
OUTPUT SIGNALS:
        HIGH_TRUE   F1
        HIGH_TRUE   F2

   REDUCED EQUATIONS:

F1.EQN   = /X*/Y
                     + /Z*/X ;   "(2 terms)

F2.EQN   = /Z*/Y ;   "(1 term)

SIMULATION OF DCS_TT   -   Report Level 2

| | | | | F | F | |
Time ns	X	Y	Z	1	2	Messages
0	0	0	0	1	1	
25	0	0	0	1	1	
50	0	0	1	1	0	
75	0	1	0	1	0	
100	0	1	1	0	0	
125	1	0	0	0	1	
150	1	0	1	0	0	
175	1	1	0	0	0	
200	1	1	1	0	0	

Design Example 7
Page 209 in *Modern Digital Design*

   Obtain the *design file* for the Boolean functions
SUM1(C_IN1,A1,B1) = $\Sigma m(1,2,4,7)$ and C_OUT2(C_IN1,A1,B1) =
$\Sigma m(3,5,6,7)$ represented by the full adder truth table on
page 209 using the Truth Table entry method. For simulation
use the test language constructs SET and CLOCKF to specify
test vectors just for the Boolean function SUM1. Choose the
CONFIGURATION tool and select the Espresso equation
reduction level then compile the *design file* by running the
COMPILE DESIGN tool. Obtain the reduced equations and the
simulation results contained in the Documentation [.DOC]
file.

Solution

```
"Design file
TITLE                       Full adder FA;
COMMENT                     Truth Table entry format;

FUNCTION FA_TT;

INPUT C_IN1, A1, B1;
OUTPUT SUM1, C_OUT2;

     TRUTH_TABLE

        C_IN1, A1, B1 :: SUM1, C_OUT2;
        0,      0, 0 :: 0,     0;
        0,      0, 1 :: 1,     0;
        0,      1, 0 :: 1,     0;
        0,      1, 1 :: 0,     1;
        1,      0, 0 :: 1,     0;
        1,      0, 1 :: 0,     1;
        1,      1, 0 :: 0,     1;
        1,      1, 1 :: 1,     1;

     END; "Truth table

     SIMULATION "Simulation section is optional
     STEP 25ns;
     TRACE C_IN1,A1,B1,SUM1;

        SET [C_IN1, A1, B1] = 0, SUM1 = 0;
        CLOCKF;
        SET [C_IN1, A1, B1] = 3;
        CLOCKF;
        SET [C_IN1, A1, B1] = 5;
        CLOCKF;
        SET [C_IN1, A1, B1] = 6;
        CLOCKF;
        SET [C_IN1, A1, B1] = 1, SUM1 = 1;
        CLOCKF;
        SET [C_IN1, A1, B1] = 2;
        CLOCKF;
        SET [C_IN1, A1, B1] = 4;
        CLOCKF;
        SET [C_IN1, A1, B1] = 7;
        CLOCKF;

     END; "Simulation
END FA_TT; "Function
```

The reduced equations provided in the documentation
file are shown below. In this case the Espresso equation
reduction level provided minimum SOP solutions. The
simulation results are also shown below in truth table

format. The simulation ran successfully since there are no errors in the simulation output truth table for SUM1.

EQUATIONS FOR FUNCTION FA_TT

INPUT SIGNALS:
        HIGH_TRUE   C_IN1
        HIGH_TRUE   A1
        HIGH_TRUE   B1
OUTPUT SIGNALS:
        HIGH_TRUE   SUM1
        HIGH_TRUE   C_OUT2

REDUCED EQUATIONS:

SUM1.EQN   = B1*/A1*/C_IN1
                   + /B1*A1*/C_IN1
                   + /B1*/A1*C_IN1
                   + B1*A1*C_IN1 ;   "(4 terms)

C_OUT2.EQN   = C_IN1*B1
                   + A1*B1
                   + A1*C_IN1 ;   "(3 terms)

SIMULATION OF FA_TT   -   Report Level 2

	C_IN1	A1	B1	SUM1	
Time ns	B	B	B	B	Messages
0	0	0	0	0	
25	0	0	0	0	
50	0	1	1	0	
75	1	0	1	0	
100	1	1	0	0	
125	0	0	1	1	
150	0	1	0	1	
175	1	0	0	1	
200	1	1	1	1	

The Keyword TRACE was used in the source file for FA_TT (the *design file* for FA_TT) to specify all the inputs and only output SUM1. Since C_OUT2 was not requested it does not appear in the simulation results. The Bs indicate the default base for the signals in the TRACE statement. Other bases that may be used are decimal (DEC), hexadecimal (HEX), and octal (OCT).

19

Design Example 8
Page 232 in *Modern Digital Design*

Obtain the *design file* for the Demultiplexer
represented by the truth table on page 232 using the Truth
Table entry method. For simulation use the test language
constructs SET and CLOCKF to specify input combinations for
the input variables when EN = 1 and let the outputs be
supplied by the simulator. Let this be followed by a truth
table format for the input variables when En = 0. Choose
the CONFIGURATION tool and select the Espresso equation
reduction level then compile the *design file* by running the
COMPILE DESIGN tool. Obtain the reduced equations and
simulation results contained in the Documentation [.DOC]
file.

Solution

```
"Design file
TITLE                   Demultiplexer DMUX;
COMMENT                 Truth Table entry format;

FUNCTION DMUX_TT;

INPUT EN, X, Y;
OUTPUT D0, D1, D2, D3;

    TRUTH_TABLE

        EN, X, Y :: D0, D1, D2, D3;
        0,  X, X :: 0,  0,  0,  0;
        1,  0, 0 :: 1,  0,  0,  0;
        1,  0, 1 :: 0,  1,  0,  0;
        1,  1, 0 :: 0,  0,  1,  0;
        1,  1, 1 :: 0,  0,  0,  1;

    END; "Truth table

    SIMULATION "Simulation section is optional
    STEP 25ns;

        SET [EN,X,Y] = 4;
        CLOCKF;
        SET [EN,X,Y] = 5;
        CLOCKF;
        SET [EN,X,Y] = 6;
        CLOCKF;
        SET [EN,X,Y] = 7;
        CLOCKF;
        EN, X, Y :: D0, D1, D2, D3;
        0,  X, X :: S,  S,  S,  S;

    END; "Simulation
END DMUX_TT; "Function
```

The reduced equations provided in the documentation file are shown below. In this case the Espresso equation reduction level provided minimum SOP solutions. The simulation results are also shown below in truth table format. The simulation ran successfully since no errors are shown in the simulation results. Notice that the default is simulator insert (if an output signal is not specified then the output signal is automatically set to S; however, if an input is not specified then the input signal is automatically set to don't care).

EQUATIONS FOR FUNCTION DMUX_TT

```
INPUT SIGNALS:
        HIGH_TRUE   EN
        HIGH_TRUE   X
        HIGH_TRUE   Y
OUTPUT SIGNALS:
        HIGH_TRUE   D0
        HIGH_TRUE   D1
        HIGH_TRUE   D2
        HIGH_TRUE   D3
```

REDUCED EQUATIONS:

D0.EQN   = /Y*/X*EN ;   "(1 term)

D1.EQN   = Y*/X*EN ;   "(1 term)

D2.EQN   = /Y*X*EN ;   "(1 term)

D3.EQN   = Y*X*EN ;   "(1 term)

SIMULATION OF DMUX_TT   -   Report Level 2

Time ns	E N	X	Y	D 0	D 1	D 2	D 3	Messages
0	1	0	0	1	0	0	0	
25	1	0	0	1	0	0	0	
50	1	0	1	0	1	0	0	
75	1	1	0	0	0	1	0	
100	1	1	1	0	0	0	1	
125	0	X	X	0	0	0	0	

Design Example 9
Page 238 in *Modern Digital Design*

Obtain the ***design file*** for the Multiplexer represented by the truth table on page 238 using the Logic Equation entry method. For simulation use the test language

constructs SET and CLOCKF to specify test vectors for the
Boolean function to check your equation entry, i.e., run a
functional simulation using test vectors to compare the
output against your original design. Choose the
CONFIGURATION tool and select the Espresso equation
reduction level then compile the *design file* by running the
COMPILE DESIGN tool. Obtain the reduced equation and the
simulation results contained in the Documentation [.DOC]
file.

Solution

```
"Design file
TITLE                   Multiplexer MUX;
COMMENT                 Logic Equation entry format;

FUNCTION MUX_LE;

INPUT EN, DI0, DI1, DI2, DI3, SI1, SI0;
OUTPUT F;

    F = /EN*DI0*/SI1*/SI0  +  /EN*DI1*/SI1*SI0
      + /EN*DI2*SI1*/SI0  +  /EN*DI3*SI1*SI0;

    SIMULATION "Simulation section is optional
    STEP 25ns;

        SET EN=1,F=0;
        CLOCKF;
        SET EN=0,DI0=0,SI1=0,SI0=0,F=0;
        CLOCKF;
        SET DI0=1,F=1;
        CLOCKF;
        SET DI0=.X.,DI1=0,SI0=1,F=0;
        CLOCKF;
        SET DI1=1,F=1;
        CLOCKF;
        SET DI1=.X.,DI2=0,SI1=1,SI0=0,F=0;
        CLOCKF;
        SET DI2=1,F=1;
        CLOCKF;
        SET DI2=.X.,DI3=0,SI1=1,SI0=1,F=0;
        CLOCKF;
        SET DI3=1,F=1;
        CLOCKF;

    END; "Simulation
END MUX_LE; "Function
```

The equation provided in the documentation file is
shown below. In this case the Espresso equation reduction
level provided a minimum SOP solution since further
reduction is not possible. The simulation results are also
shown below in truth table format. The simulation ran

22

successfully since there were no errors in the simulation results.

EQUATIONS FOR FUNCTION MUX_LE

INPUT SIGNALS:
        HIGH_TRUE   EN
        HIGH_TRUE   DI0
        HIGH_TRUE   DI1
        HIGH_TRUE   DI2
        HIGH_TRUE   DI3
        HIGH_TRUE   SI1
        HIGH_TRUE   SI0
OUTPUT SIGNALS:
        HIGH_TRUE   F

REDUCED EQUATIONS:

F.EQN   = SI0*/SI1*/EN*DI1
                   + /SI0*SI1*/EN*DI2
                   + SI0*SI1*/EN*DI3
                   + /SI0*/SI1*DI0*/EN ;   "(4 terms)

SIMULATION OF MUX_LE - Report Level 2

```
              D  D  D  D  S  S
           E  I  I  I  I  I  I
Time ns    N  0  1  2  3  1  0  F    Messages
------------------------------------------------
0          1  X  X  X  X  X  X  0
25         1  X  X  X  X  X  X  0
50         0  0  X  X  X  0  0  0
75         0  1  X  X  X  0  0  1
100        0  X  0  X  X  0  1  0
125        0  X  1  X  X  0  1  1
150        0  X  X  0  X  1  0  0
175        0  X  X  1  X  1  0  1
200        0  X  X  X  0  1  1  0
225        0  X  X  X  1  1  1  1
```

Design Example 10
Page 252 in *Modern Digital Design*

Obtain the **design file** for the Binary to Hexadecimal character generator represented by the truth table in Fig. E5-17a on page 252 using the Truth Table entry method.   For simulation use the test language constructs SET and CLOCKF to specify the input combinations for the input variables and let the outputs be supplied by the simulator.   Choose the CONFIGURATION tool and select the Espresso equation reduction level then compile the **design file** by running the COMPILE DESIGN tool.   Obtain the reduced equations and

simulation results contained in the Documentation [.DOC]
file.

Solution

```
"Design file
TITLE                    Binary to Hex Char Gen BTH;
COMMENT                  Truth Table entry method;

FUNCTION BTH_TT;

INPUT D,C,B,A;
OUTPUT OA,OB,OC,OD,OE,OF,OG;

    TRUTH_TABLE

        D, C, B, A :: OA, OB, OC, OD, OE, OF, OG;
        0, 0, 0, 0 :: 1,  1,  1,  1,  1,  1,  0;
        0, 0, 0, 1 :: 0,  1,  1,  0,  0,  0,  0;
        0, 0, 1, 0 :: 1,  1,  0,  1,  1,  0,  1;
        0, 0, 1, 1 :: 1,  1,  1,  1,  0,  0,  1;
        0, 1, 0, 0 :: 0,  1,  1,  0,  0,  1,  1;
        0, 1, 0, 1 :: 1,  0,  1,  1,  0,  1,  1;
        0, 1, 1, 0 :: 1,  0,  1,  1,  1,  1,  1;
        0, 1, 1, 1 :: 1,  1,  1,  0,  0,  0,  0;
        1, 0, 0, 0 :: 1,  1,  1,  1,  1,  1,  1;
        1, 0, 0, 1 :: 1,  1,  1,  1,  0,  1,  1;
        1, 0, 1, 0 :: 1,  1,  1,  0,  1,  1,  1;
        1, 0, 1, 1 :: 0,  0,  1,  1,  1,  1,  1;
        1, 1, 0, 0 :: 1,  0,  0,  1,  1,  1,  0;
        1, 1, 0, 1 :: 0,  1,  1,  1,  1,  0,  1;
        1, 1, 1, 0 :: 1,  0,  0,  1,  1,  1,  1;
        1, 1, 1, 1 :: 1,  0,  0,  0,  1,  1,  1;

    END; "Truth table

    SIMULATION "Simulation section is optional
    STEP 25ns;

        "SET [D,C,B,A] = 00H;
        "CLOCKF;
        "SET [D,C,B,A] = 01H;
        "CLOCKF;
        "SET [D,C,B,A] = 02H;
        "CLOCKF;

        "Continuing in this manner
        "we can enter all possible values
        "up through and including

        "SET [D,C,B,A] = 0FH;
        "CLOCKF;
```

```
"This is one way, the brute force method,
"to request the simulation results.

"Another more compact method
"would be to use a FOR-DO construct
"as illustrated below.

VAR i;
FOR i = 0 to 15 DO
Begin
   SET [D,C,B,A] = i;
   CLOCKF;
END; "for-do construct

     END; "Simulation
 END BTH_TT; "Function
```

The reduced equations provided in the documentation file are shown below. In this case the Espresso equation reduction level provided a minimum SOP solution for all of the output signals except OG. The literal count for a minimum SOP solution for output signal OG is 11 while the literal count for the Espresso reduced equation for output OG is 12 (very close to a minimum SOP solution). The simulation results are also shown below in truth table format. The simulation ran successfully since no errors are shown in the simulation output. Both the brute force method and the FOR-DO construct method provide the same simulations results only the FOR-DO construct requires less typing.

```
EQUATIONS FOR FUNCTION BTH_TT

INPUT SIGNALS:
        HIGH_TRUE   D
        HIGH_TRUE   C
        HIGH_TRUE   B
        HIGH_TRUE   A
OUTPUT SIGNALS:
        HIGH_TRUE   OA
        HIGH_TRUE   OB
        HIGH_TRUE   OC
        HIGH_TRUE   OD
        HIGH_TRUE   OE
        HIGH_TRUE   OF
        HIGH_TRUE   OG
```

REDUCED EQUATIONS:

```
OA.EQN              = /D*C*A
                    + D*/A
                    + /B*D*/C
                    + B*C
                    + B*/D
                    + /C*/A ;   "(6 terms)


OB.EQN              = /B*/C
                    + A*B*/D
                    + /A*/C
                    + A*/B*D
                    + /A*/B*/D ;   "(5 terms)


OC.EQN              =./B*/C
                    + A*/C
                    + /B*A
                    + /C*D
                    + C*/D ;   "(5 terms)


OD.EQN              = A*B*/C
                    + /D*/A*/C
                    + /A*B*C
                    + D*/B
                    + A*/B*C ;   "(5 terms)


OE.EQN              = D*B
                    + /A*B
                    + D*C
                    + /C*/A ;   "(4 terms)


OF.EQN              = B*D
                    + C*/B*/D
                    + /B*/A
                    + /C*D
                    + C*/A ;   "(5 terms)


OG.EQN              = /D*C*/A
                    + /B*C*A
                    + D*B
                    + B*/C
                    + D*/C ;   "(5 terms)
```

26

SIMULATION OF BTH_TT - Report Level 2

| | | | | | O | O | O | O | O | O | O | |
Time ns	D	C	B	A	A	B	C	D	E	F	G	Messages
0	0	0	0	0	1	1	1	1	1	1	0	
25	0	0	0	0	1	1	1	1	1	1	0	
50	0	0	0	1	0	1	1	0	0	0	0	
75	0	0	1	0	1	1	0	1	1	0	1	
100	0	0	1	1	1	1	1	1	0	0	1	
125	0	1	0	0	0	1	1	0	0	1	1	
150	0	1	0	1	1	0	1	1	0	1	1	
175	0	1	1	0	1	0	1	1	1	1	1	
200	0	1	1	1	1	1	1	0	0	0	0	
225	1	0	0	0	1	1	1	1	1	1	1	
250	1	0	0	1	1	1	1	1	0	1	1	
275	1	0	1	0	1	1	1	0	1	1	1	
300	1	0	1	1	0	0	1	1	1	1	1	
325	1	1	0	0	1	0	0	1	1	1	0	
350	1	1	0	1	0	1	1	1	1	0	1	
375	1	1	1	0	1	0	0	1	1	1	1	
400	1	1	1	1	1	0	0	0	1	1	1	

Design Example 11
Page 398 in *Modern Digital Design*

Obtain the ***design file*** for the Boolean function represented by the Karnaugh map in Fig. E7-13a on page 398 using the Truth Table entry method. For simulation use the test language constructs SET and CLOCKF to specify test vectors for the Boolean function to check your truth table, i.e., run a functional simulation using test vectors to compare the output against your original design. Choose the CONFIGURATION tool and select the Espresso equation reduction level then compile the ***design file*** by running the COMPILE DESIGN tool. Obtain the reduced equation and simulation results contained in the Documentation [.DOC] file.

Solution

```
"Design file
TITLE                Combinational logic equation CLE;
COMMENT              Truth Table entry format;

FUNCTION CLE_TT;

INPUT A, B, C;
OUTPUT F;
```

```
       TRUTH_TABLE

          A, B, C :: F;
          0, 0, 0 :: 0;
          0, 0, 1 :: 0;
          0, 1, 0 :: 1;
          0, 1, 1 :: 1;
          1, 0, 0 :: 0;
          1, 0, 1 :: 0;
          1, 1, 0 :: 0;
          1, 1, 1 :: 1;

       END; "Truth table

       SIMULATION "Simulation section is optional
       STEP 25ns;

          SET [A, B, C] = 0, F = 0;
          CLOCKF;
          SET [A, B, C] = 1;
          CLOCKF;
          SET [A, B, C] = 2, F = 1;
          CLOCKF;
          SET [A, B, C] = 3;
          CLOCKF;
          SET [A, B, C] = 4, F = 0;
          CLOCKF;
          SET [A, B, C] = 5;
          CLOCKF;
          SET [A, B, C] = 6;
          CLOCKF;
          SET [A, B, C] = 7, F = 1;
          CLOCKF;


        END; "Simulation
  END CLE_TT; "Function
```

The reduced equation provided in the documentation file
is shown below. The simulation results are also shown below
in truth table format. The simulation ran successfully
since there are no errors in the simulation results.

```
EQUATIONS FOR FUNCTION CLE_TT

INPUT SIGNALS:
        HIGH_TRUE   A
        HIGH_TRUE   B
        HIGH_TRUE   C
OUTPUT SIGNALS:
        HIGH_TRUE   F

REDUCED EQUATIONS:

F.EQN              = B*/A
```

```
                 + C*B ;   "(2 terms)
```

SIMULATION OF CLE_TT - Report Level 2

```
Time ns   A  B  C  F     Messages
-----------------------------------
0         0  0  0  0
25        0  0  0  0
50        0  0  1  0
75        0  1  0  1
100       0  1  1  1
125       1  0  0  0
150       1  0  1  0
175       1  1  0  0
200       1  1  1  1
```

Design Example 12
Page 404 in *Modern Digital Design*

     Obtain the ***design file*** for the Priority Encoder represented by the truth table on page 404 using the Truth Table entry method.   For simulation use the table format (truth table) to specify input combinations for the Boolean functions and let the outputs be supplied by the simulator. Choose the CONFIGURATION tool and select the Espresso equation reduction level then compile the ***design file*** by running the COMPILE DESIGN tool.   Obtain the reduced equations and simulation results contained in the Documentation [.DOC] file.

Solution

```
"Design file
TITLE               Priority Encoder PE;
COMMENT             Truth Table entry format;

FUNCTION PE_TT;

INPUT I7, I6, I5, I4, I3, I2, I1, I0;
OUTPUT F2, F1, F0;

   TRUTH_TABLE

      I7, I6, I5, I4, I3, I2, I1, I0 :: F2, F1, F0;
      1,  X,  X,  X,  X,  X,  X,  X  :: 1,  1,  1;
      0,  1,  X,  X,  X,  X,  X,  X  :: 1,  1,  0;
      0,  0,  1,  X,  X,  X,  X,  X  :: 1,  0,  1;
      0,  0,  0,  1,  X,  X,  X,  X  :: 1,  0,  0;
      0,  0,  0,  0,  1,  X,  X,  X  :: 0,  1,  1;
      0,  0,  0,  0,  0,  1,  X,  X  :: 0,  1,  0;
      0,  0,  0,  0,  0,  0,  1,  X  :: 0,  0,  1;
      0,  0,  0,  0,  0,  0,  0,  1  :: 0,  0,  0;
```

```
      END; "Truth table

      SIMULATION "Simulation section is optional
      STEP 25ns;

          I7, I6, I5, I4, I3, I2, I1, I0 :: F2, F1, F0;
          1,  X,  X,  X,  X,  X,  X,  X  :: S,  S,  S;
          0,  1,  X,  X,  X,  X,  X,  X  :: S,  S,  S;
          0,  0,  1,  X,  X,  X,  X,  X  :: S,  S,  S;
          0,  0,  0,  1,  X,  X,  X,  X  :: S,  S,  S;
          0,  0,  0,  0,  1,  X,  X,  X  :: S,  S,  S;
          0,  0,  0,  0,  0,  1,  X,  X  :: S,  S,  S;
          0,  0,  0,  0,  0,  0,  1,  X  :: S,  S,  S;
          0,  0,  0,  0,  0,  0,  0,  1  :: S,  S,  S;

      END; "Simulation
END PE_TT; "Function
```

The reduced equations provided in the documentation
file are shown below. The simulation results are also shown
below in truth table format.

```
EQUATIONS FOR FUNCTION PE_TT

INPUT SIGNALS:
        HIGH_TRUE   I7
        HIGH_TRUE   I6
        HIGH_TRUE   I5
        HIGH_TRUE   I4
        HIGH_TRUE   I3
        HIGH_TRUE   I2
        HIGH_TRUE   I1
        HIGH_TRUE   I0
OUTPUT SIGNALS:
        HIGH_TRUE   F2
        HIGH_TRUE   F1
        HIGH_TRUE   F0

REDUCED EQUATIONS:

F2.EQN              = I4
                    + I5
                    + I6
                    + I7 ;   "(4 terms)

F1.EQN              = /I7*/I6*/I3*/I4*/I5*I2
                    + /I7*/I6*I3*/I4*/I5
                    + I6
                    + I7 ;   "(4 terms)

F0.EQN              = /I6*/I4*I1*/I2
                    + /I6*I3*/I4
                    + I5*/I6
                    + I7 ;   "(4 terms)
```

```
SIMULATION OF PE_TT   -   Report Level 2

          I  I  I  I  I  I  I  I  F  F  F
Time ns   7  6  5  4  3  2  1  0  2  1  0      Messages
-----------------------------------------------------------
0         1  X  X  X  X  X  X  X  1  1  1
25        1  X  X  X  X  X  X  X  1  1  1
50        0  1  X  X  X  X  X  X  1  1  0
75        0  0  1  X  X  X  X  X  1  0  1
100       0  0  0  1  X  X  X  X  1  0  0
125       0  0  0  0  1  X  X  X  0  1  1
150       0  0  0  0  0  1  X  X  0  1  0
175       0  0  0  0  0  0  1  X  0  0  1
200       0  0  0  0  0  0  0  1  0  0  0
```

It is interesting to note that the Espresso equation reduction level provides equations with a literal count of 27 for this example compared to a literal count of 33 using logic equation entry. Compiling the *design file* PE_TT with a commercial version of PLDesigner using the Quine-McCluskey equation reduction level provides a literal count of 22.

Design Example 13
Page 416 in *Modern Digital Design*

Obtain the *design file* for Boolean Function $F(A,B,C,D)$ = $\Sigma m(0,2,4,5,6,7,8,10,11,15)$ on page 416 using the Logic Equation entry method. For simulation use the table format (truth table) to specify test vectors for the Boolean functions to check your equation entry, i.e., run a functional simulation using test vectors to compare the output against your original design. Choose the CONFIGURATION tool and select the Sum Of Products equation reduction level then compile the *design file* by running the COMPILE DESIGN tool. Obtain the reduced equation and simulation results contained in the Documentation [.DOC] file.

Solution

```
   "Design file
   TITLE                Logic Hazard-free circuit LHFC;
   COMMENT              Logic Equation entry format;

   FUNCTION LHFC_LE;

   INPUT A, B, C, D;
   OUTPUT F;

       F = /B*/D + /A*B + A*C*D + /A*/C*/D + /A*C*/D
           + B*C*D + A*/B*C; "Use Sum Of Products equation
                             "reduction level
```

31

```
    SIMULATION "Simulation section is optional
    STEP 25ns;

        A, B, C, D :: F;
        0, 0, 0, 0 :: 1;
        0, 0, 0, 1 :: 0;
        0, 0, 1, 0 :: 1;
        0, 0, 1, 1 :: 0;
        0, 1, 0, 0 :: 1;
        0, 1, 0, 1 :: 1;
        0, 1, 1, 0 :: 1;
        0, 1, 1, 1 :: 1;
        1, 0, 0, 0 :: 1;
        1, 0, 0, 1 :: 0;
        1, 0, 1, 0 :: 1;
        1, 0, 1, 1 :: 1;
        1, 1, 0, 0 :: 0;
        1, 1, 0, 1 :: 0;
        1, 1, 1, 0 :: 0;
        1, 1, 1, 1 :: 1;

    END; "Simulation
  END LHFC_LE; "Function
```

The reduced equation provided in the documentation file is shown below. Notice that the equation has not been altered, i.e., reduced. The simulation results are also shown below in truth table format. The simulation ran successfully since there are no errors in the simulation results.

EQUATIONS FOR FUNCTION LHFC_LE

INPUT SIGNALS:
        HIGH_TRUE   A
        HIGH_TRUE   B
        HIGH_TRUE   C
        HIGH_TRUE   D
OUTPUT SIGNALS:
        HIGH_TRUE   F

REDUCED EQUATIONS:

F.EQN   = C*D*B
                   + /A*B
                   + C*A*D
                   + C*A*/B
                   + C*/A*/D
                   + /C*/A*/D
                   + /D*/B ;   "(7 terms)

32

```
SIMULATION OF LHFC_LE  -   Report Level 2

Time ns   A  B  C  D  F      Messages
-------------------------------------------
0         0  0  0  0  1
25        0  0  0  0  1
50        0  0  0  1  0
75        0  0  1  0  1
100       0  0  1  1  0
125       0  1  0  0  1
150       0  1  0  1  1
175       0  1  1  0  1
200       0  1  1  1  1
225       1  0  0  0  1
250       1  0  0  1  0
275       1  0  1  0  1
300       1  0  1  1  1
325       1  1  0  0  0
350       1  1  0  1  0
375       1  1  1  0  0
400       1  1  1  1  1
```

Design Example 14
Page 448 in *Modern Digital Design*

Obtain the **design file** for the S-R NOR latch circuit shown in Figure 8-7 on page 448 using the Logic Equation entry method. For simulation use the test language constructs SET and CLOCKF to specify the input combinations for the input variables and let the outputs be supplied by the simulator. Provide messages to help interpret the simulation results. Choose the CONFIGURATION tool and select the Espresso equation reduction level then compile the **design file** by running the COMPILE DESIGN tool. Obtain the equation and simulation results contained in the Documentation [.DOC] file. Compare the simulation results with the state diagram
on page 453.

Solution

```
"design file
TITLE                 NOR latch NL;
COMMENT               Logic Equation entry format;

FUNCTION NL_LE;

INPUT S, R;
OUTPUT Q, X;

     X = /(S + q);   " or S /+ q
     Q = /(X + R);   " or X /+ R
```

33

```
SIMULATION "Simulation section is optional
STEP 25ns;
    INITIAL X = 1;
    INITIAL Q = 0;

    MESSAGE ('Do not set or reset the latch');
    SET S = 0;
    SET R = 0;
    CLOCKF;
    MESSAGE ('Reset the latch');
    SET S = 0;
    SET R = 1;
    CLOCKF;
    MESSAGE ('Set & reset the latch at same time');
    SET S = 1;
    SET R = 1;
    CLOCKF;
    MESSAGE ('Set the latch');
    SET S = 1;
    SET R = 0;
    CLOCKF;
    MESSAGE ('Do not set or reset the latch');
    SET S = 0;
    SET R = 0;
    CLOCKF;
    MESSAGE ('Set the latch');
    SET S = 1;
    SET R = 0;
    CLOCKF;
    MESSAGE ('Set & reset the latch at same time');
    SET S = 1;
    SET R = 1;
    CLOCKF;
    MESSAGE ('Do not set or reset the latch');
    SET S = 0;
    SET R = 0;
    CLOCKF;
    MESSAGE ('Set the latch');
    SET S = 1;
    SET R = 0;
    CLOCKF;
    MESSAGE ('Do not set or reset the latch');
    SET S = 0;
    SET R = 0;
    CLOCKF;
    MESSAGE ('Reset the latch');
    SET S = 0;
    SET R = 1;
    CLOCKF;
    MESSAGE ('Do not set or reset the latch');
    SET S = 0;
    SET R = 0;
    CLOCKF;
```

34

```
      END; "Simulation
END NL_LE; "Function
```

The equations provided in the documentation file are shown below. The simulation results are also shown below in truth table format. The Q (next state) output produced by the simulator for the input combinations specified can be seen to follow the state diagram for the NOR latch illustrated on page 453 in *Modern Digital Design*.

EQUATIONS FOR FUNCTION NL_LE

```
INPUT SIGNALS:
      HIGH_TRUE  S
      HIGH_TRUE  R
OUTPUT SIGNALS:
      HIGH_TRUE  Q
      HIGH_TRUE  X
```

REDUCED EQUATIONS:

Q.EQN   = /R*/X ;   "(1 term)

X.EQN   = /Q*/S ;   "(1 term)

SIMULATION OF NL_LE   -   Report Level 2

Time ns	S	R	Q	X		Messages
0	0	0	0	1		
25	0	0	0	1		Do not set or reset the latch
50	0	1	0	1		Reset the latch
75	1	1	0	0		Set & reset latch at same time
100	1	0	1	0		Set the latch
125	0	0	1	0		Do not set or reset the latch
150	1	0	1	0		Set the latch
175	1	1	0	0		Set & reset latch at same time
200	0	0	0	0	U	Do not set or reset the latch
225	1	0	1	0		Set the latch
250	0	0	1	0		Do not set or reset the latch
275	0	1	0	1		Reset the latch
300	0	0	0	1		Do not set or reset the latch

NOTE: UNSTABLE SIMULATION STATES INDICATED BY TRAILING "U"

Notice in this case the simulator flagged the race condition that occurs when S and R change at the same time as indicated by the trailing U shown on the line where this input condition occurred in the simulation truth table output.

Design Example 15
Page 471 in *Modern Digital Design*

Obtain the **design file** for the Gated D latch circuit shown in Figure 8-20 on page 470 and represented by the Karnaugh map on page 471 using the Truth Table entry method. For simulation use the table format (truth table) to specify input combinations for the Boolean functions and let the outputs be supplied by the simulator. Provide messages to help interpret the simulation results. Choose the CONFIGURATION tool and select the Espresso equation reduction level then compile the **design file** by running the COMPILE DESIGN tool. Obtain the reduced equation and simulation results contained in the Documentation [.DOC] file. Compare the simulation results with the state diagram on page 471.

Solution

```
"design file
TITLE                      Gated D latch GDL;
COMMENT                    Truth Table entry format;

FUNCTION GDL_TT;

INPUT C, D;
OUTPUT Q;

    TRUTH_TABLE

    q, C, D :: Q;
    0, 0, 0 :: 0;
    0, 0, 1 :: 0;
    0, 1, 0 :: 0;
    0, 1, 1 :: 1;
    1, 0, 0 :: 1;
    1, 0, 1 :: 1;
    1, 1, 0 :: 0;
    1, 1, 1 :: 1;

    END; "Truth table

    SIMULATION "Simulation section is optional
    STEP 25ns;

        INITIAL Q = 1;
        MESSAGE ('C = 0, D has no effect');
        C, D :: Q;
        0, 0 :: S;
        MESSAGE ('C = 0, D has no effect');
        C, D :: Q;
        0, 1 :: S;
```

36

```
        MESSAGE ('Set the latch');
        C, D :: Q;
        1, 1 :: S;
        MESSAGE ('Reset the latch');
        C, D :: Q;
        1, 0 :: S;
        MESSAGE ('C = 0, D has no effect');
        C, D :: Q;
        0, 0 :: S;
        MESSAGE ('C = 0, D has no effect');
        C, D :: Q;
        0, 1 :: S;
        MESSAGE ('C = 0, D has no effect');
        C, D :: Q;
        0, 0 :: S;
        MESSAGE ('Reset the latch');
        C, D :: Q;
        1, 0 :: S;
        MESSAGE ('Set the latch');
        C, D :: Q;
        1, 1 :: S;
        MESSAGE ('Reset the latch');
        C, D :: Q;
        1, 0 :: S;
        MESSAGE ('Reset the latch');
        C, D :: Q;
        1, 0 :: S;
        MESSAGE ('Set the latch');
        C, D :: Q;
        1, 1 :: S;
        MESSAGE ('C =  0, D has no effect');
        C, D :: Q;
        0, 1 :: S;
        MESSAGE ('Set the latch');
        C, D :: Q;
        1, 1 :: S;
        MESSAGE ('Reset the latch');
        C, D :: Q;
        1, 0 :: S;
        MESSAGE ('C = 0, D has no effect');
        C, D :: Q;
        0, 0 :: S;
        MESSAGE ('Reset the latch');
        C, D :: Q;
        1, 0 :: S;
        MESSAGE ('Set the latch');
        C, D :: Q;
        1, 1 :: S;

    END; "Simulation
END GDL_TT; "Function
```

The reduced equations provided in the documentation
file are shown below. The simulation results are also shown

below in truth table format. The Q (next state) output produced by the simulator for the input combinations specified can be seen to follow the state diagram for the Gated D latch illustrated on page 471 in *Modern Digital Design*.

EQUATIONS FOR FUNCTION GDL_TT

```
INPUT SIGNALS:
      HIGH_TRUE  C
      HIGH_TRUE  D
OUTPUT SIGNALS:
      HIGH_TRUE  Q
```

REDUCED EQUATIONS:

```
Q.EQN              = Q*/C
                   + D*C ;   "(2 terms)
```

SIMULATION OF GDL_TT - Report Level 2

Time ns	C	D	Q	Messages
0	0	0	1	
25	0	0	1	C = 0, D has no effect
50	0	1	1	C = 0, D has no effect
75	1	1	1	Set the latch
100	1	0	0	Reset the latch
125	0	0	0	C = 0, D has no effect
150	0	1	0	C = 0, D has no effect
175	0	0	0	C = 0, D has no effect
200	1	0	0	Reset the latch
225	1	1	1	Set the latch
250	1	0	0	Reset the latch
275	1	0	0	Reset the latch
300	1	1	1	Set the latch
325	0	1	1	C = 0, D has no effect
350	1	1	1	Set the latch
375	1	0	0	Reset the latch
400	0	0	0	C = 0, D has no effect
425	1	0	0	Reset the latch
450	1	1	1	Set the latch

Design Example 16
Page 473 in *Modern Digital Design*

Obtain the *design file* for the Gated J-K latch circuit shown in Figure 8-23 on page 473 and represented by the Characteristic equation on page 474 using the Logic Equation entry method. Use the test language constructs SET and CLOCKF to specify the input combinations for the input variables and let the outputs be supplied by the simulator. Provide messages to help interpret the simulation results.

38

Choose the CONFIGURATION tool and select the Espresso
equation reduction level then compile the *design file* by
running the COMPILE DESIGN tool. Obtain the reduced
equation and simulation results contained in the
Documentation [.DOC] file. Compare the simulation results
with the state diagram on page 475.

Solution

```
"design file
TITLE                   Gated J-K latch GJKL;
COMMENT                 Logic Equation entry format;

FUNCTION GJKL_LE;

INPUT C, J, K;
OUTPUT Q;

     Q = /q*C*J + q*/C + q*/K;

     SIMULATION "Simulation section is optional
     STEP 25ns;

         INITIAL Q = 1;

         SET C = 0;

         MESSAGE ('C = 0, J and K have no effect');
         SET J = 0;
         SET K = 0;
         CLOCKF;
         MESSAGE ('C = 0, J and K have no effect');
         SET J = 1;
         SET K = 0;
         CLOCKF;
         MESSAGE ('C = 0, J and K have no effect');
         SET J = 0;
         SET K = 0;
         CLOCKF;
         MESSAGE ('C = 0, J and K have no effect');
         SET J = 0;
         SET K = 1;
         CLOCKF;
         MESSAGE ('C = 0, J and K have no effect');
         SET J = 0;
         SET K = 0;
         CLOCKF;
         MESSAGE ('C = 0, J and K have no effect');
         SET J = 1;
         SET K = 1;
         CLOCKF;
         MESSAGE ('C = 0, J and K have no effect');
         SET J = 1;
         SET K = 0;
```

```
        CLOCKF;
        SET C = 1;
        MESSAGE ('Do not set or reset the latch');
        SET J = 0;
        SET K = 0;
        CLOCKF;
        MESSAGE ('Set the latch');
        SET J = 1;
        SET K = 0;
        CLOCKF;
        MESSAGE ('Do not set or reset the latch');
        SET J = 0;
        SET K = 0;
        CLOCKF;
        MESSAGE ('Reset the latch');
        SET J = 0;
        SET K = 1;
        CLOCKF;
        MESSAGE ('Reset the latch');
        SET J = 0;
        SET K = 1;
        CLOCKF;
        MESSAGE ('Set & reset latch, at same time');
        SET J = 1;
        SET K = 1;
        CLOCKF;
        MESSAGE ('Set the latch');
        SET J = 1;
        SET K = 0;
        CLOCKF;

    END; "Simulation
  END GJKL_LE; "Function
```

The reduced equation provided in the documentation file is shown below (notice that the equation will not reduce). The simulation results are also shown below in truth table format.   The Q (next state) output produced by the simulator for the input combinations specified can be seen to follow the state diagram for the Gated J-K latch illustrated on page 475 in *Modern Digital Design*.

```
EQUATIONS FOR FUNCTION GJKL_LE

INPUT SIGNALS:
        HIGH_TRUE   C
        HIGH_TRUE   J
        HIGH_TRUE   K
OUTPUT SIGNALS:
        HIGH_TRUE   Q
```

REDUCED EQUATIONS:

```
Q.EQN            = /Q*C*J
                 + Q*/C
                 + /K*Q ;   "(3 terms)
```

SIMULATION OF GJKL_LE - Report Level 2

Time ns	C	J	K	Q		Messages
0	0	0	0	1		
25	0	0	0	1		C = 0, J and K have no effect
50	0	1	0	1		C = 0, J and K have no effect
75	0	0	0	1		C = 0, J and K have no effect
100	0	0	1	1		C = 0, J and K have no effect
125	0	0	0	1		C = 0, J and K have no effect
150	0	1	1	1		C = 0, J and K have no effect
175	0	1	0	1		C = 0, J and K have no effect
200	1	0	0	1		Do not set or reset the latch
225	1	1	0	1		Set the latch
250	1	0	0	1		Do not set or reset the latch
275	1	0	1	0		Reset the latch
300	1	0	1	0		Reset the latch
325	1	1	1	0	U	Set & reset latch, at same time
350	1	1	0	1		Set the latch

NOTE: UNSTABLE SIMULATION STATES INDICATED BY TRAILING "U"

Notice in this case the simulator flagged the instability in the circuit for the input conditions C J K = 111 as indicated by the trailing U shown on the line where this condition occurs in the simulation truth table output.

Our goal has been to provide design and analysis examples to assist first time users of the software package PLDesigner. By presenting a variety of design examples that the Student Version of PLDesigner V1.5 can easily handle, we hope to encourage you to use PLDesigner in your digital design courses. Synchronous sequential design examples using PLDesigner are presented in Chapter 9 of *Modern Digital Design*. The simulation section was not presented in the synchronous sequential design examples in the text but can be added if desired.